Encyclopedia of Alternative and Renewable Energy: Dye Sensitized Solar Cells

Volume 26

Encyclopedia of Alternative and Renewable Energy: Dye Sensitized Solar Cells

Volume 26

Edited by **Terence Maran and David McCartney**

New York

Published by Callisto Reference,
106 Park Avenue, Suite 200,
New York, NY 10016, USA
www.callistoreference.com

Encyclopedia of Alternative and Renewable Energy:
Dye Sensitized Solar Cells
Volume 26
Edited by Terence Maran and David McCartney

International Standard Book Number: 978-1-63239-200-8 (Hardback)

Printed in the United States of America.

Contents

Permissions

List of Contributors

Preface

The world is advancing at a fast pace like never before. Therefore, the need is to keep up with the latest developments. This book was an idea that came to fruition when the specialists in the area realized the need to coordinate together and document essential themes in the subject. That's when I was requested to be the editor. Editing this book has been an honour as it brings together diverse authors researching on different streams of the field. The book collates essential materials contributed by veterans in the area which can be utilized by students and researchers alike.

This book focuses on dye-sensitized solar cells (DSSCs), which are considered to be highly promising as they are made of low cost materials with simple and economical manufacturing procedures and can be molded into malleable sheets. These cells have emerged as a new class of energy conversion tools, which represent the third generation of solar technology. The method of converting solar energy into electricity in these devices is quite distinctive. The energy conversion efficiency produced by the DSSCs is low, but, it has upgraded quickly in the last years. It is assumed that the DSSCs will take an admirable place in the large scale production for the future. This book covers various aspects related to DSSCs and examines investigations of dyes for dye sensitized solar cells: Ruthenium-Complex, Metal-Free, Metal-Complex Porphyrins and Natural Dyes, and also discusses its future possibilities in fabrication, doping and characterization of polyaniline and metal oxides.

Each chapter is a sole-standing publication that reflects each author's interpretation. Thus, the book displays a multi-facetted picture of our current understanding of application, resources and aspects of the field. I would like to thank the contributors of this book and my family for their endless support.

Editor

Dye Sensitized Solar Cells Principles and New Design

Yang Jiao, Fan Zhang and Sheng Meng
*Beijing National Laboratory for Condensed Matter Physics and
Institute of Physics, Chinese Academy of Sciences, Beijing
China*

1. Introduction

It is generally believed that fossil fuels, the current primary but limited energy resources, will be replaced by cleaner and cheaper renewable energy sources for compelling environmental and economic challenges in the 21st century. Solar energy with its unlimited quantity is expected to be one of the most promising alternative energy sources in the future. Devices with low manufacturing cost and high efficiency are therefore a necessity for sunlight capture and light-to-energy conversion.

The dye-sensitized solar cell (DSSC), invented by Professor M. Grätzel in 1991 (O'Regan & Grätzel, 1991), is a most promising inexpensive route toward sunlight harvesting. DSSC uses dye molecules adsorbed on the nanocrystalline oxide semiconductors such as TiO_2 to collect sunlight. Therefore the light absorption (by dyes) and charge collection processes (by semiconductors) are separated, mimicking the natural light harvest in photosynthesis. It enables us to use very cheap, wide band-gap oxide semiconductors in solar cells, instead of expensive Si or III-V group semiconductors. As a result, much cheaper solar energy at $1 or less per peak Watt ($1/pW) can be achieved. For comparison, the dominant crystalline or thin-film Si solar cells have a price of >$4-5/pW presently and are suffering from the world-wide Si shortage. The fabrication energy for a DSSC is also significantly lower, 40% of that for a Si cell.

In this book chapter, we will present the principles of DSSC and detail the materials employed in a DSSC device in section 2. In section 3, the fabrication processes are shown. Then we discuss the energy conversion mechanism at the microscopic level in section 4. After this we try to give new design of the dye molecule and adsorption anchoring configurations to give hints on improving the energy conversion efficiency and making more stable devices in section 5. At last we present our conclusion and perspectives.

2. Principles of dye sensitized solar cells

2.1 Components

The current DSSC design involves a set of different layers of components stacked in serial, including glass substrate, transparent conducting layer, TiO_2 nanoparticles, dyes, electrolyte, and counter electrode covered with sealing gasket. The typical configuration is shown in Fig. 1.

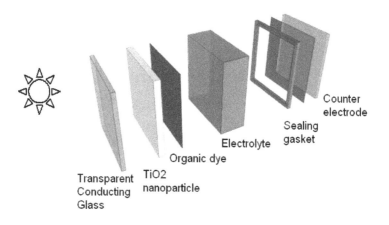

Fig. 1. Typical configuration of a DSSC.

2.1.1 Transparent conducting glass

In the front of the DSSC there is a layer of glass substrate, on top of which covers a thin layer of transparent conducting layer. This layer is crucial since it allows sunlight penetrating into the cell while conducting electron carriers to outer circuit. Transparent Conductive Oxide (TCO) substrates are adopted, including F-doped or In-doped tin oxide (FTO or ITO) and Aluminum-doped zinc oxide (AZO), which satisfy both requirements. ITO performs best among all TCO substrates. However, because ITO contains rare, toxic and expensive metal materials, some research groups replace ITO with FTO. AZO thin films are also widely studied because the materials are cheap, nontoxic and easy to obtain. The properties of typical types of ITO and FTO from some renowned manufacturers are shown in Table 1.

Conductive glass	Company	Light transmittance	Conductivity (Ohm/sq)	Thickness (mm)	Size (cm×cm)
ITO	Nanocs	>85%	5	1.1	1x3
ITO	PG&O	85%	4.5	1.1	2×3
FTO	NSG	>84%	<7	3	100×100

Table 1. Properties of a few types of commercial ITO and FTO materials.

2.1.2 TiO₂ nanoparticles

DSSC has a low efficiency less than 1% until Professor Grätzel employs porous TiO_2 as the anode material. Usually a layer of negatively doped TiO_2 (n-TiO_2) nanoparticles is used. The advantages of TiO_2 include high photosensitivity, high structure stability under solar irradiation and in solutions, and low cost. The typical particle size is 8-30 nm in diameter, and the TiO_2 films thickness is 2-20 μm, with the maximum efficiency located at a thickness of 12-14μm depending on dyes and electrolyte chosen (Ito et al., 2008). However, as a wide bandgap semiconductor (~3.2 eV), TiO_2 absorbs only UV light, which comprises only a small fraction (~5%) of solar spectrum. As a result, dye molecules are employed for visible light capture. Only nanocrystalline TiO_2 provides high light capture efficiency, with external quantum efficiency (incident photon-to-charge efficiency) typically in the range of 60-90% using nanocrystal forms in comparison with <0.13% using the monocrystal form (Grätzel, 2005). The reason lies in the high surface-to-volume ratios for porous nanocrystal materials.

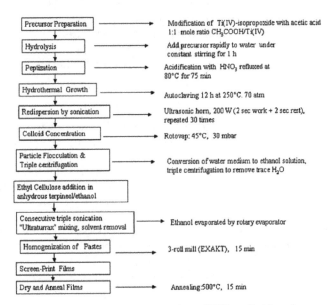

Scheme 1. Flow diagram depicting preparation of TiO$_2$ colloid and paste used in screen-printing technique for DSSC production. Adopted from (Ito et al., 2008). Copyright: 2007 Elsevier B. V.

2.1.3 Dyes

Dye molecules are the key component of a DSSC to have an increased efficiency through their abilities to absorb visible light photons. Early DSSC designs involved transition metal coordinated compounds (e.g., ruthenium polypyridyl complexes) as sensitizers because of their strong visible absorption, long excitation lifetime and efficient metal-to-ligand charge transfer (O'Regan & Grätzel, 1991; Grätzel, 2005; Ito et al., 2008). However, high cost of Ru dyes (>$1,000/g) is one important factor hindering the large-scale implementation of DSSC. Although highly effective, with current maximum efficiency of 11% (Grätzel, 2005), the costly synthesis and undesired environmental impact of those prototypes call for cheaper, simpler, and safer dyes as alternatives.

Organic dyes, including natural pigments and synthetic organic dyes, have a donor-acceptor structure called as push-pull architecture, thus improving short circuit current density by improving the absorption in red and infrared region. Natural pigments, like chlorophyll, carotene, and anthocyanin, are freely available in plant leaves, flowers, and fruits and fulfill these requirements. Experimentally, natural-dye sensitized TiO$_2$ solar cells have reached an efficiency of 7.1% and high stability (Campbell et al., 2007).

Even more promising is the synthetic organic dyes. Various types have recently been developed, including indolic dyes (D102, D149) (Konno et al., 2007), and cyanoacrylic acids (JK, C209). The same as some natural dyes, they are not associated with any metal ions, being environmental benign and easily synthesized from abundant resources on a large scale. The efficiency has reached a high level of 10.0-10.3% (Zeng et al., 2010). They are relatively cheap, at the cost of one-tenth of corresponding Ru dyes. Light soaking experiments have confirmed they possess long-time stability: 80% efficiency has been maintained after 1,200 hours of light-soaking at 60 °C (~5 million turnovers). The commercialized production of these synthetic dyes has been established in China this year. A single dye usually has a limited adsorption spectrum, so some research groups use several kinds of dyes to relay energy transfer and compensate each other and have achieved good results (Hardin et al., 2010).

2.1.4 Electrolyte

Currently three different kinds of electrolytes have been used in real DSSCs with pros and cons of each kind: (i) the most common electrolyte is I^-/I_3^- in organic solvents, such as acetonitrile. Sometimes lithium ion is added to facilitate electron transport. This kind of electrolyte is good for ion diffusion and infiltrate well with TiO_2 film, keeping highest efficiency of all DSSCs. But limited long-term stability due to volatilization of liquid hinders its wide use. (ii) Inorganic ionic liquids made of salts or salt mixture. It looks like solid while it has properties of liquid and it performs well in conductivity. But after a long period of time, its efficiency declines. (iii) Solid electrolyte, such as spiro-MeOTAD or CuI (Konno et al., 2007). For CuI, its instability and crystallization makes it hard to fill in the porous TiO_2 films. The problem can be solved by adding ionic liquid into the electrolyte. Spiro-MeOTAD is a typical kind of organic hole conductor, which has been developed for years and the DSSC based on this kind of electrolyte has reached the efficiency of 5% (Yu et al., 2009).

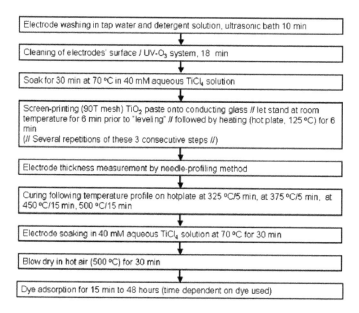

Scheme 2. Schematic representation for fabrication of dye-sensitized-TiO_2 electrodes. Adopted from (Ito et al., 2008). Copyright: 2007 Elsevier B. V.

2.1.5 Counter electrode

On the back of the DSSC there presents another glass substrate covered with a thin layer of Pt used as the catalyst to regenerate I⁻ and as the cathode material. Pt is the best material to make efficient devices technically. But considering high expenses, carbon cathode has been an ideal substitute, such as carbon black, carbon nanotubes etc. In 2006, Grätzel's group employs carbon black as the material of counter electrode, and reaches an efficiency of 9.1%, which is 83% of that using Pt (Yu et al., 2009).

Conducting polymers can also be used. Polyaniline film on stainless steel by electrochemical polymerization bas been reported as a counter electrode of DSSC (Qin et al., 2010). It is cheap and non-fragile.

2.2 Fabrication

In this section, we introduce Grätzel's new fabrication technologies for DSSCs having a conversion efficiency of solar light to electricity power over 10% (Ito et al., 2008). It consists of pre-treatment of the working photoelectrode by $TiCl_4$, variations in layer thickness of the transparent nanocrystalline-TiO_2 and applying a topcoat light-scattering layer as well as the adhesion of an anti-reflecting film to the electrode's surface.

First, one prepares glass substrate with a transparent conducting layer. Then, the DSSC working electrodes are prepared as shown in Scheme 1 & 2. Scheme 1 depicts procedures to produce paste (A) containing 20 nm nanocrystalline TiO_2 particles. For the paste used in the light-scattering layers (paste B), 10nm TiO_2 particles which were obtained following the peptization step and in a procedure analogous to those of 20 nm TiO_2 outlined in Scheme 1, were mixed with 400 nm TiO_2 colloidal solution. Paste A and B are coated on FTO, forming the TiO_2 "double layer" film. It is treated with $TiCl_4$ before sintering. After cooling, the electrode is immersed in dye solutions.

Finishing fabrication of all parts, the dye-covered TiO_2 electrode and Pt-counter electrode are assembled into a sandwich type cell and sealed with a hot-melt gasket. For the counter electrode, a hole (1-mm diameter) is drilled in the FTO glass. The hole is made to let the electrolyte in via vacuum backfilling. After the injection of electrolyte, the hole is sealed using a hot-melt ionomer film and a cover glass as shown in Fig. 2.

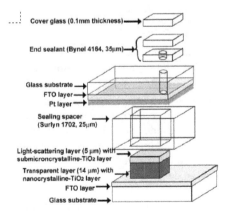

Fig. 2. Fabrication of the dye sensitized solar cells. Adopted from (Ito et al., 2008). Copyright: 2007 Elsevier B. V.

3. Processes

The way that DSSC works is quite simple, shown by the scheme in Fig. 3. The central idea is to separate the light absorption process from the charge collection process, mimicking natural light harvesting procedures in photosynthesis, by combining dye sensitizers with semiconductors. For most DSSC devices, the I^-/I_3^- couple is often used as a redox shuttle.

3.1 Sunlight absorption and electronic excitation
Photons of different energy in sunlight strike on the cell, penetrating into the dye layer since both the ITO layer with glass substrate and the TiO_2 nanocrystals are transparent to visible light. If photon energy is close to the energy gap of the dye molecule, namely, the energy difference between the highest occupied molecular orbital (HOMO) and lowest unoccupied molecular orbital (LUMO), it will be absorbed by the dye, promoting one electron from HOMO to LUMO. To be effective, usually it requires HOMO of the dye to reside in the bandgap of the semiconductor, and the LUMO to lie within the conduction band of the semiconductor. But this is not always true. Meng and collaborators have shown that electrons can be injected from cyanin dyes to the TiO_2 nanowire within 50 fs after excitation, although the dye LUMO is originally lower than the TiO_2 conduction band minimum (CBM) by 0.1-0.3 eV (Meng et al., 2008). This extends the current understanding of the mechanisms of DSSC operation and thus can enhance the ability to optimize their design and efficiency.

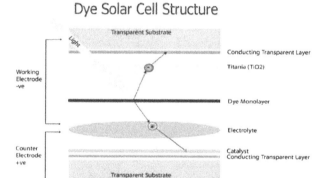

Fig. 3. Schematic illustration of the operation principle for molecular photovoltaic cell. Adopted from (Grätzel, 2009) Copyright: 1991 Nature Publishing Group.

3.2 Electron-hole separation, electron injection and collection
The excited electron will then inject into the conduction band of semiconductor through the interfacial bonds between the dye and the TiO_2, and then be transported and finally collected by the ITO electrode (the anode). On the other hand, the hole generated by photon excitation remains on the molecule during this process, since the HOMO of dye is separated from all other energy levels of the device.

3.3 Redox reaction at dye/electrolyte interface, and at counter electrode
There is no energy channels for the hole to diffuse into TiO_2 photoanode. As a result, the hole is eventually filled up by electrons from electrolyte ions, which conduct current

between the cathode and the dye molecule. At the same time, reduction of oxidized dye by iodide produces triiodide. The triiodide diffuses to counter electrode and accepts electrons from external load, regenerating the iodides. The overall process will provide electron flow from the ITO side (anode) to the outer circuit, with the potential difference equals to the incident photon energy if no voltage loss occurs during electron injection, diffusion and the neutral dye regeneration. In reality, however, these processes are always present. So the anode-cathode potential difference, namely, the open-circuit voltage V_{oc}, is mainly determined by the difference between conduction band bottom and electrolyte anion energy level, typically around 0.6-0.9 V.

3.4 Recombination

In all solar cells, the light generated carriers have a probability to meet their conjugate carriers and recombine. Recombination reduces the output of electrical power in DSSC. The photo injected electrons in the TiO_2 have two possible recombination pathways: direct recombination with the oxidized dyes or with the I_3^- in the electrolyte (Haque et al., 2005), which are shown in Fig. 4. The latter process is dominant and has been thoroughly studied. It can be reduced by adding a thin TiO_2 under-layer between the FTO and the nanocrystalline porous TiO_2 layer, because the connection between the FTO and the electrolyte is an important recombination route (Liu et al., 2006).

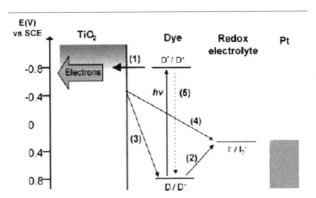

Fig. 4. Different electron transfer processes in the solar cell: (1) electron injection from dye excited state into the conduction band of TiO_2 semiconductor; (2) regeneration of the dye cation by electron transfer from the redox couple; (3) charge recombination to the cation of the dye; (4) recombination to the redox couple; and (5) excited state decay to the ground state. Adopted from (Haque et al., 2005), Copyright: 2005 American Chemical Society.

4. Electron dynamics in DSSC

A DSSC benefits from its imitation of natural photosynthesis in that it separates sunlight absorption--which requires a large space--from electron collection processes which need highly pure materials and being most efficient on a small lengthscale. The biggest challenge to develop DSSCs is to realize both functions in the same system and to improve efficiency on both sides. By combining dye sensitizers with oxide semiconductor nanoparticles, DSSCs resolve this conflict. Visible light absorption efficiency is improved by >1000 times on nanoparticles compared to that of single-crystal surfaces, due to high surface/volume ratio of the former (Grätzel, 2005). Nevertheless, the mechanism and detailed dynamics of electron-hole separation at the dye/semiconductor interface, especially on a microscopic scale, remains elusive.

Experimentally it is observed that at various dye/TiO$_2$ interface, the timescale of excited electron injection ranges from the shortest 3 fs for biisonicontinic acid in vacuum to 100 ps for the triplet state injection of Ru-complex N719 in devices (see, for example, the summary in Meng & Kaxiras, 2010). The huge time span suggests rich physical factors may play a role. Understanding this process will help us tune charge transfer timescale and improve sunlight-to-electricity conversion efficiency. For instance, electron injection will be 3.3 times slower with the addition of a CH$_2$ group inserted between the dye molecule and the semiconductor, predicted from exponential decay of tunneling electron density when increasing separation distance in a non-adiabatic process. This is indeed observed in experiments using Re dyes (ReC1A-ReC3A) (Asbury et al., 2000). However, this pronounced time increase is not observed in experiments on Zn-porphyrins with one or four oligo(phenylethylny) bridges (Chang et al., 2009).

Meng et al. examined the influence of various factors on the electron injection efficiency using state-of-the-art first-principles calculations within the framework of time-dependent density functional theory (TDDFT). From TDDFT simulations, they found that among the three organic dyes they investigated longer molecules do involve a longer injection time, which is consistent with intuition (Meng & Kaxiras, 2010). Furthermore, the time elongation is only 1.2 times (by inserting a (CH)$_2$ group) or 1.3 times (inserting a thiophene group). This indicates that adiabatic processes play a major role in these cases, which also explains the Zn-porphyrin experiment. Figure 5 shows the structure of the three organic dyes and their absorption intensity in neutral and deprotonated states.

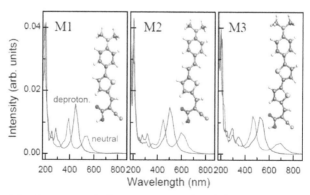

Fig. 5. Structure of the three organic dyes investigated and their absorption spectra. Blue lines: neutral dye; Red: deprotonated dyes. Adopted from (Meng & Kaxiras, 2010). Copyright: 2010 American Chemical Society.

On the basis of a previous work which illustrates an ultrafast electron-hole separation process at the anthocyanin/TiO$_2$ interface (Meng et al., 2008), they extended to demonstrate further that various factors including dye molecular size, binding geometry, and point defects on the TiO$_2$ surface will greatly influence electron collection efficiency. To analyze quantitatively the charge transfer and electron-hole separation process, they use the integral of excited electron (hole) density projected onto the TiO$_2$ orbitals, χ, as a function of time after photon absorption, with χ defined as

$$\chi = \int dr\,|\psi(r)|^2 \qquad\qquad \psi(r) = \sum_{j \in TiO_2} c_j \varphi_j(r) \qquad\qquad (1)$$

where c_j are the coefficients of atomic basis states φ_j in the Kohn-Sham orbital $\psi_{KS}(r)$ of interest (either the excited electron or the hole)

$$\psi_{KS}(r) = \sum_j c_j \varphi_j(r) \qquad\qquad (2)$$

The curves in Fig. 6 show the injection probability, χ, of the excited electron and hole as a function of simulation time.

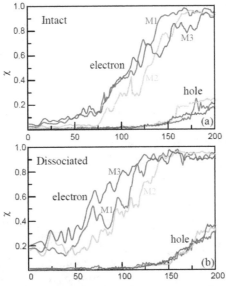

Fig. 6. Electron-hole dynamics from TDDFT simulations. Vertical axis demonstrates electrons injected into TiO$_2$, horizontal axis denotes time simulated. Adopted from (Meng & Kaxiras, 2010). Copyright: 2010 American Chemical Society

They also found that dye adsorption configurations significantly affect electron injection. By comparing to measured spectroscopy, intact and dissociative dye adsorptions are identified. The former is 30-50 fs slower than the latter. Different adsorption configurations of intact dyes result in injection time varies by three folds. The difference is mainly caused by the interface dipole moments. A positive dipole at the interface introduces an upshift of CBM, which will suppress excited electron transfer from the dye molecule to semiconductor conduction bands.

The semiconductor surface also imposes a fundamental influence on electron-hole dynamics. Dye adsorption on surface oxygen vacancies is very stable; it leads to a strong electronic coupling between the dye and the surface resulting in an electron injection time of ~50fs, 2-3 times faster than that on defect-free surfaces. But this improvement is at the cost of fast electron-hole recombination, which will reduce device efficiency. These simulations could explain well the two injection times at 40 fs and 200 fs observed in experiments, which would correspond to adsorption on defects and on clean surfaces, respectively.

They also studied electron back transfer after the excited electron has been completely injected into the TiO_2. All the research results indicate that electrons and holes are separated in space at a time ~200 fs, to assure DSSCs work well.

5. New design

Recent developments of effective all-organic dyes have introduced a novel Donor-π-Acceptor (D-π-A) structure, with the donor and acceptor connected via a molecular bridge conductor (π-group) and the acceptor directly bound to semiconductor surface. Upon excitation, electrons in the HOMOs, originally distributed around the donor group, are promoted to the LUMOs, which are centred around the acceptor. This excitation process effectively pushes extra electrons to the acceptor part of the dyes and leaves holes on the donor part. Importantly, the proximity between the acceptor and the semiconductor surface, mutually bound via covalent and/or hydrogen bonds, allows ultrafast transfer of excited electrons to the conduction bands of the semiconductor (Meng et al., 2008; Duncan & Prezhdo, 2007). Consequently, this leads to efficient electron-hole separation and ultimately electricity generation.

Laboratory development of D-π-A dyes often invokes a trial-and-error approach, which requires extensive chemical synthesis and expensive materials processing with a slow progress. In this regard, theoretical screening of potential organic dyes using state-of-the-art first principles computation shows a great promise, significantly reducing the cost to develop efficient dyes and expediting discovery of new ones.

5.1 Enhanced light absorption

Based on first principles calculations on the electronic structure and optical absorption of organic molecules, we design a set of Donor-π-Acceptor (D-π-A) type of dyes with inserting and changing intermediate spacer groups or electron acceptor groups, to be used in DSSCs for sunlight harvesting and energy conversion. We found that with these modifications, the electronic levels and corresponding optical absorption properties of organic dyes can be gradually tuned, with dyes having novel 1,4-cyclohexadiene groups as promising candidates for red light absorption and achieving high extinction coefficients. This study opens a way for material design of new dyes with target properties for further optimizing the performance of dye solar cells.

Our calculations are performed within the framework of DFT (Kohn & Sham, 1965) (for structure optimization) and TDDFT (Runge & Gross, 1984) (for excited states and optical absorption), using SIESTA (Soler et al., 2002) and Gaussian (Frisch et al., 2009) packages. It is well known that popular functionals (such as PBE and B3LYP) systematically underestimate charge-transfer excitation energies and result optical band splitting, leading to problematic predictions on optical excitations. Most recently, significant progresses have been made by including self-interaction correlations at the long range in hybrid fuctionals, to properly account for the long-range charge transfer excitation while retain the reliable description of short-range correlation interactions. Our previous work shows that the long-range corrected (LC) functionals ωB97X (Chai & Head-Gordon, 2008) and CAM-B3LYP (Yanai et al., 2004)) yield better comparison with experiment. Since ωB97X has less parameters and shows better agreement with experimental results (after solvent effect correction of 0.1-0.3 eV (Pastore et al., 2010)), we use this functional in the following for predicting optical properties of new dyes.

For structure optimization, we use pseudopotentials of the Troullier-Martins type to model the atomic cores, the PBE form of the exchange-correlation functional, and a local basis set of double-ζ polarized orbitals. An auxiliary real space grid equivalent to a planewave cutoff of 680 eV is used for the calculation of the electrostatic (Hartree) term. A molecular structure is considered fully relaxed when the magnitude of forces on the atoms is smaller than 0.03eV/Å. Optical absorption are extracted from TDDFT simulations within the linear response regime at fixed dye geometry in gas phase. We use the 6-31G(d) basis set throughout this book chapter, which has been shown to yield negligible difference in electron density and energy accuracy compared with those including additional diffuse functions (Pastore et al., 2010). Within TDDFT, different functionals based on adiabatic approximation are used, including B3LYP and ωB97X. The latter one, belonging to LC functionals, are employed to correct excitation energies for charge-transfer excitations, which prevail in these dyes. The spectrum is obtained using the following formula based on calculated excitation energies and oscillator strengths,

$$\varepsilon(\omega) = 2.174 \times 10^8 \sum_I \frac{f_I}{\Delta} \exp\left[-2.773 \frac{(\omega - \omega_I)^2}{\Delta^2}\right], \tag{3}$$

where ε is the molar extinction coefficient given in unit of M⁻¹cm⁻¹, energies and ω_I are in unit of cm⁻¹ and f_I is the corresponding oscillator strength. Such a formula will satisfy the well-known relationship

$$4.32 \times 10^{-9} \int \varepsilon(\omega) d\omega = \sum_I f_I. \tag{4}$$

We use a gaussian full-bandwidth of $\Delta = 3000 \, \text{cm}^{-1}$ at half-height for optical band broadening to mimic thermodynamic oscillations in dye structures and excitations at room temperature in vacuum.

Organic dyes with the cyanoacrylic acid anchoring and electron with-drawing group have been very successful in real devices. We choose Y-1, Y-2 and Y-3 molecules (left column, Fig. 7) as the model dyes because they have a simple structure and have been successfully synthesized for DSSC applications (for example, as a1, b1, b2 dyes in reference Kitamura et al., 2004). Then the model dyes are modified by inserting a cyclohexadene in the C=C double bond on the backbone. The resulting dyes are named as dyes Y-1ben, Y-2ben, Y-3ben hereafter for convenience. Another model dye Y-1ben2 is constructed by inserting an additional cyclohexadene ring in the Y-1ben (right column, Fig. 7). All these dyes have a strong structural similarity with each other; we hope upon these variations in the dye structures, various aspects of the atomic structure, electronic structure, absorption character and therefore the structure-property relationship of organic dyes can be investigated.

The calculated optical absorption spectra are shown in Fig. 8. We found that the absorption spectra will be red shifted by prolonging the oligoene backbone (compare the green and red lines) and inserting cyclohexadiene moiety (compare the lines in the same colour). Attaching benzene rings to the amide nitrogen could enhance the absorption intensity a little but barely changes the position of maximum absorption (compare blue and green lines). On the other hand, inserting cyclohexadiene group could modify the spectra significantly, both in peak positions and intensity (compare the lines in Fig. 8 in the same colour). This would make dyes with cyclohexadiene group attractive candidates for future development of DSSC devices, especially for high extinction, long wavelength light absorption.

We analyse further the electronic energy level calculated using B3LYP/6-31G(d) and the first excitation energy from ωB97X/6-31G(d) (Table 1). It is known that the solvent effects will lower dye absorption energy by 0.1-0.3 eV (Pastore et al., 2010); therefore experimental values quoted in Table 2 which are measured in solution would have been corrected by 0.1-

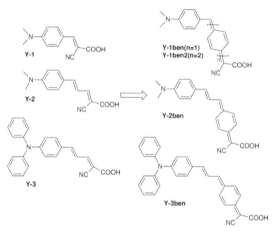

Fig. 7. Chemical structure of model dyes. The left column from up to bottom depicts Y-1, Y-2, Y-3, and the right column shows Y-1ben (Y-1ben2), Y-2ben, Y-3ben.

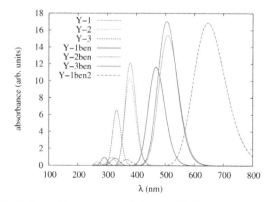

Fig. 8. Calculated optical absorption spectra of the model dyes.

0.3 eV, this leads to a better agreement with the results from LC-TDDFT for dyes in vacuum. For Y-1 dye the LC seems not necessary, maybe due to its short length.

Intuitively, the energy levels of the whole dye molecule are modified by the electric field introduced by the presence of chemical groups at the acceptor end. They would change according to the electronegativity of these chemical groups. The higher negativity, the lower the LUMO level, the smaller the energy gap would be. Our results proved this simple rule by showing that inserting electron withdrawing groups C=C or cyclohexadiene will down shift the LUMO level and producing a smaller energy gap (Table 1).

Following this way we obtain a small energy gap of 1.41 eV for Y-1ben2. In particular, we emphasize that the LUMO level is still higher by at least 0.7 eV than the conduction band edge (CBE) of anatase TiO_2, located at -4.0 eV (Kavan et al., 1996). This would provide enough driving force for ultrafast excited state electron injection. Another requisite in interface energy level alignment for DSSCs to work properly is that the dye HOMO has to be lower than the I^-/I_3^- redox potential at about -4.8 eV. The potential level is comparable to cyclohexadiene dye HOMOs in Table 2. However, we note here that by comparing calculated energy levels with experimental measured potentials, the HOMO and LUMO positions are generally overestimated by ~0.6 eV and ~0.3 eV, respectively. We expect the same trend would occur for dyes discussed in the present work; taking this correction into account, the designed new dyes would have a perfect energy alignment with respect to the CBE of TiO2 (more than 0.4 eV lower than LUMOs) and the I^-/I_3^- redox potential (~0.5-1 eV higher than HOMOs) favouring DSSC applications.

Dyes	Y-1	Y-2	Y-3	Y-1ben	Y-2ben	Y-3ben	Y-1ben2
LUMO	-2.22	-2.49	-2.63	-2.92	-3.05	-3.09	-3.30
HOMO	-5.64	-5.36	-5.38	-5.15	-5.02	-5.06	-4.71
gap	3.42	2.87	2.75	2.23	1.97	1.97	1.41
ωB97X	3.74	3.28	3.28	2.65	2.44	2.46	1.92
Exp.[a]	3.28	3.01	2.98				

Table 2. Electronic and optical properties of dyes predicted from first-principles calculations. The LOMO, HOMO and gap are results using B3LYP exchange-correlation potential. ωB97X indicates the first excitation energy using the ωB97X long correction.
[a] The data are from reference (Kitamura et al., 2004). The first absorption peak observed at 1.6×10^{-5} mol • dm^{-3} in ethanol. Energy levels are in unit of eV.

To further demonstrate the electronic properties for these promising dyes, we show the wavefunction plots for the molecular orbitals HOMO and LUMO of dye Y-1ben. The contour reveals that the HOMO and LUMO extend over the entire molecule. Other modified dyes have a similar characteristic.

We conclude that with the insertion of electron with-drawing groups C=C and cyclohexadiene in the backbone of oligoene dyes, we can tune the dye electronic levels relative to TiO_2 conduction bands and the corresponding optical absorption properties as shown in Fig. 8 and Table 2, for optical performance when used in real DSSC devices. In particular, we propose that the model dyes with cyclohexadiene group might be promising candidates for red to infrared light absorption, which may offer improved sunlight-to-electricity conversion efficiency when used alone or in combination with other dyes. The anchoring geometry will also influence the energy level alignment and energy conversion efficiency, which will be discussed in the next subsection.

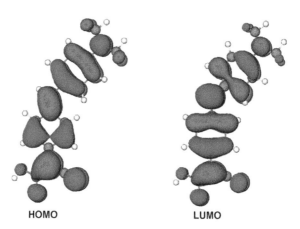

HOMO   **LUMO**

Fig. 9. Wavefunction plots for the representative dye Y-1ben. HOMO and LUMO orbitals are shown.

5.2 Stable anchoring

Following the similar principles, we have extended our study to the design of purely organic dyes with a novel acceptor, since the acceptor group connects the dye to the semiconductor and plays a critical role for dye anchoring and electron transfer processes. Our strategy is to test systematically the influence of chemical group substitution on the electronic and optical properties of the dyes.

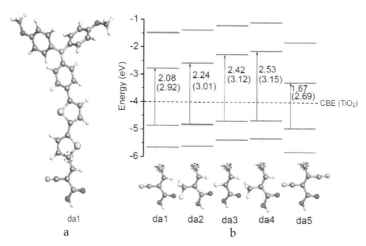

Fig. 10. (a) Optimized structure for the model dye da1. (b) The structure of acceptor groups by design and corresponding electronic level. The dashed circle marks the carbon atom connecting the acceptor part and the π bridge. Adapted from (Meng et al., 2011). Copyright: 2011 American Chemical Society.

Chemical groups of different electronegativity, size and shape and at different sites have been investigated. As an example, we consider here a specific modification on existing dye structures: replacement of the -CN group of cyanoacrylic side with other elements or groups. This gives a group of new dyes which we label da-n (with n = 1-5, an index). Organic dyes with the carboxylate-cyanoacrylic anchoring group have been very successful in real devices. From the point of view of electronic structure and optical absorption, it is possible that the side cyano group (-CN) has a positive influence on light absorption and anchoring to the TiO_2 surface (Meng et al., 2010). Accordingly, we consider the possibility of replacing -CN by other chemical groups and examine the dependence of dye performance on these groups. In Fig. 10 we show the set of dye acceptor structures we have investigated. We have replaced the cyano -CN group in model dye da1 by –CF_3, -F, and –CH_3 groups, which are labelled da2, da3, and da4, respectively. Model dye da1, shown in Fig. 10, has a very similar structure to that of D21L6 dye synthesized experimentally (Yum et al., 2009), except that the hexyl tails at the donor end are replaced by methyl groups. The electronic energy levels of these modified dyes in the ground-state are also shown in Fig. 10, as calculated using B3LYP/6-31G(d). Compared to the relatively small gap of 2.08 eV for da1, the energy gap is increased by all these modifications. We have also tried many other groups, such as –BH_2, -SiH_3, etc., for substituting the -CN group. All these changes give a larger energy gap. This may explain the optimal performance in experiment of the cyano CN group as a part of the molecular anchor, which yields the lowest excitation energy favouring enhanced visible light absorption. The changes in electronic structure introduced by substitution of the -CN group by other groups are a result of cooperative effects of both electronegativity and the size and shape of the substituted chemical groups.

We did not find any single-group substitution for the -CN that produces a lower energy gap. Therefore we extended our investigation to consider other possibilities. We found that with the substitution of the -H on the next site of the backbone by another -CN group, the ground-state energy gap is reduced to 1.67 eV (see Fig. 10, dye da5). The LUMO level is higher than the conduction band edge of anatase TiO_2 (dashed line, Fig. 10) and HOMO level is lower than the redox potential of tri-iodide, providing enough driving force for fast and efficient electron-hole separation at the dye/TiO_2 interface. With the above systematic modifications of the dye acceptor group, the electronic levels can be gradually tuned and as a consequence dyes with desirable electronic and optical properties can be identified. Influence of these acceptor groups on excited state electron injection can be studied in the same way as that in Section 4.

We also strive to design dye acceptors that render a higher stability of the dye/TiO_2 interface when used in outdoors applications. It was previously found that some organic dyes do not bind to the TiO_2 photoanode strongly enough, and will come off during intensive light-soaking experiments, while other dyes show higher stability in such tests (Xu et al., 2009). Dye anchors with desirable binding abilities will contribute greatly to the stability of interface. A type of dye anchor under design is a cyano-benzoic acid group, whose binding configuration to the anatase TiO_2 (101) surface is shown in Fig. 11. A particular advantage of cyano-benzoic acid as a dye acceptor is that, it strongly enhances dye binding onto TiO_2 surfaces. We investigate the binding geometries of this organic dye on TiO_2 (101) using DFT. Among several stable binding configurations, the one with a bidentate bond and a hydrogen bond between -CN and surface hydroxyl (originating from dissociated carboxyl acid upon adsorption), is the most stable with a binding energy of 1.52 eV. The bond lengths are d_{Ti1-O1}=2.146 Å, d_{Ti2-O2}=2.162 Å, and $d_{CN...HO}$=1.80 Å. Since there are three bonds formed, the dye is strongly stabilized on TiO_2. Experimentally dyes with cyano-benzoic acid anchors have been successfully synthesized and the corresponding stability is under test (Katono et al., submitted).

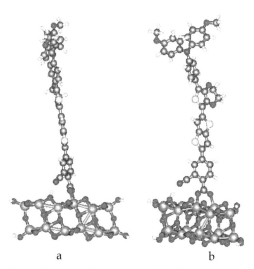

a   b

Fig. 11. (a) Side and (b) front views of adsorption of the dye with a cyano-benzoic acid anchor on a TiO$_2$ nanoparticle.

6. Conclusion

In summary, after a brief introduction of the principles of DSSCs, including their major components, the fabrication procedures and recent developments, we try to focus on the atomic mechanisms of dye adsorption and electron transport in DSSCs as obtained from accurate quantum mechanical simulations. We discovered that electron injection dynamics is strongly influenced by various factors, such as dye species, molecular size, binding group, and surface defects; more importantly, the time scale for injection can be tuned by changing these parameters. Based on the knowledge about the interface electronic structure and dynamics at the molecular level, we strive to design new dye molecules and anchoring configurations employing state-of-the-art first principles calculations. Our results show that upon systematic modifications on the existing dye structures, the optical absorption and energy levels could be gradually tuned. In particular, we propose that by inserting cyclohexadiene groups into spacing C=C double bonds, simple organic dyes could be promising candidates for enhanced red to infrared light absorption. This study opens a way for material design of new dyes with target properties to advance the performance of organic dye solar cells.

7. Acknowledgment

We thank our collaborators Professor Efthimios Kaxiras and Professor Michael Grätzel. This work is financially supported by NSFC (No. 11074287), and the hundred-talent program and knowledge innovation project of CAS.

8. References

Asbury, J. B.; Hao, E.; Wang, Y. & Lian, T. (2000). Bridge Length-Dependent Ultrafast Electron Transfer from Re Polypyridyl Complexes to Nanocrystalline TiO2 Thin

Films Studied by Femtosecond Infrared Spectroscopy. *Journal of Physical Chemistry B* Vol 104, pp. 11957-11964, ISSN 1520-6106

Campbell, W. M.; Jolley, K. W.; Wagner, P.; Wagner, K.; Walsh, P. J.; Gordon, K. C.; Schmidt-Mende, L.; Nazeeruddin, M. K.; Wang, Q.; Grätzel, M. & Officer, D. L. (2007). Highly Efficient Porphyrin Sensitizers for Dye-Sensitized Solar Cells, *Journal of Physical Chemistry C*, Vol. 111, No.32, (August 2007), pp. 11760-11762, ISSN 1932-7447

Chai, J.-D.; Head-Gordon, M. (2008). Systematic Optimization of Long-Range Corrected Hybrid Density Functionals, *Journal of Chemical Physics*, Vol. 128, No.8, (February 2008) pp.084106, ISSN 0021-9606

Chang, C. W.; Luo, L.; Chou, C. K.; Lo, C. F.; Lin, C. Y.; Hung, C. S.; Lee, Y. P. & Diau, E. W. (2009). Femtosecond Transient Absorption of Zinc Porphyrins with Oligo(phenylethylnyl) Linkers in Solution and on TiO2 Films. *Journal of Physical Chemistry C* Vol 113, pp. 11524-11531, ISSN 1932-7447

Duncan, W. R.; Colleen, F. C.; Oleg, V. P. (2007). Time-Domain Ab Initio Study of Charge Relaxation and Recombination in Dye-Sensitized TiO$_2$, *Journal of the American Chemical Society*, Vol.129, No.27, (June 2007) pp.8528-8543, ISSN 0002-7863

Frisch, M. J. et al. (2009). Gaussian 09, Revision A.1, Gaussian Inc.: Wallingford, CT, 2009

Grätzel, M. (2005). Mesoscopic Solar Cells for Electricity and Hydrogen Production from Sunlight, *Chemistry Letters*, Vol.34, No.1, (January 2005), pp. 8-13, ISSN 0366-7022

Grätzel, M. (2009). Recent Advances in Sensitized Mesoscopic Solar Cells, *Accounts of Chemical Research*, Vol.42, No.11, (November 2009), pp. 1788-1798, ISSN 0001-4842

Haque, S. A.; Palomares, E.; Cho, B. M.; Green, A. N. M.; Hirata, N.; Klug, D. R. & Durrant, J. R. (2005). Charge Separation Versus Recombination in Dye-Sensitized Nanocrystalline Solar Cells: the Minimization of Kinetic Redundancy, *Journal of the American Chemical Society*, Vol.127, No.10, (March 2005), pp.3456-3462, ISSN 0002-7863

Hardin, B.E.; Yum, J.-H.; Hoke, E.T.; Jun, Y.C.; Pechy, P.; Torres, T.; Brongersma, M.L.; Nazeeruddin, M.K.; Graetzel, M.; & McGehee, M.D. (2010). High Excitation Transfer Efficiency from Energy Relay Dyes in Dye-Sensitized Solar Cells, *Nano Letters*, Vol. 10, pp. 3077-3083, ISSN 1530-6984

Ito, S.; Murakami, T. N.; Comte, P.; Liska, P.; Grätzel, C.; Nazeeruddin, M. K. & Grätzel, M. (2008). Fabrication of Thin Film Dye Sensitized Solar Cells With Solar to Electric Power Conversion Efficiency over 10%, *Thin Solid Films*, Vol.516, No.14, (May 2008), pp.4613-4619, ISSN 0040-6090

Katono, M.; Bessho, T.; Meng, S.; Zakeeruddin, S. M.; Kaxiras, E.; Grätzel, M. (2011). Submitted.

Kavan, L.; Grätzel, M.; Gilbert, S. E.; Klemenz, C.; Scheel, H. J. (1996). Electrochemical and Photoelectrochemical Investigation of Single-Crystal Anatase, *Journal of American Chemical Society*. Vol. 118, No.28 (July 1996) pp. 6716-6723, ISSN 0002-7863

Kitamura, T.; Ikeda, M.; Shigaki, K.; Inoue, T.; Anderson, N. A.; Ai, X.; Lian, T. & Yanagida, S. (2004). Phenyl-Conjugated Oligoene Sensitizers for TiO2 Solar Cells, *Chemcal Materials*, Vol.16, No.9, (February 2004) pp. 1806-1812, ISSN 0897-4756

Kohn, W.; Sham, L. J. (1965). Self-Consistent Equations Including Exchange and Correlation Effects. *Physical Review*, Vol.140, No.4A, (1965) pp. 1133-1138, ISSN 0031-899X

Konno, A.; Kumara, G. R. A.; Kaneko, S. (2007). Solid-state solar cells sensitized with indoline dye, *Chemistry Letters*, Vol.36, No.6, (June 2007), pp.716-717, ISSN 0366-7022

Liu, X.; Huang, Z.;Meng, Q. et al. (2006). Recombination Reduction in Dye-Sensitized Solar Cells by Screen-Printed TiO_2 Underlayers, *Chinese Physics letters*, Vol.23, No.9, (June 2006), pp.2606-2608, ISSN 0256-307X

Meng, S.; Ren, J. & Kaxiras, E. (2008). Natural Dyes Adsorbed on TiO2 Nanowire for Photovoltaic Applicaitons: Enhanced Light Absorption and Ultrafast Electron Injection, *Nano Letters*, Vol.8, No.10, (September 2008), pp.3266-3272, ISSN 1530-6984

Meng, S. & Kaxiras, E. (2010). Electron and Hole Dynamics in Dye-Sensitized Solar Cells : Influencing Factors and Systematic Trends, *Nano Letters*, Vol. 10, No.4 (April 2010), pp 1238-1247, ISSN 1530-6984

Meng, S.; Kaxiras, E; Nazeeruddin, Md. K. & Grätzel, M. (2011). Design of Dye Acceptors for Photovoltaics from First-principles Calculations, *Journal of Physical Chemistry C*, in press, ISSN 1932-7447

O'Regan, B. & Grätzel, M. (1991). A Low-Cost, High-Efficientcy Solar-Cell Based on Dye-Sensitized Colloidal TiO_2 Films, *Nature*, Vol.353, No.6346, (October 1991), pp. 737-740, ISSN 0028-0836

Pastore, M.; Mosconi, E,; De Angelis, F. & Grätzel, M. (2010). A Computational Investigation of Organic Dyes for Dye-Sensitized Solar Cells: Benchmark, Strategies, and Open Issues, *Journal of Physical Chemistry C*, Vol.114, No.15 (April 2010) pp. 7205-7212, ISSN 1932-7447

Qin, Q.; Tao, J. & Yang, Y. (2010). Preparation and Characterization of Polyaniline Film on Stainless Steel by Electrochemical Polymerization as a Counter Electrode of DSSC, *Synthetic Metals*, Vol.160, No.11-12, (June 2010) pp.1167-1172, ISSN 0379-6779

Runge, E.; Gross, E. K. U. (1984). Density-Functional Theory for Time-Dependent Systems. *Physical Review Letter*, Vol.52, No.12, (1984) pp.997-1000, ISSN 0031-9007

Soler, J. M.; Artacho, E.; Gale, J. D.; Garcia, A.; Junquera, J.; Ordejon, P. & Sanchez-Portal, D. (2002). The SIESTA Method for Ab Initio Order-N Materials Simulation, *Journal of Physics: Condensed Matter*, Vol.14, No.11, (March 2002) pp. 2745-2779, ISSN 0953-8984

Xu, M.; Wenger, S.; Bala, H.; Shi, D.; Li, R.; Zhou, Y.; Zakeeruddin, S. M.; Grätzel, M. & Wang, P. (2009). Tuning the Energy Level of Organic Sensitizers for High-Performance Dye-Sensitized Solar Cells, *Journal of Physical Chemistry C*, Vol. 113, No.7 (February 2009) pp.2966-2973, ISSN 1932-7447

Yanai, T.; Tew, D. P. & Handy, N. C. (2004). A New Hybrid Exchange-Correlation Functional Using the Coulomb-Attenuating Method (CAM-B3LYP), *Chemical Physics Letter*, Vol. 393, No.1-3 (July 2004) pp. 51-57, ISSN 0009-2614

Yum, J. H.; Hagberg, D. P.; Moon, S. J.; Karlsson, K. M.; Marinado, T.; Sun, L.; Hagfeldt, A.; Nazeeruddin, M. K & Grätzel, M. (2009). Panchromatic Response in Solid-State Dye-Sensitized Solar Cells Containing Phosphorescent Energy Relay Dyes, *Angewandte Chemie-International Edition*, Vol. 48, No.49 (2009) pp. 1576-1580, ISSN1433-7851

Yu, Z.; Li, D.; Qin, D.; Sun, H.; Zhang, Y.; Luo, Y. & Meng, Q. (2009). Research and Development of Dye-Sensitized Solar Cells, *Materials China*, Vol.28, No.7-8, (August 2009), pp. 7-15, ISSN 1674-3962

Zeng, W.; Cao, Y.; Bai, Y.; Wang, Y.; Shi, Y.; Zhang, M.; Wang, F.; Pan, C. & Wang, P. (2010). Efficient Dye-Sensitized Solar Cells with an Organic Photosensitizer Ferturing Orderly Conjugated Ethylenedioxythiophene and Dithienosilole Blocks. *Chemistry of Materials*, Vol.22, No.5, (March 2010) pp. 1915-1925, ISSN 0897-4756

2

Investigation of Dyes for Dye-Sensitized Solar Cells: Ruthenium-Complex Dyes, Metal-Free Dyes, Metal-Complex Porphyrin Dyes and Natural Dyes

Seigo Ito
Department of Electrical Engineering and Computer Sciences,
Graduate School of Engineering, University of Hyogo, Hyogo
Japan

1. Introduction

Following the first report on dye-sensitized solar cells (DSCs) by Prof. Grätzel in 1991, thousands of papers have been published with the aim of making DSCs commercially viable (Fig. 1). They are attractive because of their low-cost materials and convenient fabrication by a non-vacuum, high-speed printing process. One of the key materials in DSCs is the sensitizer dye. Ruthenium-complex dyes are used to make DSCs with conversion efficiencies of over 10%; recently, the Grätzel group reported a DSC using ruthenium dye (Z991) which achieved a conversion efficiency of 12.3%. Research into synthetic ruthenium-free dyes, including metal-free organic dyes and metal-complex porphyrin dyes, has intensified because of the high cost of ruthenium. Indoline dyes and oligothiophene dyes are used to make DSCs with conversion efficiencies greater than 9% and 10%, respectively. A zinc-porphyrin dye produced a conversion efficiency of 11.4%.

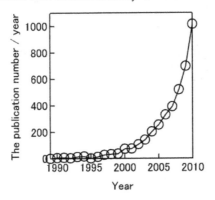

Fig. 1. Number of publications on DSCs in each year from 1989 to 2010. The data was obtained by searching online databases (Scopus, Elsevier). The keywords were "dye", "solar" and "cell", and the document type was limited to "article".

Natural dyes have also been studied for use in DSCs because they are cheaper than synthetic dyes, and exhibit moderate energy conversion efficiency. Natural chlorophyll dyes show energy conversion efficiencies of over 4% and Monascus yellow dye yielded a conversion efficiency of 2.3%. Despite the wide variety of natural dyes available, other natural dyes do not yield energy conversion efficiencies of over 2%, although some natural dyes derivatives are capable of higher energy conversion efficiencies. In this review, the principles and fabrication methods of DSCs are explained, and recent research on sensitizing dyes, including ruthenium-complex dyes, metal-free organic dyes, metal-complex porphyrin dyes, and natural dyes, is reviewed.

2. Ruthenium-complex dyes

Ruthenium dye DSCs were first reported in 1991 by O'Regan and Grätzel in Nature [1]. These first ruthenium dye DSCs achieved a 7.1% conversion efficiency (Fig. 2). However, the structure of the ruthenium dye was complicated and contained three ruthenium metal centers. In 1993, Nazeeruzzin et al. published DSCs with 10.3% conversion [2], using a ruthenium dye sensitizer (N3, Fig. 3: [cis-di(thiocyanato)bis(2,2-bipyridine-4,4-dicarboxylate)ruthenium]]), which contained one ruthenium center and was thus simpler than the ruthenium dye reported in 1991. At the end of the 1990s, Solaronix SA (Switzerland) began selling the materials for constructing DSCs: ruthenium dyes, electrodes, electrolytes, TiO_2 paste, fluorine-doped tin oxide (FTO)/glass plates, and sealing materials. This led to a blossoming of DSC research, using the N3 dye (Fig. 3) and nanocrystalline-TiO_2 electrodes made using doctor-blading methods, which resulted in the development of sandwich-type solar cells (Fig. 4).

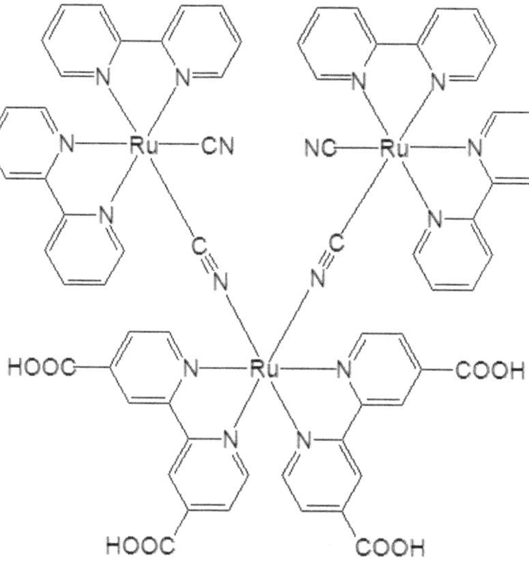

Fig. 2. A ruthenium dye reported in *Nature* (1991) by Dr. O'Regan and Prof. Grätzel [1].

Investigation of Dyes for Dye-Sensitized Solar Cells: Ruthenium-Complex Dyes, Metal-Free Dyes, Metal-Complex
Porphyrin Dyes and Natural Dyes

21

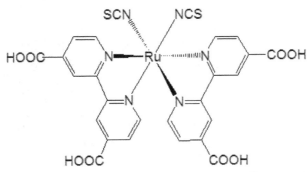

Fig. 3. The ruthenium dye N3, which achieved 10% conversion efficiencies in DSCs, reported by Dr. Nazeeruddin of the Grätzel group [2].

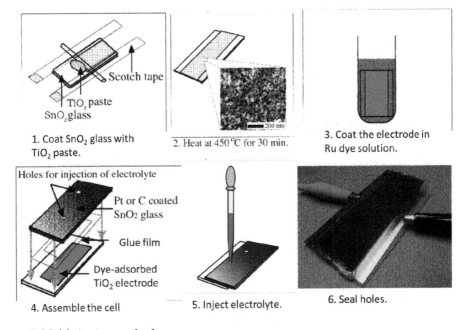

Fig. 4. DSC fabrication methods.

The high performance of the N3 sensitizer adsorbed on to nanocrystalline-TiO_2 films (Fig. 5) brought a significant advance in DSC technology. The sensitizer adsorbed on to the TiO_2 surface absorbs a photon to produce an excited state, which efficiently transfers one electron to the TiO_2 conduction band (Fig. 6). The oxidized dye is subsequently reduced by electron donation from an electrolyte containing the iodide/triiodide redox system. The injected electron flows through the semiconductor network to arrive at the back contact then through the external load to the counter electrode, which is made of platinum sputtered conducting glass. The circuit is completed by the reduction of triiodide at the counter electrode, which regenerates iodide.

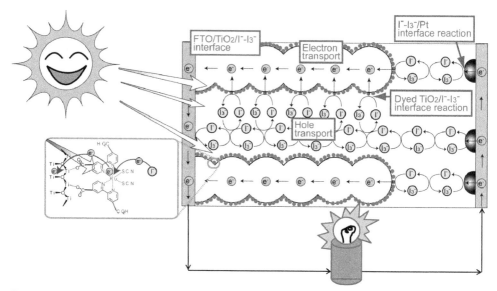

Fig. 5. Structure and electron movement in DSCs.

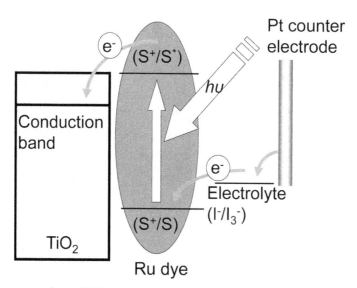

Fig. 6. Electron transfer in DSCs.

The high performance achieved by dye-sensitized TiO_2 solar cell devices depends on several factors, such as the broad range of visible light absorbed by the dye and the dye's relatively long-lived excited states with energies near those of the TiO_2 conduction band. Moreover, the presence of terminal carboxylic acid groups allows the sensitizer to be stably anchored to the semiconductor surface, ensuring high electronic coupling between the dye and the semiconductor, which is required for efficient charge injection.

Investigation of Dyes for Dye-Sensitized Solar Cells: Ruthenium-Complex Dyes, Metal-Free Dyes, Metal-Complex
Porphyrin Dyes and Natural Dyes

23

However, it was difficult to reproduce the 10% efficiency observed in the first published DSCs. In 2001, Nazeeruzzin *et al.* reported DSCs with 10.4% efficiency using a ruthenium dye called 'black dye' in the *Journal of the American Chemical Society* (Fig. 7) [3]. Although black dye looks green in solvent, on a porous nanocrystalline-TiO_2 electrode the DSC looks black, because its wide absorption band covers the entire visible range of wavelengths. The conversion efficiency was confirmed by the National Renewable Energy Laboratory in the United States. Subsequently, using black dye, Wang *et al.* (AIST, Japan) reported a 10.5% efficiency [4], and Chiba *et al.* (Sharp Co. Ltd., Japan) reported an 11.1% efficiency, confirmed by AIST [5]. In 2006, Nazeeruzzin *et al.* reported a new dye, N179, which was similar to N3, but which achieved an 11.2% conversion efficiency in DSCs (Fig. 8) [6]. N3 has four H^+ counterions, whereas N719 has three TBA^+ and one H^+ counterions (Fig. 9). The change in the counterions alters the speed of adsorption onto the porous TiO_2 electrode; N3 is fast (3 h) whereas N719 is slow (24 h), thus N719 gives a higher conversion efficiency than N3.

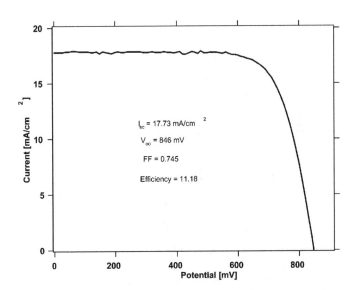

Fig. 7. Structure of black dye for DSCs [3]. TBA^+: tetrabutylammonium.

Fig. 8. Photo I-V curve for DSC with 11.1% conversion efficiency, using N719 [6].

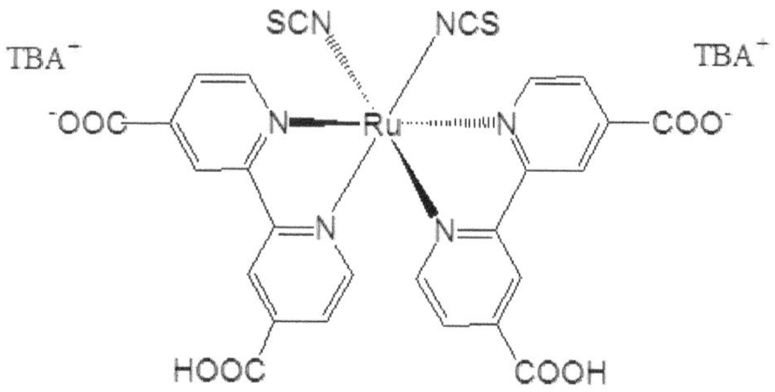

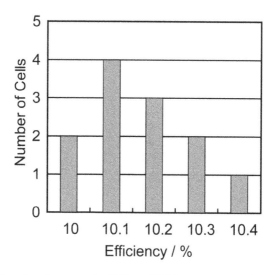

Fig. 9. A ruthenium dye (N719) for DSCs [6]. TBA$^+$: tetrabutylammonium.

In 2008, Ito *et al.* detailed DSC fabrication methods [7] that improved the reproducibility of the results for DSC conversion efficiencies (Fig. 10), and many papers have subsequently reported conversion efficiencies close to 10%. The components of high efficiency DSCs, such as the TiCl$_4$-treated photoelectrodes, the thickness of the transparent nanocrystalline-TiO$_2$-layer, the light-scattering layer (Fig. 11), and the anti-reflective film on transparent conducting oxide (TCO) substrates, have been optimized. These components have a significant effect on the conversion efficiency. TiCl$_4$ treatment is necessary for improving the mechanical strength of the TiO$_2$ layer. The thickness of the TiO$_2$ layer affects the photocurrent and the photovoltage of the devices. Furthermore, the photocurrent can also be increased by using an anti-reflective film. The combination of both transparent and light-scattering layers in a double layer system (Fig. 12) and an anti-reflective film creates a photon-trapping effect, which has been used to enhance the quantum efficiency, known as the incident photon-to-electricity conversion efficiency (IPCE).

Fig. 10. Histogram showing the reproducibility of DSC conversion efficiencies. Reported values for 12 DSC devices produced over a 24 h period. [7].

In this section, the fabrication method and the influence of different procedures on the photovoltaic performance of high-efficiency DSCs is described [7]. Two types of TiO_2 paste containing nanocrystalline-TiO_2 (20 nm) and macrocrystalline-TiO_2 (400 nm) particles (Fig. 11) were prepared, which gave transparent and light-scattering layers, respectively (Fig. 12) [8]. The synthesis of *cis*-di(thiocyanato)-N,N'-bis(2,2'-bipyridyl-4-carboxylic acid-4'-tetrabutylammoniumcarboxylate)ruthenium(II) (N-719, Fig. 9) has previously been reported [6]. The purification of N-719 was carried out by repeating the following method three times. The N719 complex was dissolved in water containing 2 equiv of tetrabutylammonium hydroxide. The concentrated solution was filtered through a sintered glass crucible, applied to a water-equilibrated Sephadex LH-20 column, and then the adsorbed complex was eluted using water. The main band was collected and the pH of the solution was lowered to 4.3 using 0.02 M HNO_3. The titration was carried out slowly over a period of 3 h and then the solution was kept at –20 °C for 15 h. After allowing the flask to warm to 25 °C, the precipitated complex was collected on a glass frit and air-dried.

(a)

(b)

Fig. 11. SEM images of the surface of TiO_2 submicrometer particles (400C, JGC-CCIC) (upper) and a mixture of TiO_2 submicrometer particles and nanoparticles (PST-400C, JGC-CCIC) (lower). Images were acquired at 50,000× magnification [8].

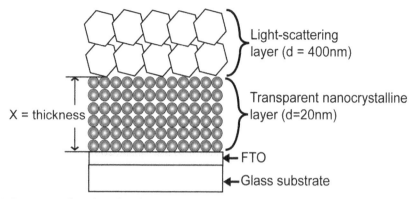

Fig. 12. Structure of DSC with a double layer of nanocrystalline-TiO$_2$ electrodes. X in Figure shows the thickness of the nanocrystalline-TiO$_2$ electrodes that was varied during the optimization of high-efficiency DSC [7, 8].

FTO glass was used as a current collector (4 mm thick, Solar, Nippon Sheet Glass). In order to prepare the DSC working electrodes, the FTO glass was first cleaned in a detergent solution using an ultrasonic bath for 15 min, and then rinsed with tap water, pure water, and ethanol. After treatment in a UV-O$_3$ system for 18 min, the FTO glass plates were immersed in a 40 mM aqueous TiCl$_4$ solution at 70 °C for 30 min, washed with pure water and ethanol and then dried. The FTO glass plate was coated with a layer of nanocrystalline-TiO$_2$ paste (anatase, d = 20 nm) by screen printing. The screen-printing procedure was repeated to get an appropriate thickness for the working electrode. After the nanocrystalline-TiO$_2$ paste was dried at 125 °C, two layers of macrocrystalline-TiO$_2$ paste (anatase, d = 400 nm) were deposited by screen printing to form a light-scattering TiO$_2$ film, 4-5 μm thick. The TiO$_2$-coated electrodes were gradually heated under an air flow at 325 °C for 5 min, at 375 °C for 5 min, at 450 °C for 15 min and at 500 °C for 15 min.

The sintered TiO$_2$ film was treated again with a 40 mM TiCl$_4$ solution as described above, rinsed with pure water and ethanol, and sintered again at 500 °C for 30 min. The TiO$_2$ electrode was allowed to cool to 80 °C, and was then immersed in a 0.5 mM acetonitrile/*tert*-butyl alcohol (1:1) solution of N-719 dye for 20-24 h at room temperature to ensure complete uptake of the sensitizer dye. The dye uptake time must be optimized for each dye, for example: indoline dyes, 4 h [9, 10]; porphyrin dyes, 1 h [11]; and natural dyes, 15 min [12].

To prepare the counter electrode, a hole was drilled in the FTO glass (2.2 mm thick, TEC 15 Nippon Sheet Glass) by sandblasting. The perforated sheet was washed with H$_2$O, and with a 0.1 M HCl solution in ethanol, and then cleaned by ultrasound in an acetone bath for 10 min. Residual organic contaminants were removed by heating in air for 15 min at 400 °C; then, the Pt catalyst was deposited on the FTO glass by coating the glass with a drop of H$_2$PtCl$_6$ solution (2 mg Pt in 1 mL ethanol) and repeating the heat treatment at 400 °C for 15 min.

The dye-covered TiO$_2$ electrode and Pt-counter electrode were assembled into a sandwich cell (Fig. 13) and sealed on a heating stage with a hot-melt gasket, made of an ionomer (25 μm thick, Surlyn 1702, DuPont). A drop of the electrolyte; a 0.60 M solution of butylmethylimidazolium iodide, 0.03 M I$_2$, 0.10 M guanidinium thiocyanate and 0.50 M 4-*tert*-butylpyridine in acetonitrile/valeronitrile (85:15 v/v) was introduced into the cell via vacuum backfilling. The cell was placed in a vacuum, and subsequent exposure to ambient pressure pushed the electrolyte into the cell. Finally, the hole was covered by a hot-melt ionomer film (35 μm thick, Bynel 4164, Du-Pont) and a cover glass (0.1 mm thick), and sealed with a hot soldering iron.

The electrolyte must be optimized for each dye: black dye is used with 0.6 M dimethyl propyl imidazolium iodide, 0.1 M lithium iodide, 0.05 M iodine, and 0.5 M *tert*-butylpryidine in acetone [5]; D149, D205, and Monascus yellow (a natural dye) are used with 0.10 M lithium iodide, 0.60 M butylmethylimidazolium iodide, 0.05 M I_2, and 0.05 M 4-*tert*-butylpyridine in acetonitrile/valeronitrile (85:15) [9, 10, 12]; YD-2 (a porphyrin dye) is used with 1.0 M 1,3-dimethylimidazolium iodide, 0.03 M iodine, 0.5 M *tert*-butylpyridine, 0.05 M LiI, 0.1 M guanidinium thiocyanate, in acetonitrile/valeronitrile (85:15 v/v) [13].

Photovoltaic measurements were taken using an AM 1.5 solar simulator (100 mW cm^{-2}). The power of the simulated light was calibrated using a reference Si photodiode equipped with an infrared (IR) cut-off filter in order to reduce the mismatch between the simulated light and the AM 1.5 spectrum in the 350-750 nm region to less than 2% [14]. The current-voltage (I-V) curves were obtained by applying an external bias to the cell and measuring the photocurrent with a digital source meter.

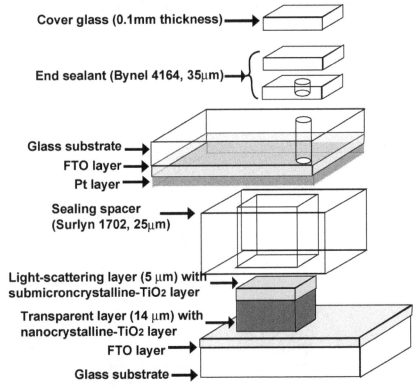

Fig. 13. Structure of DSCs.

In order to demonstrate the effect of the sensitizing dye on the photovoltaic performance, two types of TiO_2 working electrode were prepared on the FTO/glass substrate (Fig. 14)

[15]. The first type of working electrode, nano-TiO_2, is a double layer of mesoporous TiO_2 (TiO_2 nanoparticles d = 20 nm; nanocrystalline TiO_2 layer, 14 µm; microcrystalline TiO_2 layer, 4 µm) screen-printed onto the FTO. The second type of electrode, UL/nano-TiO_2, used a compact TiO_2 underlayer (UL) deposited by spray pyrolysis between the porous TiO_2 layer and the FTO. Figure 14 shows the dark I-V characteristics of the two types of mesoscopic-TiO_2 electrodes, with and without ruthenium dye. The onset of the dark current in the nano-TiO_2 electrode occurred at a low forward bias. The compact TiO_2 UL suppresses the dark current, shifting its onset by several hundred millivolts. This indicates that the triiodide reduction in the exposed part of the FTO layer is responsible for the high dark current observed in the nanocrystalline TiO_2 film alone. Adsorption of the N-719 dye onto the nano-TiO_2 electrode also suppresses the dark current (Fig. 15), indicating that the ruthenium sensitizer itself worked as an effective blocking layer on the FTO layer. In contrast, the dark-current curves of UL/nano-TiO_2 were shifted to slightly lower voltages by the adsorption of the N-719 dye (Fig. 15) indicating that the sensitizer increases the dark current on electrodes where the FTO surface is already blocked. This can be attributed to the TiO_2 band shifting to positive values by surface protonation; the protons can be supplied by the ruthenium dye. The photovoltaic results are shown in Figure 16 and confirm the trends observed in the dark currents. The dye loaded nanocrystalline TiO_2 film alone gave a lower conversion efficiency (Fig. 16, nano-TiO_2/Ru-dye). Introducing the compact TiO_2 UL in the nano-TiO_2/Ru-dye electrode increased the open-circuit photovoltage (V_{OC}) by 27 mV and the short-circuit photocurrent density (J_{SC}) by 1 mA cm^{-2}. The difference between the nano-TiO_2/Ru-dye and UL/nano-TiO_2/Ru-dye electrodes arose from the UL suppressing the charge recombination at the FTO layer. Mathematical modeling of charge-recombination carried out by Ferber et $al.$ (Fig. 17) [16] shows good agreement with these I-V curves. Therefore, the observed improvement of the V_{OC} and J_{SC} from using a UL on the FTO layer agrees with the theoretical calculations. The suppression of dark current is enhanced by introducing a compact layer between the FTO and the TiO_2 nanocrystals, and leads to an increase in the V_{OC}. However, the performance of DSCs with spray-pyrolyzed TiO_2 ULs is less reproducible; to avoid this problem, $TiCl_4$ treatment between the FTO and nanocrystalline-TiO_2 layers is used instead (Fig. 10) [7].

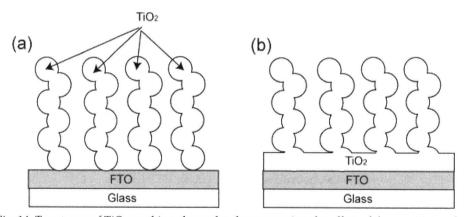

Fig. 14. Two types of TiO_2 working electrodes demonstrating the effect of the sensitizing dye on the photovoltaic results: (a) nano-TiO_2 and (b) UL/nano-TiO_2 [15].

Investigation of Dyes for Dye-Sensitized Solar Cells: Ruthenium-Complex Dyes, Metal-Free Dyes, Metal-Complex
Porphyrin Dyes and Natural Dyes

29

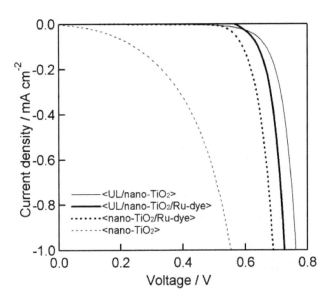

Fig. 15. Dark current-voltage characteristics of the mesoscopic TiO₂ electrodes shown in Fig.
14 in sandwich cells, with and without adsorbed ruthenium dye. The counter electrode was
Pt-coated FTO [15].

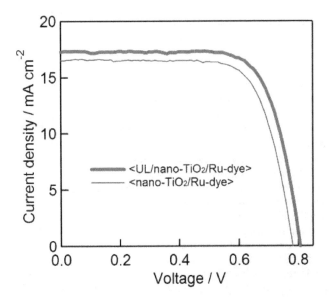

Fig. 16. Photovoltage-current curves for DSCs with two types of electrodes (shown in Fig.
14) under a solar simulator (AM 1.5, 100 mW cm⁻²) [15].

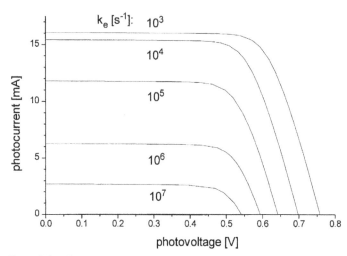

Fig. 17. The effect of the electron recapture rate constant k_e on the I-V curve of the modeled DSC. Base case parameters [16].

It was found that the V_{OC} was shifted by modifying the number of protons in the ruthenium dye (Fig. 18) [17]. When dyes that contain protonated carboxylic groups are adsorbed, the anchoring groups transfer most of their protons to the TiO_2 surface, and the positive charge of the surface shifts the Fermi level in a positive direction. The electric field, which is associated with the surface dipole generated by the positive charge, enhances the dye adsorption and assists electron injection from the sensitizer's excited state into the TiO_2 conduction band, thus increasing photocurrents. However, the positive shift in the Fermi level decreases the gap between the iodide/triiodide redox couple and the Fermi level, resulting in a lower open-circuit potential. In contrast, adsorption of a sensitizer that contains no protons shifts the Fermi level in a negative direction, leading to a higher value for the open-circuit potential, while the value of the short circuit current is low. Therefore, there is an optimum degree of protonation for the sensitizer, where the product of the short circuit photocurrent and the open circuit potential is high, thus maximizing the power conversion efficiency of the cell. Varying the degree of protonation of the sensitizer, however, also changes its electronic structure; therefore it is important to investigate how the energy and composition of the excited states change as a function of the protonation of the terminal carboxylic acid groups.

In order to enhance the photocurrent of DSCs, the mesoscopic surface area of the nanocrystalline-TiO_2 photoelectrode has been improved by chemical bath deposition of TiO_2 from $TiCl_4$ (Fig. 19) [7]. BET measurements confirmed that the surface area was increased by 20%. Moreover, a photon-trapping system has been applied to porous TiO_2 electrodes using double-layer system consisting of transparent and light-scattering layers (Fig. 12) [7, 8]. Figure 20 shows the photovoltaic characteristics of the DSC were improved by the $TiCl_4$ treatment and the double layer system (J_{SC} = 18.2 mA cm^{-2}, V_{OC} = 789 mV, FF = 0.704, and η = 10.1%). It has previously been reported that the light-scattering layer is important not only for the photon-trapping system, but also for photovoltaic generation. DSCs with the dye-sensitized light-scattering-TiO_2 layer, but without the transparent nanocrystalline-TiO_2 layer, gave a conversion efficiency of 5% [18].

Investigation of Dyes for Dye-Sensitized Solar Cells: Ruthenium-Complex Dyes, Metal-Free Dyes, Metal-Complex
Porphyrin Dyes and Natural Dyes

31

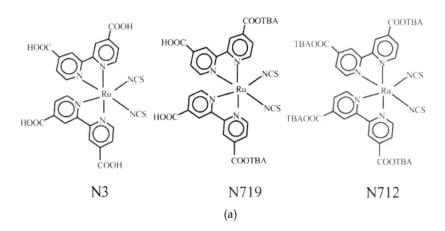

(a)

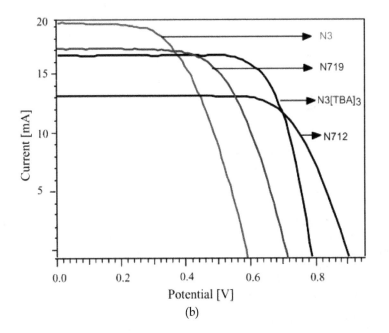

(b)

Fig. 18. (a) Structures of ruthenium dyes, and (b) the effect of dye protonation on photocurrent-voltage characteristics of nanocrystalline TiO_2 cell sensitized with N3 (4 protons), N719 (2 protons), N3[TBA]$_3$ (1 proton), and N712 (0 protons) dyes, measured under AM 1.5 sun using 1 cm^2 TiO_2 electrodes with an I$^-$/I$_3^-$ redox couple in methoxyacetonitrile [17].

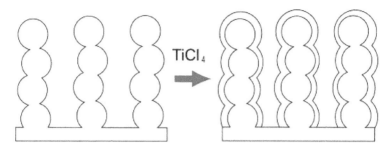

Fig. 19. Effect of the TiCl$_4$ treatment. An additional TiO$_2$ layer (1 nm thick) was coated on the surface of the nanocrystalline TiO$_2$ porous film [7].

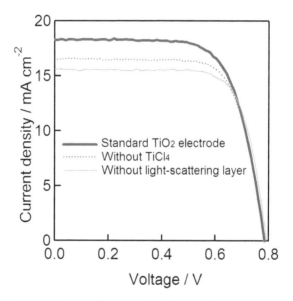

Fig. 20. Differences in the I-V curves from standard TiO$_2$ electrodes, electrodes with no TiCl$_4$ treatment and electrodes with no light-scattering layer. The transparent and light-scattering layers were 14 μm and 5 μm thick, respectively. Photovoltaic characteristics: standard TiO$_2$ electrode, J_{SC} = 18.2 mA cm^{-2}, V_{OC} = 789 mV, FF = 0.704 and η = 10.1%; without TiCl$_4$, J_{SC} = 16.6 mA cm^{-2}, V_{OC} = 778 mV, FF = 0.731 and η = 9.40%; without light-scattering layer, J_{SC} = 15.6 mA cm^{-2}, V_{OC} = 791 mV, FF = 0.740 and η = 9.12% [7].

Photoconversion happens on the surface of the dye-covered TiO$_2$ layer and the surface area can be calculated from the thickness of the porous layer. Therefore, in order to optimize the photovoltaic performance of DSCs, it is important to understand the relationship between the thickness of the nanocrystalline-TiO$_2$ layer and the conversion efficiency of the DSC (Fig. 21) [7, 11]. A thickness of around 14 μm was confirmed as the optimum for DSCs using N719 dye. Hence, the total optimum thickness of the TiO$_2$ layer, consisting of the transparent layer (14 μm) and the light-scattering layer (5 μm), was around 19 μm for N719.

However, a total thickness of 32 μm was necessary for DSCs using black dye, because it has
a lower photo-absorbance coefficient than N719 [4].

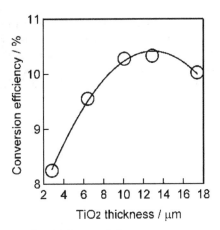

Fig. 21. Relationship between the thickness of the transparent nanocrystalline-TiO₂ layer (in
Fig. 12) and the conversion efficiency of DSCs with anti-reflective films. Each point is the
average of four cells [7].

Because glass substrates reflect 8-10% of the incident light, an anti-reflective film is
necessary to enhance the photovoltaic performance of DSCs. The light-reflecting losses were
eliminated by a self-adhesive fluorinated polymer film (Arktop, Asahi Glass) that also
served as a 380 nm UV cut-off filter. Masks made of black plastic tape were attached to the
Arktop filter to reduce scattered light. Figure 22 shows the IPCE of an electrode with anti-
reflective film compared to the electrode without. The anti-reflective film enhances the IPCE
from 87% to 94%, increasing the conversion efficiency by 5%.

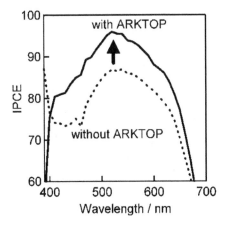

Fig. 22. Effect of anti-reflective film (Arktop) on the IPCE of DSC. A double-layer electrode
(14 μm thick transparent and 5 μm light-scattering TiO₂ layers) was used.[7].

Optimizing the thickness of the nanocrystalline-TiO$_2$ layer, TiCl$_4$ treatments, and the anti-reflective film have allowed DSCs with conversion efficiencies of over 10% to be fabricated. Figure 10 shows the photoconversion efficiency of DSCs made at the same time; among the 12 devices, the reproducibility of DSCs with a conversion efficiency of over 10% was 100%. The error in the experimental results falls within the measurement error for a solar simulator, and therefore conversion efficiencies of 10.2 ± 0.2% is highly reproducible. Using the above technique, DSCs with efficiencies of 11.3% [19] and 12.3% [20] containing the dyes C101 (Fig. 23) and Z991 (Fig. 24), respectively, have recently been published by the Grätzel group, which are the highest conversion efficiencies published to date.

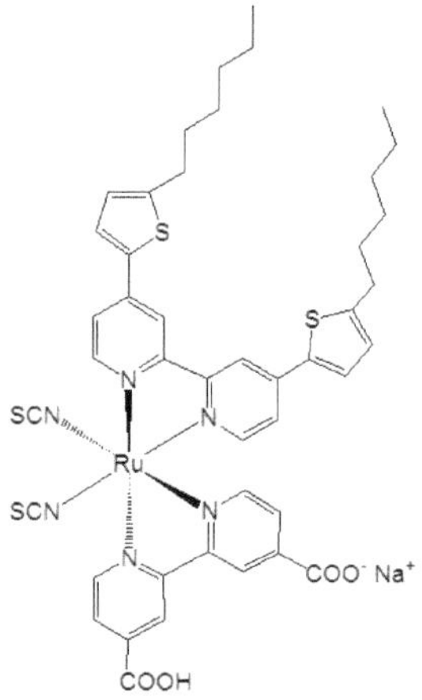

Fig. 23. The structure of the ruthenium dye C101, which achieved an 11.3% conversion efficiency in DSCs [19].

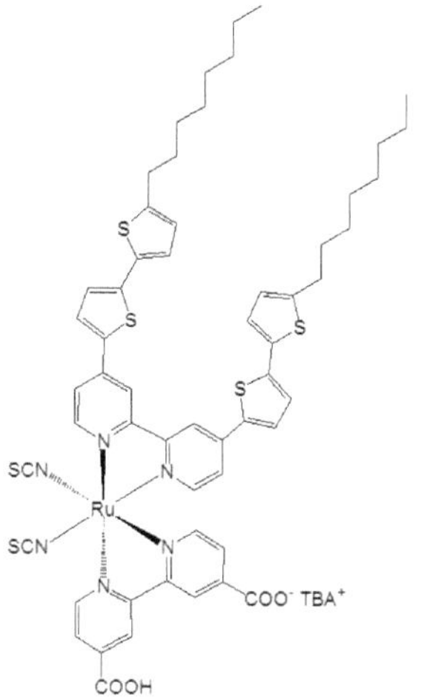

Fig. 24. The structure of the ruthenium dye Z991 which achieved a conversion efficiency of 12.3% in DSCs [20].

3. Metal-free dyes

Ruthenium complex dyes are capable of delivering DSCs with high photoenergy-conversion efficiencies. However, because ruthenium is a rare and expensive metal ruthenium dyes are not suitable for cost-effective, environmentally friendly photovoltaic systems. This limits the range of applications for these complexes, and makes the development of DSCs that use metal-free, organic dyes essential for their practical use. Recently, numerous organic dyes for high-efficiency DSCs have been reported. New organic dyes with efficiencies over 5% include hemicyanine dye (Fig. 25) (η = 5.1%) [21], polyene-diphenylaniline dye (Fig. 26) (η = 5.1%) [22], thienylfluorene dye (Fig. 27) (η = 5.23%) [23], phenothiazine dye (Fig. 28) (η = 5.5%) [24], thienothiophene-thiophene-derived dye (Fig. 29) (η = 6.23%) [25], phenyl-conjugated polyene dye (η = 6.6%) (Fig. 30) [26], N,N-dimethylaniline-cyanoacetic acid (Fig. 31) (η = 6.8%) [27, 28], oligothiophene dye (Fig. 32) (η = 7.7%) [29], coumarin dye (Fig. 33) (η = 8.2%) [30], indoline dye (Fig. 34) (η = 9.03%) [9, 31] and oligo-phenylenevinylene-unit dye (Fig. 35) (η = 9.1%) [32].

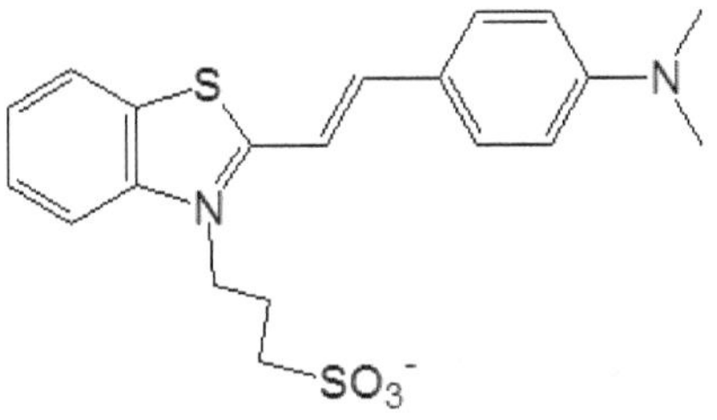

Fig. 25. Hemicyanine dye [21].

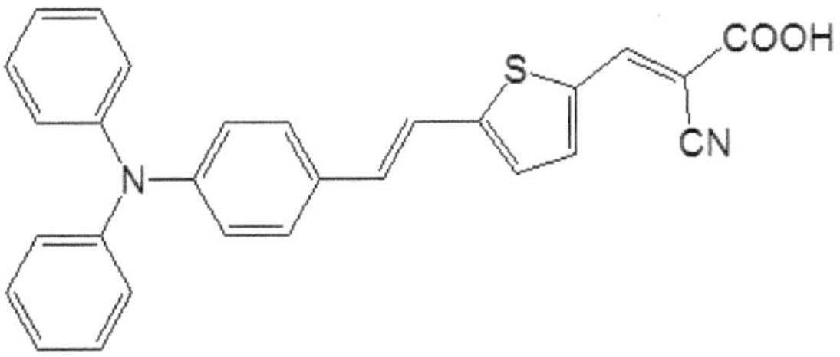

Fig. 26. Polyene-diphenylaniline dye [22].

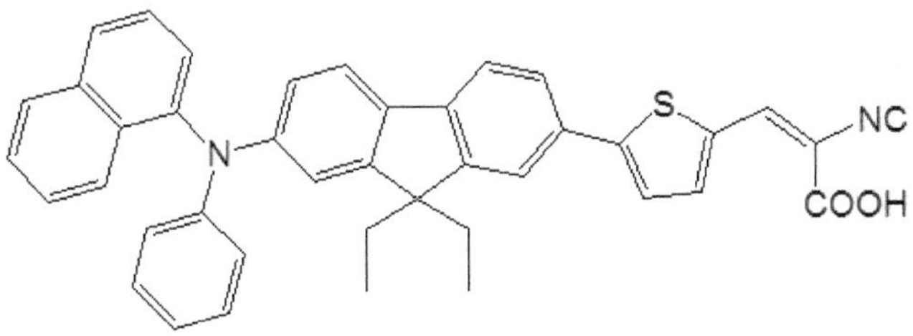

Fig. 27. Thienylfluorene dye [23].

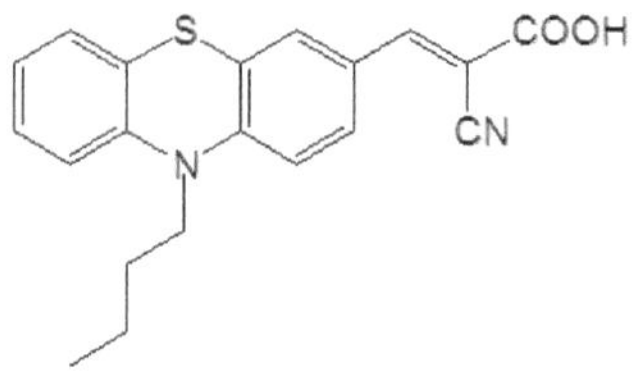

Fig. 28. Phenothiazine dye [24].

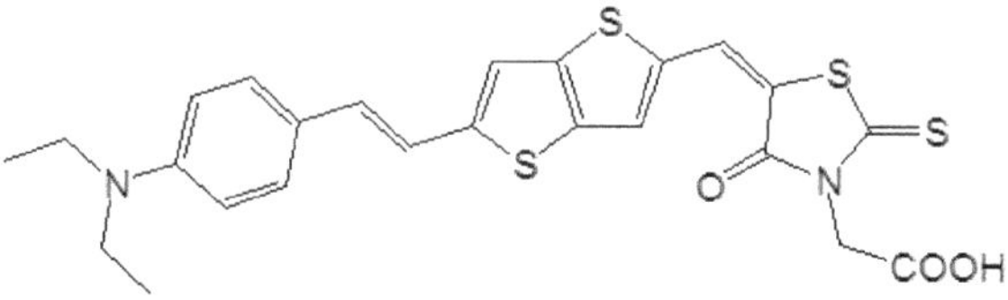

Fig. 29. Thienothiophene-thiophene-derived dye [25].

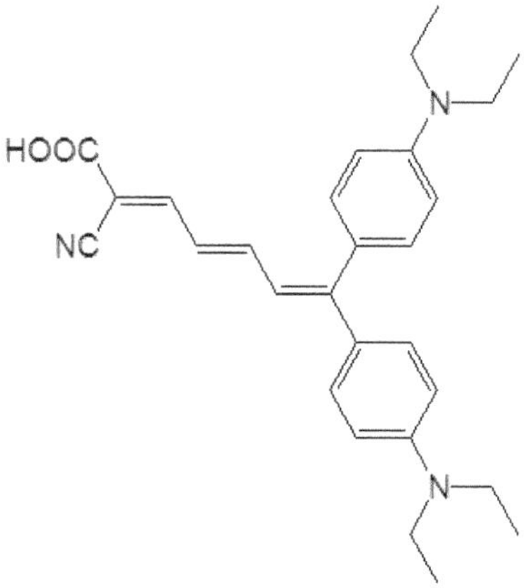

Fig. 30. Phenyl-conjugated polyene dye [26].

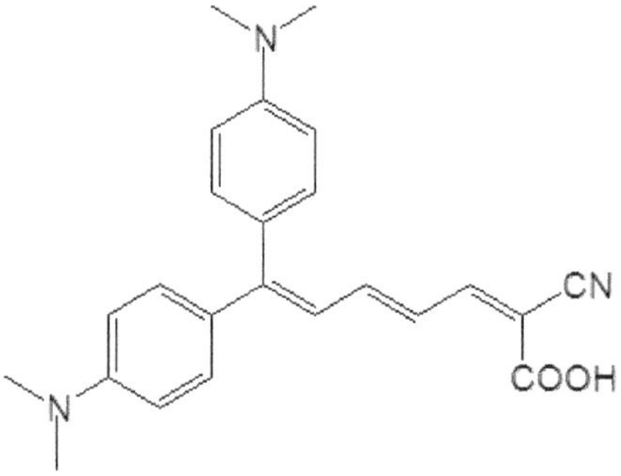

Fig. 31. *N,N*-dimethylaniline-cyanoacetic acid [27, 28].

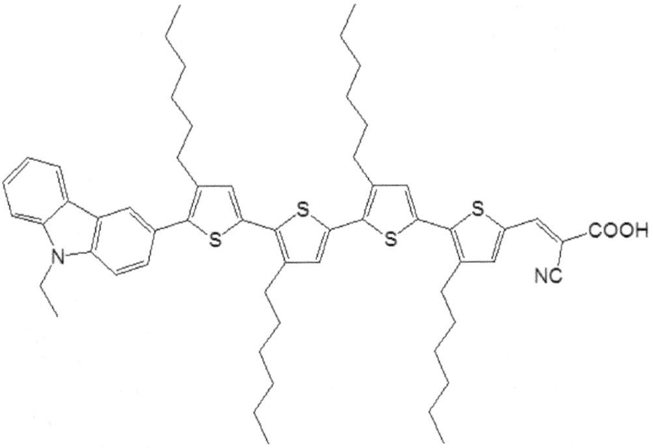

Investigation of Dyes for Dye-Sensitized Solar Cells: Ruthenium-Complex Dyes, Metal-Free Dyes, Metal-Complex
Porphyrin Dyes and Natural Dyes

37

Fig. 32. Oligothiophene dye [29].

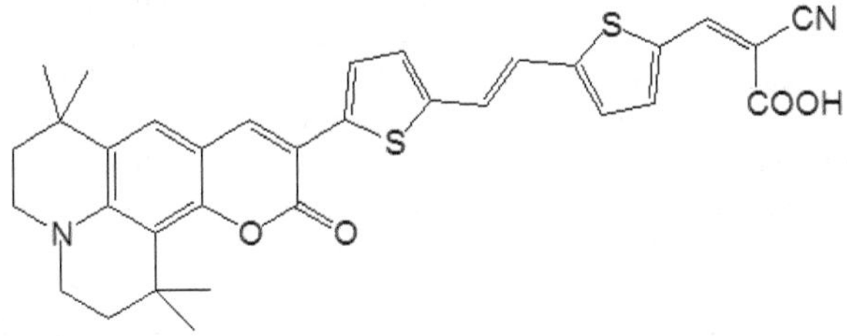

Fig. 33. Coumarin dye [30],

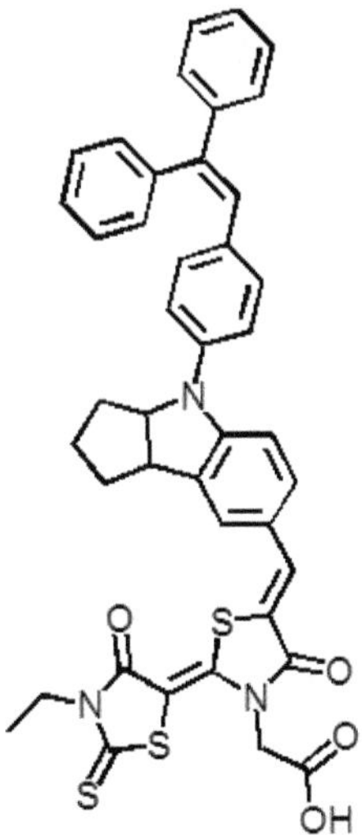

Fig. 34. Indoline dye (D149) [31].

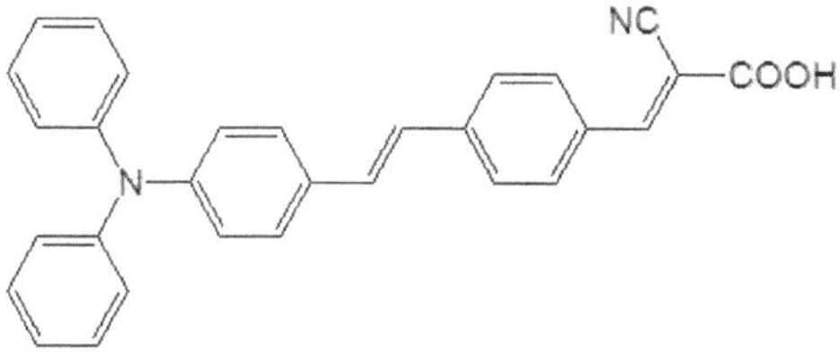

Fig. 35. Oligo-phenylenevinylene-unit dye [32].

In order to improve the conversion efficiency values, the structure of organic dye photosensitizers needed to be altered. For example, controlling the aggregation of dye molecules improves the photocurrent generation; π-stacked aggregation (D and H aggregation) on the nanocrystalline-TiO$_2$ electrodes should normally be avoided. Aggregation may lead to intermolecular quenching and the presence of molecules that are not functionally attached to the TiO$_2$ surface which act as filters. Some ruthenium complexes (black dye and N719) have shown their best results using chenodeoxycholic acid (CDCA), which functions as an anti-aggregation reagent and improves the photovoltaic effect. However, indoline dyes and coumarin dyes form photoactive aggregates on nanocrystalline-TiO$_2$ electrodes for DSCs, in a process known as J-aggregation.

In order to control the aggregation between dye molecules, an indoline dye with an n-octyl substituent on the rhodanine ring of D149 (Fig. 34) was synthesized, to give dye D205 (Fig. 36) [33]. Figure 37 shows the photovoltaic characteristics of DSCs using D149 and D205. Table 1 shows that n-octyl substitution increased the V_{OC} regardless of whether CDCA was present. CDCA increased the V_{OC} of D205 by approximately 0.054 V, but had little effect on D149, which only showed an increase of 0.006 V. The combination of CDCA and the n-octyl chain (D205) significantly improved the V_{OC} by up to 0.710 V, which is 0.066 V higher (by 10.2%) than that of D149 with CDCA. Kroeze et al. [34] showed that the alkyl substitution of dyes improved the V_{OC}, because of the blocking effect on the charge recombination between triiodide and electrons injected in the nanocrystalline-TiO$_2$ electrodes. Therefore, the V_{OC} variation observed in Figure 37 indicates that the charge recombination was impeded by the blocking effect, arising from the combination of the n-octyl chain and CDCA.

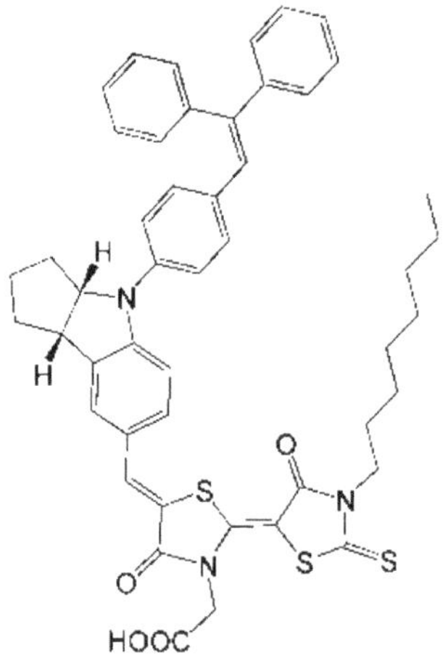

Fig. 36. Indoline dye (D205) [33].

Investigation of Dyes for Dye-Sensitized Solar Cells: Ruthenium-Complex Dyes, Metal-Free Dyes, Metal-Complex
Porphyrin Dyes and Natural Dyes

39

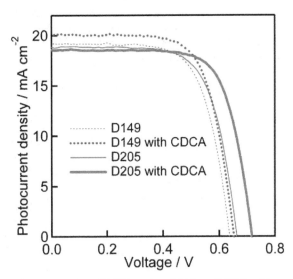

Fig. 37. Photovoltaic characteristics of DSCs using D149 and D205 with or without CDCA
[33].

Photovoltaic parameter	Without CDCA		With CDCA	
	D149	D205	D149	D205
Jsc / mA cm^{-2}	19.08±0.26	18.99±0.19	19.86±0.10	18.68±0.08
Voc / V	0.638±0.05	0.656±0.11	0.644±0.13	0.710±0.07
FF	0.682±0.06	0.678±0.09	0.694±0.06	0.707±0.09
η / %	8.26±0.09	8.43±0.16	8.85±0.18	9.40±0.12

Table 1. Photovoltaic characteristics of DSCs with indoline dyes shown in Figure 34 and 36.
Each result was obtained from three DSCs [33].

Without CDCA, the variation in J_{SC} arising from n-octyl substitution on the rhodanine ring
was small (0.5% of J_{SC}). However, in the presence of CDCA, the effect of the n-octyl chain
was significant; the substitution of the n-octyl chain (from D149 to D205) with CDCA
decreased J_{SC} by 5.9%. The effect of n-octyl substitution and CDCA on the FF was similarly
small; without CDCA, the n-octyl substitution decreased the FF by 0.6%, and with CDCA,
the n-octyl substitution increased the FF by 1.9%.

Without CDCA, the increase in conversion efficiency from D149 to D205 was only by 2.1%,
and in the presence of CDCA, the increase was by 6.2%. The resulting average conversion
efficiency for D205 with CDCA was an outstanding 9.40% (Table 2). The highest conversion
efficiency value of 9.52% was achieved with a DSC based on D205 (J_{SC}: 18.56 mA cm^{-2}, V_{OC}:
0.717 V, and FF: 0.716). Reproducible efficiencies from 9.3% to 9.5% were obtained with the
D205 solar cell.

Recently, the group of Prof. Peng Wang has reported the synthesis of organic dyes with
conversion efficiencies in DSCs that rival those of ruthenium dyes; the highest value for a

ruthenium dye is 12.3%, under the same measurement conditions [20]. Dye C217 (Figure 38) produced DSCs with 9.8% conversion efficiency, [35] and dye C219 (Figure 39) achieved a 10.1% conversion efficiency [36]. These results strongly suggest that utilizing organic dye photosensitizers is a promising approach for producing high-performance, low cost, recyclable DSCs.

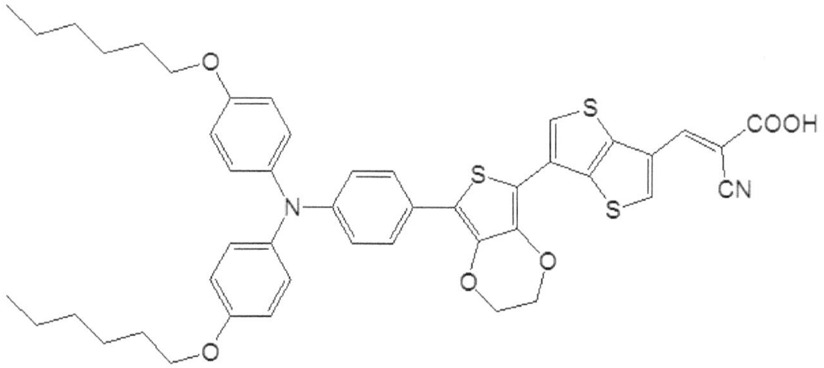

Fig. 38. C217 dye [35].

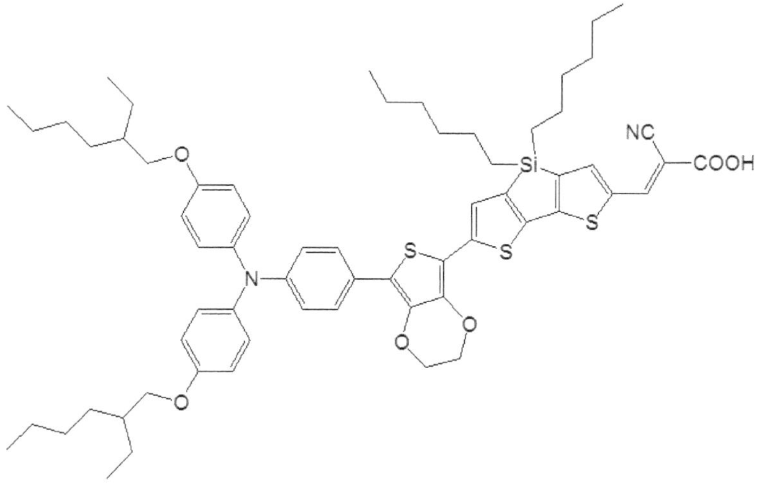

Fig. 39. C219 dye [36].

4. Metal-complex porphyrin dye

A further strategy for avoiding the use of expensive ruthenium in DSC dyes is to use complexes containing inexpensive metals. Large π-aromatic molecules, such as porphyrins and phthalocyanines, are attractive potential candidates for thin, low-cost, efficient DSCs, because of their photostability and high light-harvesting capability. Porphyrins show strong absorption and emission in the visible region, as well as tunable redox potentials. These

Investigation of Dyes for Dye-Sensitized Solar Cells: Ruthenium-Complex Dyes, Metal-Free Dyes, Metal-Complex
Porphyrin Dyes and Natural Dyes

41

properties mean they have many potential applications, in areas such as optoelectronics, catalysis, and chemosensing. Self-assembled porphyrin molecular structures play a key role in solar energy research, because the photosynthetic systems of bacteria and plants contain chromophores based on porphyrins, which efficiently collect and convert solar energy into chemical energy. Various artificial photosynthetic model systems have been designed and synthesized in order to elucidate the factors that control the photoinduced electron-transfer reaction. Inspired by efficient energy transfer in naturally occurring photosynthetic reaction centers, numerous porphyrins and phthalocyanines have been synthesized and tested in DSCs.

Campbell et al. have reported zinc porphyrin dyes (Fig. 40) which have conversion efficiencies of 7.1% [37]. A recently reported series of zinc porphyrin dyes with donor-acceptor (D–A) substituents exhibit promising photovoltaic properties with a conversion efficiency of 6.8% (YD-2, Fig. 41) [38]. Bessho et al. optimized the fabrication method for YD-2-sensitized DSCs, resulting in the achievement of an 11% solar-to-electric power conversion efficiency under standard conditions (AM 1.5G, 100 mW cm^{-2} intensity) (Fig. 42) [13], which is the highest conversion efficiency for a DSC using a ruthenium-free dye so far.

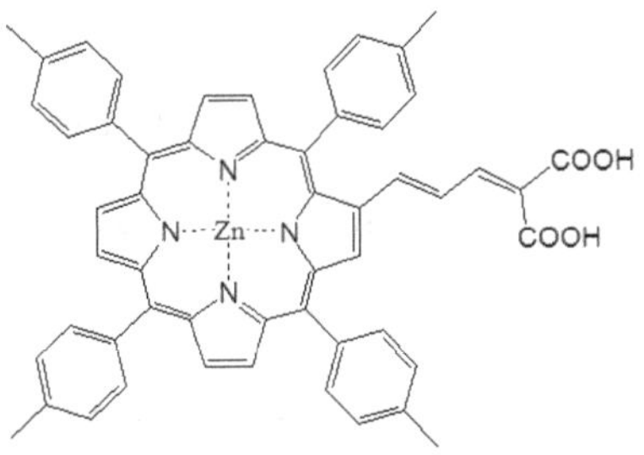

Fig. 40. Zinc porphyrin dyes by synthesized by Campbell et al. [37]

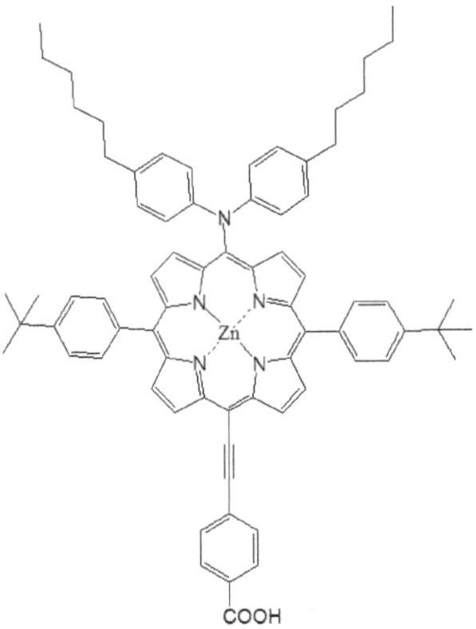

Fig. 41. Structure of porphyrin dye YD-2 [38].

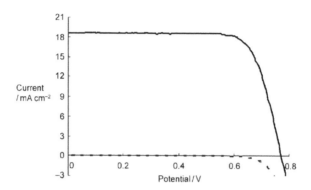

Fig. 42. Photocurrent density–voltage (J–V) characteristics of a device using YD-2 as sensitizer under AM 1.5G illumination (100 mWcm²). Values for dark current (solid line) and 100% sun (dotted line) are shown [13].

5. Natural dyes

Although high performance synthetic ruthenium-free dyes have been developed, their synthesis is time-consuming and laborious. Furthermore, they must be tested for toxicity before they can be used commercially. These problems could be solved if inexpensive, non-toxic, natural dyes, such as pigments used in food coloring, could be used in DSCs. Natural dyes are easily and safely extracted from plants, which means they are cheap and widely available, and do not require complex synthesis or toxicity testing. Therefore, the use of natural dyes is important for the development of cheap, commercially available DSCs. Natural dyes have shown moderate energy conversion efficiencies in DSCs [39-45]; natural chlorophyll dyes achieved energy conversion efficiencies of over 4% (Fig. 43) [39, 40]. However, despite the huge range of natural dyes, most of the other dyes tested yielded energy conversion efficiencies below 2%, although some derivatives synthesized from natural dyes have produced energy conversion efficiencies of over 2% [46, 47]).

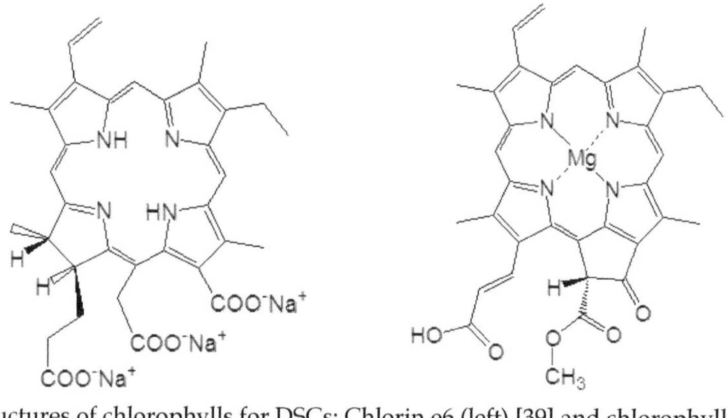

Fig. 43. Structures of chlorophylls for DSCs: Chlorin e6 (left) [39] and chlorophyll c (right) [40].

Investigation of Dyes for Dye-Sensitized Solar Cells: Ruthenium-Complex Dyes, Metal-Free Dyes, Metal-Complex Porphyrin Dyes and Natural Dyes

43

The natural dye Monascus yellow produces DSCs with over 2% conversion efficiency [12]. It is extracted from Monascus (red yeast rice), which is the product of *Monascus purpureus* fermentations. Monascus is a dietary staple in some Asian countries, and is traditionally made by inoculating rice soaked in water with *Monascus purpureus* spores. The mixture is incubated at room temperature for 3–6 days, and its core turns bright red and the outside turns reddish purple. Because of the low cost of cultivation, some producers have extracted the highly colored fermentation products to use as food pigment dyes. Monascus red, which is one of the red dyes in Monascus fermentations, has been used as a sensitizing dye for DSCs [41], which gave a 0.33% conversion efficiency. Monascus yellow, which is also extracted from Monascus fermentations (Fig. 44), was used as a novel sensitizer. The DSC with Monascus yellow achieved a photovoltaic performance where $J_{SC} = 6.1$ mA cm^{-2}, $V_{OC} = 0.57$ V, $FF = 0.66$, and $\eta = 2.3\%$ (Fig. 45).

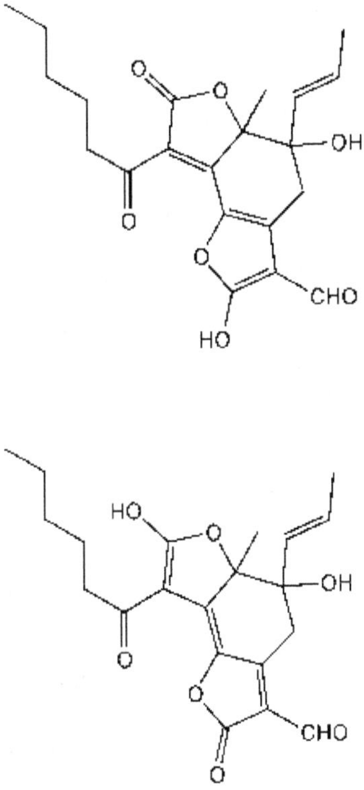

Fig. 44. Structures of dye molecules in Monascus yellow, supplied as a mixture of two isomers in an extract from Monascus fermentations [12].

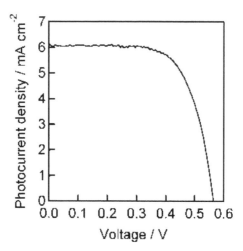

Fig. 45. I-V curve for DSC using Monascus yellow dye [12].

Figure 46 shows the UV–VIS absorption spectra and IPCE spectra of Monascus yellow on transparent nanocrystalline-TiO_2 electrodes (prepared using a TiO_2 paste, PST-18NR, CCIC, Japan) [12]. Without acetic acid treatment, the absorption peak at 426 nm was just 8% and the IPCE value was close to zero. After acetic acid treatment, the absorption peak increased to 69%, resulting in the IPCE value increasing to 47% at 450 nm. The action spectrum of the cell sensitized by Monascus yellow largely agrees with the absorption spectrum of the dye adsorbed to the TiO_2 film. The IPCE peak is red-shifted by 24 nm relative to the absorption peak. The strong absorption of blue light of the iodide/triiodide electrolyte is thought to decrease the IPCE value in the short wavelength region, resulting in the red shift of the peak in the action spectrum. The remarkable improvement in the IPCE value following treatment with acetic acid may be because the acetic acid promotes bonding between the Monascus yellow hydroxyl groups and the surface of the nanocrystalline-TiO_2 film. In addition, sensitizing dyes adsorbed onto TiO_2 surfaces are known to be desorbed by addition of bases such as NaOH and NH_3. Thus, the improvement of the photovoltaic performance upon addition of acetic acid can be attributed to a chemical bonding adsorption mode, in contrast to a physical (van der Waals) adsorption mode without acetic acid. When the dye is physically adsorbed onto the TiO_2 electrode, it cannot inject photoexcited electrons, resulting in a small absorption peak. In contrast, the dye adsorbed onto the TiO_2 surface with protons strengthening the chemical bonding, can efficiently inject photoexcited electrons from the dye into the electrode.

The effect of the nanocrystalline-TiO_2 layers, and the solvents used for rinsing them following dye uptake, were examined (Table 2) [12]. After rinsing with water, the conversion efficiency of the P25-based cell was higher than that of the PST-18NR-based cell. However, after rinsing with other solvents, the PST-18NR-based cell yielded higher conversion efficiencies than the P25-based cell, which arose from the difference in J_{SC}. This effect can be attributed to the surface area and pore size, which is related to the particle size. Generally, small TiO_2 particles fused together form a large surface area, which allows more dye to be loaded onto the surface. However, the small pore size prohibits the smooth transportation of redox species in the electrolyte solution, which fills the pore space. The

Investigation of Dyes for Dye-Sensitized Solar Cells: Ruthenium-Complex Dyes, Metal-Free Dyes, Metal-Complex
Porphyrin Dyes and Natural Dyes

45

TiO$_2$ particle size in PST-18NR is smaller than that of P25, thus the amount of adsorbed molecules onto the PST-18NR surface should be higher than that on the P-25 surface. This rationalizes the higher photocurrent density from the PST-18NR-based cell compared with the P25-based cell. The TiO$_2$ particles of PST-18NR have a smaller pore size compared with P25. Therefore, more water molecules, which have a high boiling point, may remain inside the smaller pores of the PST-18NR-based electrode and inhibit the electrolyte diffusion into the pores. In contrast, the other volatile solvents evaporate easily after rinsing, leading to the high photocurrent density and thus high energy conversion efficiency. The values of the V_{OC} and FF did not show significant differences between the two electrodes.

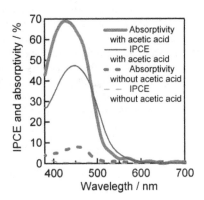

Fig. 46. UV-Vis absorption and IPCE spectra of Monascus yellow-sensitized TiO$_2$ electrodes showing the effect of acetic acid in the dye solution. The absorption spectra were measured using a transparent nanocrystalline-TiO$_2$ electrode. IPCE spectra were measured using a double-layered nanocrystalline-TiO$_2$ electrode. Each electrode was dipped in Monascus yellow solution with or without acetic acid, for 15 min. [12].

TiO$_2$ material	Rinsing solvent	J_{SC} / mA cm^{-2}	V_{OC} / V	FF	η / %
P25	water	3.50±0.30	0.537±0.021	0.697±0.018	1.29±0.04
	ethanol	3.40±0.00	0.504±0.009	0.653±0.011	1.14±0.01
	acetonitrile	3.12±0.04	0.535±0.019	0.714±0.009	1.19±0.01
	methanol	4.00±0.04	0.522±0.000	0.624±0.009	1.34±0.03
	acetone	3.72±0.04	0.504±0.009	0.692±0.009	1.39±0.02
PST-18NR	water	2.65±0.05	0.514±0.020	0.725±0.019	0.99±0.00
	ethanol	5.50±0.14	0.524±0.009	0.666±0.006	1.93±0.05
	acetonitrile	4.60±0.20	0.506±0.016	0.694±0.014	1.58±0.02
	methanol	5.54±0.22	0.547±0.002	0.633±0.002	1.92±0.00
	acetone	4.32±0.07	0.522±0.014	0.666±0.026	1.51±0.01

Table 2. Photovoltaic characteristics of DSCs using Monascus with different TiO$_2$ electrodes and rinsing solvents; short-circuit photocurrent density: J_{SC}, open-circuit photovoltage: V_{OC}, fill factor: FF, and energy conversion efficiency: η. Data were obtained using three independent measurements [12].

6. Conclusions and outlook

The main factors that affect the operation of DSCs have been discussed. The acidity of dye solution can be a critical factor for the uptake of the dye; for ruthenium dyes, the addition of TBA-OH was important for controlling the dye-uptake speed, whereas for natural dyes, the addition of acetic acid was necessary for adsorption onto nanocrystalline-TiO_2 electrodes. The resulting dye mono-layer on the nanocrystalline-TiO_2 electrodes can act as a blocking layer that maintains charge separation. Dye aggregation, which is a key factor in obtaining high-efficiency organic DSCs, can be controlled by combining substitution of the alkyl chain on the dye with the addition of CDCA. The thickness of the nanocrystalline-TiO_2 electrodes also needs to be optimized for each dye, because of the difference in the light-absorption coefficients. Screen printing methods are the most suitable for controlling the thickness of nanocrystalline-TiO_2 electrodes; the layers can be easily positioned and built up to the desired thickness. An anti-reflective film and a light-scattering TiO_2 layer on the transparent TiO_2 layer allows the incident light to be absorbed effectively, which enhances the photocurrent. The highest conversion efficiency achieved by DSCs was 12.3%, and used a ruthenium dye (Z991).

Replacing ruthenium complexes with fully organic sensitizers or complexes containing inexpensive metals is an attractive strategy for producing low cost, environmentally friendly DSCs. Ruthenium-free dyes are already producing excellent conversion efficiencies, which indicate that they are promising candidates for photosensitizers in DSCs. However, the mechanisms of dye aggregation in DSC photovoltaics are still not fully understood; in order to match the performance of ruthenium complexes, further research in this field is necessary.

Monascus yellow was found to be one of the best natural dye photosensitizers for DSCs, with a conversion efficiency of 2.3%. The chlorophyll dyes were the only natural dyes with a higher conversion efficiency of 4%. Natural food dyes are better for human health than synthetic dyes, thus Monascus yellow could be used in an educational kit for students studying DSCs.

In summary, DSCs offer a low cost, non-toxic option for the commercial production of high-performance solar cells.

7. References

[1] B. O'Regan and M. Grätzel, *Nature,* 1991, 353, 737.

[2] M. K. Nazzerruddin, A. Kay, I. Podicio, R. Humphy-Baker, E. Müller, P. Liska, N. Vlachopoulos and M. Grätzel, *J. Am. Chem. Soc.,* 1993, 115, 6382-6390.

[3] M. K. Nazeerudding, P. Péchy, T. Renouard, S. M. Zakeeruddin, R. Humphry-Baker, P. Comte, P. Liska, L.Cevey, E. Costa, V. Shklover, L. Spiccia, G. B. Deacon, C. A. Bignozzi and M. Grätzel, *J. Am. Chem. Soc.,* 123 (2001) 1613.

[4] Z.-S. Wang, T. Yamaguchi, H. Sugihara and H. Arakawa, *Langmuir,* 2005, 21, 4272-4276.

[5] Chiba, A. Islam, Y. Watanabe, R. Komiya, N. Koide and L. Han, *J. J. Appl. Phys.,* 2006, 45, L638.

[6] Md. K. Nazeeruddin, F. De Angelis, S. Fantacci, A. Selloni, G. Viscardi, P. Liska, S. Ito, B. Takeru and M. Grätzel, *J. Am. Chem. Soc.,* 2005, 127, 16835.

[7] S. Ito, T. N. Murakami, P. Comte, P. Liska, C. Grätzel, Md. K. Nazeeruddin and M. Grätzel, *Thin Soild Films,* 2008, 516, 4613.

[8] S. Ito, M. K. Nazeeruddin, S. M. Zakeeruddin, P. Péchy, P. Comte, M. Grätzel, T. Mizuno, A. Tanaka, T. Koyanagi, International Journal of Photoenergy 2009, Article ID 517609, doi:10.1155/2009/517609.

[9] S. Ito, S. M. Zakeeruddin, R. Humphry-Baker, P. Liska, R. Charvet, P. Comte, M. K. Nazeeruddin, P. Péchy, M. Takata, H. Miura, S. Uchida and M. Grätzel, Adv. Mater, 2006, 18, 1202-1205.

[10] S. Ito, H. Miura, S. Uchida, M. Takata, K. Sumioka, P. Liska, P. Comte, P. Péchy and M. Grätzel, Chem. Commun., 2008, 5194-5196.

[11] H. Imahori, Y. Matsubara, H. Iijima, T. Umeyama, Y. Matano, S. Ito, M. Niemi, N. V. Tkachenko, H. Lemmetyinen, J. Phys. Chem. C, 20010, 114, 10656.

[12] S. Ito, T. Saitou, H. Imahori, H. Uehara, N. Hasegawa, Energy Environ. Sci., 2010, 3, 905.

[13] T. Bessho, S. M. Zakeeruddin, C.-Y. Yeh, E. W.-G. Diau, M Grätzel, Angew. Chem. Int. Ed., 2010, 49, 6646.

[14] S. Ito, H. Matsui, K. Okada, S. Kusano, T. Kitamura, Y. Wada, S. Yanagida, Sol. Energy Mater. Sol. Cells 82 (2004) 421.

[15] S. Ito, P. Liska, R. Charvet , P. Comte, P. Péchy, Md. K. Nazeeruddin, S. M. Zakeeruddin and M. Grätzel, Chem. Commun., (2005) 4351.

[16] J. Ferber, R. Stangl, J. Luther, Sol. Energy Mater. Sol. Cells, 53 (1998) 29.

[17] M. K. Nazeeruddin, R. Humphry-Baker, P. Liska, M. Grätzel, J. Phys. Chem. B 2003, 107, 8981.

[18] Z. Zhang, S. Ito, B. O'Regan, D. Kunag, S.M. Zakeeruddin, P. Liska, R. Charvet, P. Comte, Md. K. Nazeeruddin, P. Péchy, R. Humphry-Baker, T. Koyanagi, T. Mizuno, M. Grätzel, Z. Phys. Chem. 221 (2007) 319.

[19] F. Gao, Y. Wang, D. Shi, J. Zhang, M. Wang, X. Jing, R. Humphry-Baker, P. Wangt, S. M. Zakeeruddin, M. Grätzel, J. Am. Chem. Soc., 2008, 130, 10720.

[20] M. Grätzel, Acc. Chem. Res., 42 (2009) 1788; the 12.3% conversion efficiency was presented by Professor Grätzel in "International Symposium on Innovative Solar Cells 2009" (The University of Tokyo, Japan, 2nd-3rd, March, 2009).

[21] Z.-S. Wang, F.-Y. Li and C.-H. Huang, J. Phys. Chem. B, 2001, 105, 9210.

[22] D. P. Hagberg, T. Edvinsson, T. Marinado, G. Boschloo, A. Hagfeldt and L. Sun, Chem. Commun., 2006, 2245.

[23] K. R. J. Thomas, J. T. Lin, Y.-C. Hsuc and K.-C. Ho, Chem. Commun., 2005, 4098.

[24] H. Tian, X. Yang, R. Chen, Y. Pan, L. Li and A. Hagfeldt, L. Sun, Chem. Commun., 2007, 3741.

[25] S.-L. Li, K.-J. Jiang, K.-F. Shao and L.-M. Yang, Chem. Commun., 2006, 2792.

[26] T. Kitamura, M. Ikeda, K. Shigaki, T. Inoue, N. A. Anderson, X. Ai, T. Lian and S. Yanagida, Chem. Mater., 2004, 16, 1806.

[27] K. Hara, M. Kurashige, S. Ito, A. Shinpo, S. Suga, K. Sayama and H. Arakawa, Chem. Commun., 2003, 252.

[28] K. Hara, T. Sato, R. Katoh, A. Furube, T. Yoshihara, M. Murai, M. Kurashige, S. Ito, A. Shinpo, S. Suga and H. Arakawa, Adv. Funct. Mater., 2005, 15, 246.

[29] N. Koumura, Z.-S. Wang, S. Mori, M. Miyashita, E. Suzuki and K. Hara, J. Am. Chem. Soc., 2006, 128, 14256.

[30] Z.-S. Wang, Y. Cui, Y. Dan-oh, C. Kasada, A. Shinpo and K. Hara, J. Phys. Chem. C, 2007, 111, 7224.

[31] T. Horiuchi, H. Miura, K. Sumioka and S Uchida, J. Am. Chem. Soc., 2004, 126, 12218.

[32] S. Hwang, J. H. Lee, C. Park, H. Lee, C. Kim, C. Park, M.-H. Lee, W. Lee, J. Park, K. Kim, N.-G. Park and C. Kim, *Chem. Commun.*, 2007, 4887.

[33] S. Ito, H. Miura, S. Uchida, M. Takata, K. Sumioka, P. Liska, P. Comte, P. Péchy, M. Grätzel, *Chem. Commun.*, 2008, 5194.

[34] J. E. Kroeze, N. Hirata, S. Koops, Md. K. Nazeeruddin, L. Schmidt-Mende, M. Grätzel and J. R. Durrant, *J. Am. Chem. Soc.* 2006, 128, 16376.

[35] G. Zhang, H. Bala, Y. Cheng, D. Shi, X. Lv, Q. Yu, P. Wang, Chem. Commun., 2009, 2198.

[36] W. Zeng, Y. Cao, Y. Bai, Y. Wang, Y. Shi, M. Zhang, F. Wang, C. Pan, P. Wang, Chem. Mater. 2010, 22, 1915.

[37] W. M. Campbell, K. W. Jolley, P. Wagner, K. Wagner, P. J. Walsh, K. C. Gordon, L. Schmidt-Mende, Md. K. Nazeeruddin, Q. Wang, M. Grätzel and D. L. Officer, *J. Phys. Chem. C*, 2007, 11, 11760.

[38] H.-P. Lu, C.-Y. Tsai, W.-N. Yen, C.-P. Hsieh, C.-W. Lee, C.-Y. Yeh, E. W.-G. Diau, *J. Phys. Chem. C* 2009, *113*, 20990.

[39] M. Ikegami, M. Ozeki, Y. Kijitori and T. Miyasaka, *Electrochemistry*, 2008, 76, 140-143.

[40] X.-F. Wang, C.-H. Zhan, T. Maoka, Y. Wada and Y. Koyama, *Chem. Phys. Lett.*, 2007, 447, 79-85.

[41] D. Zhang, N. Yamamoto, T. Yoshida and H. Minoura, *Trans. Mater. Res. Soc. Jpn.*, 2002, 27, 811-814.

[42] K. Wongcharee, V. Meeyoo and S. Chavadej, *Solar Energy Mater. Solar Cells*, 2007, 91, 566-571.

[43] A. S. Polo and N. Y. M. Iha, *Solar Energy Mater. Solar Cells*, 2006, 90, 1936-1944.

[44] G. R. A. Kumara, S. Kaneko, M. Okuya, B. Onwona-Agyeman, A. Konno and K. Tennakone, *Solar Energy Mater. Solar Cells*, 2006, 90, 1220-1226.

[45] S. Hao, J. Wu, Y. Huang and J. Lin, *Solar Energy*, 2006, 80, 209–214.

[46] A. Kay and M. Grätzel, *J. Phys. Chem.*, 1993, 97, 6272-6277.

[47] X.-F. Wang, R. Fujii, S. Ito, Y. Koyama, Y. Yamano, M. Ito, T. Kitamura and S. Yanagida, *Chem. Phys. Lett.*, 2005, 416, 1-6.

3

Comparative Study of Dye-Sensitized Solar Cell Based on ZnO and TiO$_2$ Nanostructures

Y. Chergui, N. Nehaoua and D. E. Mekki

Physics Department, LESIMS Laboratory,
Badji Mokhtar University, Annaba,
Algeria

1. Introduction

The dye-sensitized solar cell (DSC) is a third generation photovoltaic device that holds significant promise for the inexpensive conversion of solar energy to electrical energy, because of the use of inexpensive materials and a relatively simple fabrication process. The DSC is based on a nano-structured, meso-porous metal oxide film, sensitized to the visible light by an adsorbed molecular dye. The dye molecules absorb visible light, and inject electrons from the excited state into the metal oxide conduction band. The injected electrons travel through the nanostructured film to the current collector, and the dye is regenerated by an electron donor in the electrolyte solution. The DSC is fully regenerative, and the electron donor is again obtained by electron transfer to the electron acceptor at the counter electrode (Ito et al, 2006). The current certified efficiency recordis11.1% for small cells, and several large-scale tests have been conducted that illustrate the promise for commercial application of the DSC concept. A schematic presentation of the operating principles of the DSC is given in Figure1. At the heart of the system is a mesoporous oxide layer composed of nanometer-sized particles which have been sintered together to allow for electronic conduction to take place. The material of choice has been TiO$_2$ (anatase) although alternative wide band gap oxides such as ZnO, and Nb$_2$O$_5$ have also been investigated. Attached to the surface of the nanocrystalline film is a monolayer of the charge transfer dye (Janne, 2002).

Nanocrystalline electronic junctions compose of a network of mesoscopic oxide or chalcogenide particles, such as TiO$_2$, ZnO, Fe$_2$O$_3$, Nb$_2$O$_5$, WO$_3$, Ta$_2$O$_5$ or Cds and CdSe, which are interconnected to allow for electronic conduction to take place. The oxide material of choice for many of these systems has been TiO$_2$. Its properties are intimately linked to the material content, chemical composition, structure and surface morphology. From the point of the material content and morphology, two crystalline forms of TiO$_2$ are important, anatase and rutile (the third form, brookite, is difficult to obtain). Anatase is the low temperature stable form and gives mesoscopic films that are transparent and colorless (Fang et al, 2010).

The application of ZnO in excitonic solar cells, XSCs, (organic, dye sensitized and hybrid) has been rising over the last few years due to its similarities with the most studied semiconductor oxide, TiO$_2$. ZnO presents comparable bandgap values and conduction band position as well as higher electron mobility than TiO$_2$. It can be synthesized in a wide variety of nanoforms applying straight forward and scalable synthesis methodologies.

Particularly, the application of vertically-aligned ZnO nanostructures it is thought to improve contact between the donor and acceptor material in organic solar cells (OSCs), or improve electron injection in dye sensitized solar cells (DSCs).7 To date, DSC based on ZnO have achieved promising power conversion efficiency values of 6%.

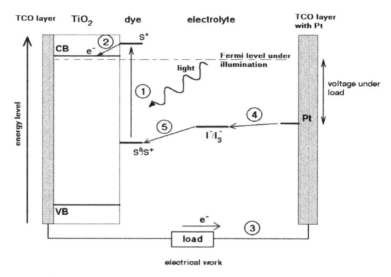

Fig. 1. Principle of operation and energy level scheme of the dye-sensitizednanocrystalline solar cell. Photo-excitation of the sensitizer (S) is followed by electron injection into the conduction band of the mesoporous oxidesemiconductor. The dye molecule is regenerated by the redox system, which itself is regenerated at the counter electrode by electrons passed through the load. Potentials are referred to the normal hydrogen electrode (NHE). The open-circuit voltage of the solar cell corresponds to the difference between the redox potential of the mediator and the Fermi level of the nanocrystallline film indicated with a dashed line (janne, 2002).

lthough most of the reported works on DSSC are based on TiO_2 porous thin films, various structures of ZnO are also being used for DSSC fabrication. The advantages of using ZnO over TiO_2 are its direct band gap (3.37 eV), higher exciton binding energy (60 meV) compared to TiO_2 (4 meV), and higher electron mobility (200 cm^2 V^{-1} s^{-1}) over TiO_2 (30 cm^{2}V^{-1} s^{-1}). However, the efficiency of the DSSC based on ZnO nanostructures is still very low (5%). Here we present a comparative study between ZnO and TiO_2 dye-sensitized solar cell (DSSC) by comparing its efficiency, fill factor, the current of short circuit I_{sc} and voltage of open circuit V_{oc} (Jingbin et al, 2010).

2. Materials

2.1 ZnO presentation
ZnO is an environment-friendly material (Myo et al, 2008), is a rogarded as one of the most promising substitution materials, and much interest has been paid to semiconductor nanostructures. Too, ZnO is one of the most important functional semi-conductor and is a very attractive material for application devices.

Zno is an environment-friendly material, is a promising candidate for exciton-related optoelctronic devices in the ultraviolet region, and it is also applicable to the devices based in reduced dimensional quantum effect because it can be grown in various crystalline form with submicron size such as whiskers, nanobelts, nanorods, nanowires, nanoplatelets, and so on. ZnO is a regarded as one of the most promising substitution materials for ITO (Introdium Thin Fims)today because of its good resistivity, high transmittance, nontoxicity, and low const. However pure ZnO thin films have a lower electric conductivity than ITO thin fims (You-Seung et al, 2008).

Recently, hollow micro/nanostructures have become of great interst because of their excellent characteristics such as density , high surface-to-volume ratio, and how coefficient sof thermal expansion and refractive index, which makes them attractive for applications ranging from catalystsupports, anti-reflection surface coatings, microwaves absorption (Lou et al, 2008; Zhong et al, 2000) , encapsulating sentive materials (Li et al, 2005; Dinsmore et al, 2002), drug delivery, and rechargeable batteries (Liang et al, 2004; Lee et al, 2003).

Rapid developments in the synthesis of hollow structures, such as CuO (Wang. et al, In press; Liu et al, 2009 Cu$_2$O (Teo et al, 2006; Chang et al , 2005), TiO$_2$ (Yu et al, 2007), SnO$_2$ (Cao et al, 2006; Wang et al, 2006), Fe$_2$O$_3$(Liu et al, 2009), Co$_3$O$_4$ (Park et al, 2009; Zhao et al, 2008), β-Ni(OH)$_2$ (Wang et al, 2005), α-MnO$_2$ (Li et al, 2006), CuS (Liu et al, 2007; Yu et 2000), Sb$_2$S$_3$ (Cao et al, 2006), ZnO (Zhou et al. 2007; Lin, et al, 2009), CdMoO$_4$(Wang et al, 2009; Wang et al, 2006), and ZnWO$_4$ (Huang et al, 2006), have greatly advanced our ability to tune their mechanical, optical, electrical, and chemical properties to satisfy the various needs of practical applications.

ZnO can be useful in many fields, such as in the rubber industry (Ibarra et al, 2002), photocatalysis(Hsu et al, 2005), sunthesis of ZnO (Hsu et al, 2005; Sun et al, 2007; Zhang et al, 2005). Cosmetic and pharmaceutical industries, and for therapeutic applications (Rosenthal et al, 2008), but little work have been done in these areas (Rosenthalet al, 2008; Chen et al, 2009; Pal et al, 2009). ZnO, as a wide gape semiconductor material, is becoming an increasing concern because of its biocompatibility, nontoxity, and good mechanical, optical, electrical properties. Thus, research about ZnO$_2$ and ZnO hollow spheres is of great importance and should be paid more attention (Ceng et al, 2009).

ZnO nanoparticules and quantium dots technologically important owing to their special properties and potential use in the fabrication of sensors, light emitters operating in the short-wavelegth range from blue to ultraviolet, transparent conducting oxides, and solar cells (özgör et al, 2005; Sakurai et al, 2002; Th et al, 2003).

Both high-quality p-and n-type ZnO thin films play an important role in the fabrication of optical devices. The production of ZnO with an n-type doping is simple without the need for intentional doping (Toshiya et al, 2008).

ZnO is expected to be one of the candidate host materials for impurity doping (Yamamoto et al, 2005; Ishizumi et al, 2005). However the optical properties of impurity-doped ZnO nanostructures are not well understood. very recently highly porous ZnO films have been successfully fabricated by electrodeposition using eosin Y(EY)dye molecules (T. Yoshida et al, 2003; T. Yoshida et al, 2004), and various lanthanoide ions were introduced into the films. (Pauporté et al, 2006). This fabrication method has been applied to high-efficiency dye-sensitized solar cells (Yoshida et al, 2004).

The application of ZnO inexcitonic solar cells, XSCs (organic, dye sensitized and hybrid) has been rising over the last few years due to its similarities with most studied semiconductor oxide, TiO_2. ZnO presents comparable band gap value and conduction band position as well as higher electron mobility than TiO_2 (Quintana et al, 2007; Keis et al, 2002). It can be synthesized in a wide variety of nanoforms (Wang, et al, 2004) applying strainght forward and scalable synthesized methodologies (Fan et al, 2006; Greene et al, 2003). particularly, the application of vertically-alligned ZnO NANOSTRUCTURES it is thought to improve contact between the donor and acceptor material in organic solar cells (OSCs), or improve electron injection in dye sensitized solar cells (DSCs) (Gonzalez et al, 2009) To date, DSC based on ZnO have achieved promising power conversion efficiency values of ~6%.(Keis et al, 2002; Yoshida et al, 2009).

Yet ZnO is not an easy material. It is a semiconductor oxide, the properties of which are greatly influenced by external conditions like synthesis methods (Huang et al, 2001; Wu et al, 2002; Elias et al, 2008; Yoshida et al, 2009), temperature (Huang et al, 2007; Guo et al, 2005). Testing atmosphere air, vacuum(Cantu et al, 2006 , 2007 ; Ahn et al, 2007), or illumination (Kenanakis et al, 2008; Feng et al, 2004; Norman et al, 1978), for example ,minimal changes in the shape of the ZnO (nanoparticules, nanorods, nanotips, etc), can produce different properties which in turn , affects the interface with any organic semiconductor or dye molecule.

Moreover, the modification of properties likes hydrophilicity/ hydrophobicity, or the amount of chemisorbed species on the ZnO surface, have already been reported to be affected by UV irradiation. In DSC, a major drawback is associated with the interaction between dye molecules and ZnO itself. Dye loading on ZnO must be carefully controlled in order to obtain the optimal power conversion efficiency for every system.(Chou et al, 2007; Fujishima et al, 1976; Kakiuchi et al, 2006). An excess in loading time results in the formation of aggregates made by the dissociation of the ZnO and the formation of $[Zn^{+2}$ –dye] complexes (Keis et al, 2002; 2000, Horiuchi et al, 2003).

In order to resolve these problems, research is currently focused on the study of new metal-free photosensitizers and the introduction of new anchoring groups (Guillén et al, 2008; Otsuka et al, 2008; Otsuka et al, 2008). Nevertheless, very recent reports indicate that the application of organic dyes could also be affected by factors like high concentration of Li^+ ions in the electrolyte or the photoinduced dye desorption (Quintana et al, 2009). As early as 178, V. J. Norman demonstrated that the absorption of organic dyes, like uranine and rhodamine B, on ZnO can be enhanced under light irradiation. A three- and two –fold increase on the amount of uranine and rhodamine B adsorbed on ZnO, respectively, was obseved after UV- light exposure (Norman et al, 1978). The latter presents important implications in ZnO-besed devices, like solar cells or diodes, since nanostructures layers of ZnO are increasingly being used on these devices, and some research groups have reported on the benefical effect of exposing ZnO-based devices to UV irradiation (Krebs et al, 2008; Verbakel et al, 2007,) .

2.2 TiO_2 presentation

Titanium dioxide is a fascinating material, with a very broad range of different possible properties, which leads to its use in application as different as toothpaste additive. TiO_2 thin films are synthesized for a broad range of different applications, which are summarized below (Estelle, 2002).

2.2.1 Optical coatings

Due to its high index of refraction, TiO$_2$ has been used for optical applications for more than 50 years (Hass et al, 1952). In particular, it is used as the high index of refraction material in multi-layer interference filtres, as anti reflection coating and as optical wave guides (Pierson, 1999). For many of these application, the mechanical and resistive of the layers are important in addition to the optical properties (Ottermann et al, 1997).

2.2.2 Microelectronics

In electric devices, the scaling down tendency leads to a decrease of the thickness of the gate oxide, which means that for fo the actually used SiO$_2$ layer, this thickness tends to atomic dimensions. Therefore, other materials are looked at with a higher dielectric constant such that a similar effective capacitance could be obtained with a thicker layer. TiO$_2$ is one of the promising materials (with Ta$_2$O$_5$ and the ternary titanate materials) for this application, due to its high dielectric constant (Pierson, 1999; Boyd et al, 2001).

2.2.3 Gas sensors

Titania fils are known to have sensing properties based on surface interactions of reducing or oxidizing species, which affect the conductivity of the film. Nano-crystalline material was in particular proven to exhibit a very high sensitivity.

Therefore, nano-grain TiO$_2$ under UV-irradiation is presently widely used as a photo catalyst different applications (water de-pollution (Rabani et al, 1998; Ding et al, 2000; Du et al, 2001), air de-odourization, NO$_x$ decomposition (Negishi et al, 2001), anti-bacteria treatment).

2.2.4 Solar cells

TiO$_2$ layers are used as photo-anodes in Grätzel's type solar cells,or in a solid state device (Bach et al, 1998). Additionaly, TiO$_2$ films used as passivation layers on silicon solar cells (Cardarelli, 2000).

2.2.5 Bio-compatible protective layers

Due to their relatively high corrosion resistance and good bio-compatibility, titanium and its alloys are commonly used for biomedical and dental implants. These beneficial properties are believed to be due to the formation of a native protective passive oxide layer. However, there is evidence that this natural layer does not prevent release from titanium in vivo, and therefore, studies have been devoted to deposition of denser, thicker and less oxygen deficient layers to improve the bio-compatibility (Pan et al, 1997). Additionally, TiO$_2$ was claimed to present good blood compatibility.

2.2.6 Protective and anti-corrosion coatings

Due to its hardness, TiO$_2$ is used as a protective layer on gold and precious metals (Battiston et al, 1999).

2.2.7 Membrans

TiO$_2$ coatings are used as membrane materials, with different porosities for different applications.

- Mesoporous membranes are used for ultra filtration or as supports for other Membranes.
- Micro-porous layers are prepared for nano filtration of liquids (Puhlfurss et al, 2000; Benfer et al, 2001).
- Ulra-microporous or dense layers are realized for gaz permselective membranes (Ha et al, 1996). Additionally, the photo catalytic properties of TiO$_2$ can be used in photo catalytic membrane reactors (Molinari et al, 2001).

2.2.8 As a component of ternary materials

Additionally,titanium dioxide is an important base for all the titanates materials. For instance, (Ba,Sr) TiO$_2$ which may become important for new geberation dynamic random access memories (Estelle, 2002).

TiO$_2$ is almost the only material suitable for industrial use at present and also probably in the future. This is because TiO$_2$ has the most efficient photo activity, the highest stability and the lowest cost (Kokoro et al, 2008). There are two types of photochemical reaction proceeding on a TiO$_2$ surface when irradiated with ultraviolet light. One includes the photo-induced redex reactions of adsorbed substances, and the other is the photo-induced hydrophilic conversion of TiO$_2$ itself. The former type has been known since the early part of the 20th centry, but the latter was found only at the end of the century. The combination of these two functions has opened up various novel applications of TiO$_2$, particularly in the field of building materials (Kokoro et al, 2008).

3. Experimental section

In the present section, we present an experimental comparaison between two dey-Sensitized solar cells based on ZnO nanotube and TiO$_2$ nanostructures.

First, an aligned ZnO nanotube arrays were fabricated by electrochemical deposition of ZnO anorods followed by chemical etching of the center part of the nanorods. The morphology of the nanotubes can be readily controlled by electrodeposition parameters. By employing the 5.1 μm length nanotubes as photoanodes for DSSC, an overall lightto- electricity conversion efficiency of 1.18% was achieved.

The current–voltage characteristic curves of DSSC fabricated using ZnO nanotubes with different lengths under simulated AM 1.5 light are shown in figure 2(A). The shortcircuit photocurrent densities (I_{sc}) obtained with nanotubes of 0.7, 1.5, 2.9 and 5.1 μm lengths were 0.68, 1.51, 2.50 and 3.24 mA cm−2, respectively. The highest photovoltaic performance of 1.18% (open-circuit voltage Voc = 0.68 V and fill factor FF = 0.58) was achieved for the sample of 5.1 μm length. This efficiency is attractive, taking into account that the film thickness is only 5.1 μm and no scattering layer is added. The Voc of the DSSC decreases upon increasing the length of the ZnO nanotubes, which is possibly related to the increase in the dark current which scales with the surface area of the ZnO film, in agreement with the previous reports on TiO2 nanotube-based DSSC.

The photon-current conversion efficiencies of DSSC using 0.7, 1.5, 2.9 and 5.1 μm length ZnO nanotubes were 0.32%, 0.62%, 0.83% and 1.18%, which were much higher than those of ZnO nanorod DSSCs (i.e. 0.11%, 0.20%, 0.39% and 0.59%). The photocurrent action spectra (figure 2(B)) display the wavelength distribution of the incident monochromatic photon-to-current conversion efficiency (IPCE). Themaximum of IPCE in the visible region is located at 520nm. This is approximately consistent with the expected maximum based on the accompanying absorption spectrum for the N719 dye (with local maxima at 390 and 535 nm), both corresponding to a metal-to-ligand charge transfer transition.

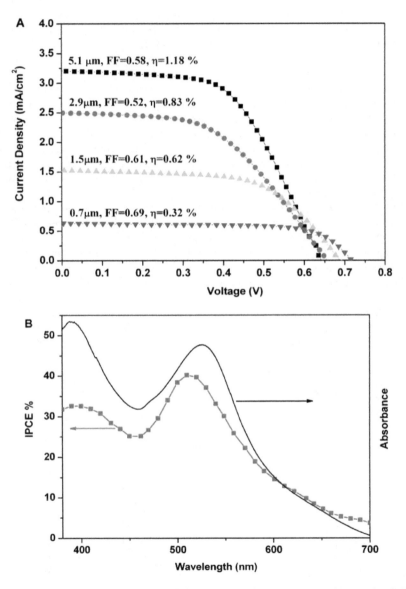

Fig. 2. Performance of the DSSC fabricated using ZnO nanotube array film under full-sun illumination: (A) current–voltage characteristic curves of DSSC with various lengths of ZnO nanotubes with Zn(OAc)2 treatment; (B) incident photon-to-current conversion efficiency (IPCE) of a ZnO nanotube-based device (square) and the absorption spectrum of the N719 dye in solution (solid line).

Second, The DSSCs based on TiO2 nanostructures grown in NaOH solution with different concentrations (0.5, 1, 3, and 10 M) are labeled as 0.5M-DSSC, 1M-DSSC, 3M-DSSC, and 10M-DSSC, respectively. To fabricate DSSCs, the substrates (FTO coated glasses) were first prepared by depositing a thin layer of nanocrystalline TiO2 paste onto FTOs using a screenprinting method. The as-prepared TiO2 membranes were then detached from the Ti plates and adhered onto the substrates as working electrodes. Besides, the DSSC comprised of commercial Degussa P25 TiO2 nanoparticles (labeled as P25-DSSC) was formed using doctor-blading method as a comparison. All of the TiO2 samples were dried under ambient conditions and annealed at 500 °C for 30 min. After cooling, they were chemically treated in a 0.2 M TiCl4 solution at 60 °C for 1 h and then annealed at 450 °C for 30 min to improve the photocurrent and photovoltaic performances. When the temperature decreased to 80 °C, the obtained samples were soaked in 0.3 mM dye solution (solvent mixture of acetonitrile and tert-butyl alcohol in volume ratio of 1:1) and kept for 24 h at room temperature. Here the cis-bis(isothiocyanato) bis (2,20-bipyridyl-4,40-dicarboxylato) ruthenium(II) bis- (tetrabutyl ammonium) (N719) was used as the sensitizer. These dye-coated electrodes were assembled into solar cells with Ptsputtered FTO counter electrodes and the electrolyte containing 0.5 M LiI, 0.05 M I2, and 0.5 M tert-butylpyridine in acetonitrile. photoinduced photocurrent density-voltage (I-V) curves of the constructed solar cells were measured on an electrochemical workstation (model CHI 660C, CH) under an AM 1.5 illumination (100 mW/cm2,model YSS-80A, Yamashita). Electrochemical impedance spectroscopic (EIS) curves of the DSSCs were also observed. The frequency range was from 0.1Hz to 100 kHz. The applied bias voltage was set to the open-circuit voltage (V_{OC}) of the DSSC, which had been determined earlier. The incident photo to current conversion efficiency (IPCE) was detected by the spectral response measuring equipment (CEP-1500, Bunkoh-Keiki. Japan). Figure 3 shows the current density voltage curves of the open cells based on different TiO2 photoelectrodes. The resultant photovoltaic parameters are summarized in Table 2.

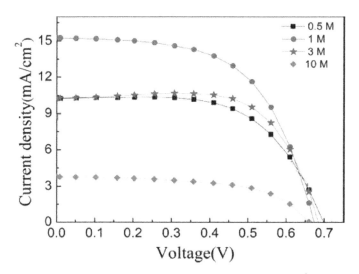

Fig. 3. I-V characteristics of dye-sensitized solar cells assembled with TiO2 films prepared with different concentrations of aqueous NaOH.

The resultant photovoltaic parameters are summarized in Table 1 for ZnO DSSC and in Table 2 for TiO2 DSSC.

Thickness(μm)	V_{oc} (V)	I_{sc} (mA cm⁻²)	FF (%)	η(%)
5.1	0.68	3.24	0.58	1.18
2.9	0.65	2.50	0.52	0.83
1.5	0.70	1.51	0.61	0.62
0.7	0.72	0.68	0.69	0.32

Table 1. Performance Characteristics of DSSCs Based onVarious ZnO nanotube.

DSSCs	V_{oc} (V)	I_{sc} (mA cm⁻²)	FF (%)	η(%)	Thickness(μm)
0.5M-DSSC	0.70	10.26	61.21	4.40	11.68
1M-DSSC	0.67	15.25	58.33	6.00	15.11
3M-DSSC	0.69	10.2	68.87	4.84	24.10
10M-DSSC	0.66	3.71	58.68	1.44	a

Table 2. Performance Characteristics of DSSCs Based on Various TiO2 Nanostructures.

From table 1 and 2, we observe the difference between the photovoltaic performance for these two type DSSC (TiO₂ Nanostructures and ZnO nanotobe), where a high photovoltaic performance is given by TiO₂ Nanostructured with different thickness and over the range of NaOH concentrations, conversion efficient increased from 4.40% at 0.5M to a maximum value of 6.00% at 1 M, which correspond to a high short-circuit photocurrent densities 15.25 mA cm⁻². Compared with ZnO nanotube DSSC, where the higher conversion efficient is 1.18% correspond to high short-circuit photocurrent densities 3.24 mA cm⁻². The cause of this difference is the based materials properties (ZnO and TiO₂), the method of fabrication and the different condition of measured I-V characteristics of temperature and illumination.

4. Simulation section

Now, to justify the experiment section, we use the computer simulation, which is an important tool for investigating the behaviour of semiconductor devices and for optimising their performance. Extraction and optimisation of semiconductor device parameters is an important area in device modelling and simulation (Chergaar. M et al, 2008; Bashahu M et al, 2007; Priyanka et al, 2007). The current–voltage characteristics of photocells, determined under illumination as well as in the dark, represent a very valuable tool for characterizing the electronic properties of solar cells. The evaluation of the physical parameters of solar cell: series resistance (R_s), ideality factor (n), saturation current (I_s), shunt resistance (R_{sh}) and photocurrent (I_{ph}) is of a vital importance for quality control and evaluation of the performance of solar cells when elaborated and during their normal use on site under different conditions. I-V characteristics of the solar cell can be presented by either a two diode or by a single diode model. Under illumination and normal operating conditions, the single diode model is however the most popular model for solar cells. In this case, the current voltage (I-V) relation of an illuminated solar cell is given by:

$$I = I_{ph} - I_d - I_p = I_{ph} - I_s \left[\exp\left(\frac{\beta}{n}(V + IR_s) \right) - 1 \right] - G_{sh}(V + IR_s) \tag{1}$$

I_{ph}, I_s, n, R_s and G_{sh} (=$1/R_{sh}$) being the photocurrent, the diode saturation current, the diode quality factor, the series resistance and the shunt conductance, respectively. I_p is the shunt current and $\beta=q/kT$ is the usual inverse thermal voltage. The circuit model of solar cell corresponding to equation (1) is presented in figure (4).

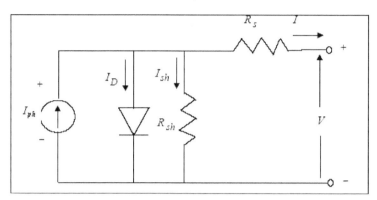

Fig. 4. Equivalent circuit model of the illuminated solar cell.

Determination of R_{sh}

The shunt resistance is considered Rsh = (1 / G_{sh}) >> R_s. the shunt conductance Gsh is evaluated from the reverse or direct bias characteristics by a simple linear fit (Nehaoua. N et al, 2010). The calculated value of G_{sh} gives the shunt current Ip = G_{sh}V.

Determination of n and R_s

Before extracting the ideality factor and the series resistance, our measured I-V characteristics are corrected considering the value of the shunt conductance as obtained from the linear fit and for V+R_sI>>kT, the current voltage relation becomes:

$$I = I_{ph} - I_s\left[\exp\left(\frac{\beta}{n}(V + IR_s)\right)\right]$$

(2)

The method concerns directly the usual measured I-V data by writing Eq. (2) in its logarithmic form:

$$\ln\left(I_{ph} - I\right) = \ln I_s + \frac{\beta}{n}(V + IR_s)$$

(3)

For a point defined by (V_0, I_0) we have:

$$\ln\left(I_{ph} - I_0\right) = \ln I_s + \frac{\beta}{n}(V_0 + I_0R_s)$$

(4)

By subtracting Eq. (3) and Eq. (4) and after a simplification we get a linear equation given by:

$$Y = \frac{\beta}{n}(R_s + X) \qquad \text{For} \quad I >> I_s$$

(5)

where:

$$Y = \frac{1}{I - I_0} \ln \frac{(I_{ph} - I)}{(I_{ph} - I_0)} \tag{6}$$

and

$$X = \frac{(V - V_0)}{(I - I_0)} \tag{7}$$

(V_0, I_0) is a point of the I-V curve.
We consider a set of I_i-V_i data giving rise to a set of X-Y values, with i varying from 1 to N. Then, we calculate X and Y values for $I_0 = I_{i0}$ and $I=I_{i0+1}$ up to $I=I_N$. This gives (N-1) pairs of X-Y data. We start again with $I_0 = I_{i0+1}$ and $I = I_{i0+2}$ up to I_N and get (N-2) additional X-Y data, and so on, up to $I_0 = I_{N-1}$. Finally, we obtain N(N-1)/2 pairs of X-Y data that means more values for the linear regression. The linear regression of equation (5) gives n and R_s.

Determination of I_{ph}

For most practical illuminated solar cells we usually consider that $I_s \ll I_{ph}$, the photocurrent can be given by the approximation $I_{sc} \approx I_{ph}$, where I_{sc} is the short-circuit current. This approximation is highly acceptable and it introduces no significant errors in subsequent calculations (Nehaoua. N et al, 2010).

Determination of I_s

The saturation current I_s was evaluated using a standard method based on the I-V data by plotting $\ln(I_{ph} - I_{cr})$ versus V_{cr} equation (8). Note that I-V data were corrected taking into account the effect of the series resistance.

$$\ln\left(I_{ph} - I_{cr}\right) = \ln\left(I_s\right) + \frac{\beta}{n} V_{cr} \tag{8}$$

When we plot $\ln(I_c)$ where $(I_c=I_{ph}-I_{cr})$ versus V_{cr}, it gives a straight line that yields I_s from the intercept with the y-axis.

4.1 Application
The method is applied on the too type of Dey-sensitized solar cell, the first one is based on TiO$_2$ nanostructures and ZnO nanotube under different condition of fabrication, illumination and temperature. The current-voltage (I-V) characteristics of TiO$_2$ nanostructures DSSC is taken from the work of (Fang Sho et al, 2010) and The current-voltage (I-V) characteristics of ZnO nanotube is taken from the work of (Jingbin Han et al, 2010). The two characteristics correspond to the higher photovoltaice performance, where η=6.00%, FF=58.33%, I_{sc}=15.25mAcm^{-2} and V_{oc}=0.67V for TiO$_2$ nanostructures, and for ZnO nanotube η=1.18%, FF=0.58%, I_{sc}=3.24mAcm^{-2} and V_{oc}=0.68V.

4.2 Results and discussion
The shunt conductance $G_{sh} = 1 / R_{sh}$ was calculated using a simple linear fit of the reverse or direct bias characteristics. The series resistance and the ideality factor were obtained from the linear regression (5) using a least square method.

In order to test the quality of the fit to the experimental data, the percentage error is calculated as follows:

$$e_i = \left(I_i - I_{i,cal}\right)\left(100 / I_i\right) \tag{9}$$

Where $I_{i,cal}$ is the current calculated for each V_i, by solving the implicit Eq.(1) with the determined set of parameters (I_{ph}, n, R_s, G_{sh}, I_s). (I_i, V_i) are respectively the measured current and voltage at the ith point among N considered measured data points avoiding the measurements close to the open-circuit condition where the current is not well-defined (Chegaar M et al, 2006).

Statistical analysis of the results has also been performed. The root mean square error (RMSE), the mean bias error (MBE) and the mean absolute error (MAE) are the fundamental measures of accuracy. Thus, RMSE, MBE and MAE are given by:

$$RMSE = \left(\sum_i |e|_i^2 / N\right)^{1/2}$$

$$MBE = \sum_i e_i / N \tag{10}$$

$$MAE = \sum_i |e|_i / N$$

N is the number of measurements data taken into account.

The extracted parameters obtained using the method proposed here for the Dey-Sensitized solar cell based on TiO2 nanostructures and ZnO nanotube are given in Table 3. Satisfactory agreement is obtained for most of the extracted parameters. good agreement is reported. Statistical indicators of accuracy for the method of this work are shown in Table 3.

	DSSC-TiO$_2$ nanostructures	DSSC-ZnO nanotube
G_{sh} (Ω^{-1})	0.001269	0.000588
R_s (Ω)	0.025923	0.383441
n	1.629251	3.560949
I_s(μA)	0.33556	0.16553
I_{ph}(mA/cm^2)	15.99	3.25
RMSE	0.850353	1.875871
MBE	0.232276	0.727544
MAE	0.757886	1.053901

Table 3. Extracted parameters for Dey-Sensitized solar cell based on TiO$_2$ nanostructures and ZnO nanotube.

Figures 5 and 6 show the plot of I-V experimental characteristics and the fitted curves derived from equation (1) with the parameters shown in Table 3 for Dey-Sensitized solar cell based on TiO$_2$ nanostructures and ZnO nanotube. The interesting point with the procedure described herein is the fact that we do not have any limitation condition on the voltage and it is reliable, straightforward, easy to use and successful for different types of solar cells.

Extracting solar cells parameters is a vital importance for the quality control and evaluation of the performance of the solar cells, this parameters are: series resistance, shunt conductance, saturation current, the diode quality factor and the photocurrent. In this work, a simple method for extracting the solar cell parameters, based on the measured current-voltage data. The method has been successfully applied to dey-Sensitized solar cell based on TiO$_2$ nanostructures and ZnO nanotube under different temperatures.

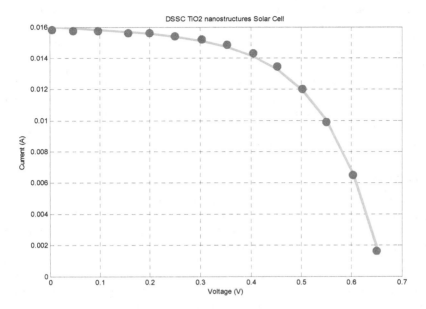

Fig. 5. Experimental (●) data and fitted curve of TiO₂ nanostructures DSSC.

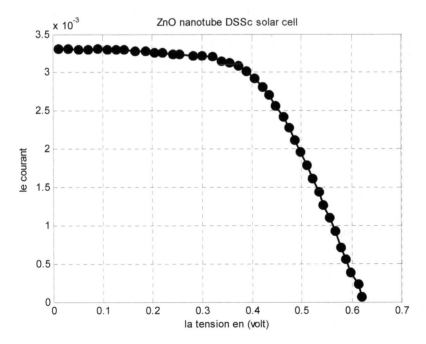

Fig. 6. Experimental (●) data and fitted curve of ZnO nanotube DSSC.

Figures 5 and 6 shows the plot of I–V experimental characteristics and the fitted curves derived from equation (1) with the parameters shown in Table 3 for the dey-Sensitized solar cell based on TiO_2 nanostructures and ZnO nanotube solar cell. Good agreement is observed for the different structure, especially for the TiO_2 nanostructures solar cells with statistical error less than 1%, and 2% for ZnO nanotube DSSC solar cells respectively, which attribute mainly to lower parasitic losses, where we can observe a low series resistance 0.025923Ω compared to $0.383441 \ \Omega$ for TiO_2 nanostructures and ZnO nanotube solar cell respectively. The interesting point with the procedure described herein is the fact that we do not have any limitation condition on the voltage and it is reliable, straightforward, easy to use and successful for different types of solar cells.

5. Conclusion

In this contribution, a simple comparative study between experimental and simulation works to improve the dey-sensitized solar cell performance of two DSSCs based on TiO_2 nanostrucures and ZnO nanotube, under differents condition of temperature. We compare the different parameters which are: the conversion efficient, the fill factor, the short-circuit photocurrent and the open-circuit voltage, where we observe a high photovoltaic performance for TiO_2 nanostrucures with maximun conversion efficient 6% compared to 1.18% for ZnO nanotube. In second time, an evaluation of the physical parameters of solar cell: series resistance (R_s), ideality factor (n), saturation current (I_s), shunt resistance (R_{sh}) and photocurrent (I_{ph}) from measured current-voltage characteristics by using a numerical method proposed by th authors. Extracting solar cells parameters is a vital importance for the quality control and evaluation of the performance of solar cells when elaborated and during their normal use on site under different conditions. Good resuts are given by the differents DSSCs, and specialy for on dey-sensitized TiO_2 nanostrucures, which justify the experimental work.

6. References

Boyd, I. W. & Zhang, J. Y. (2001), *Solid-State Electronics*, 45, 1413

Benfer, S.; Popp, U.; Richter, H. et al.(2001) *Separation and Purification Technology*, 22-23. 231.

Bashahu M, Nkundabakura P. Solar energy 2007; 81:856-863.

Bach,U. ; Lupo, D. ;Comte, P. et al.(1998), *Nature*,395 ,583.

Battiston, G., A. ; Gerbasi, R. ; Porchia, M. et al. (1999), *Chemical Vapor Deposition*,5 ,73

Chang Y.; Teo, J. J. ; Zeng, H. C. (2005), *Langmuir*. 21(3), 1074-1079, 21(3), 1074-1079

Cao, Q. H. ; Gao, Y. Q. ; Chen, X. Y. ; Mu, L. ; Yu, W. C. & Qian, Y. T.(2006). *Chem. Lett.* 35(2), 178-179.

Cao, X. B. ; Gu, L. ; Zhuge, L. ; Gao, W. J.; Wang, W. C. & Wu, S. F.(2006) *Adv. Funct. Mater.* 16(7), 896-902.

Chegaar M, Nehaoua. N, Bouhemadou. A. Energy conversion and management 2008, 49:1376-1379.

Chegaar M, Azzouzi G, Mialhe P. Solid State Electronics 2006; 50:1234-1237.

Chen, W.; Lu, Y. H. ; Wang, M. ; Kroner, L. ; Paul, H. ; Fecht, H.J.; Bednarcik, J.; Stahl, K.; Zhang, Z. L.; Wiedwald, U.; Kaiser, U.; Ziemann.,P.; Kikegawa, T.; Wu, C. D. & Jiang, J. Z. J.(2009) *Phys. Chem. C* , 113(4),1320-1324.

Cardarelli, F.(2000), *in Materials Handbook A Concise Desktop Reference*(Springler-Verlag).

Dinsmore, A. D.; Hsu, M. F.; Nikolaides, M. G.; Marquez, M.;Bausch, A. R. & Weitz, D. A.(2002) *Science*, 298(5595),1006-1009.

Ding, Z. ; Hu, X. ; Lu, G. Q. ; et al.(2000), *Langmuir*, 16 ,6216.,

Du, G., H. ; Chen, Q. ; Che, R. C. et al. (2001), *Appl. Phys. Lett.*, 79 ,3702

Estelle Wagner (2002), thesis(selective light induced chemical vapour deposition of titanium dioxide thin films), EPFL, 2650.

Engeneering Chemical Research, (199), 3381

Elias, J.; Tena-Zaera, R. & Lévy-Clément, C.(2008), *J. Phys. Chem. C*, 112, 5736.

Fujishima,A.; Iwase, T. & Honda,K. J. Am, J.(1976). *Chem. Soc.*, 98(6), 1625.

Fan, H. j.; Werner, P. & Zacharias, M.(2006) Small. 2(6),700.

Fang shao, Jing sun, lian gao, Songwang, Jianqiang luo. Physical chemestry C. Dx.doiorg/10.1021/jp110743m.

E. Guillén et al. J. Phtochem. Photobiol., A., 2008,200,364.,

I. Gonzalez-Valls and M. Lira-Cantu.(2009), *Energy Environ.* Sci. 2,1

L. E. Greene,M. Law, J. Goldberger, F. Kim, J. C. Johnson,Y. Zhang, R. J. Saykally and P. D. Yang, Angew.(2003). Chem., Int. Ed., 42(26), 3031.

Guo, M.; Diao, P. and Cai, S. M. (2005), *J. Solid State Chem.* 178, 1864

Huang, M. H.; Mao,S. ;Feick, H.;Yan, H. Q.; Wu, Y. Y.; Kind, H.; Weber, E. Russo, & Yang, P. D.(2001), *Science*, 292, 1897.,

Hass, G. Vacuum, 11,(1952),331

Huang, J. H.; Gao, L.(2006), *J. Am. Ceram. Soc.* 89(12), 3877-3880.

Hsu, C. C.; Wu, N. L. (2005), *Photochem. Phtobiol. A* 172(3), 269-274.),

Horiuchi,H.; Katoh,R.; Hara, K. Yanagida, M.; Murata, S. Arakawa, H. & Tachiya, M.(2003), *J. Phys. Chem.* B, 107, 2570.

Ha, Y. H., Nam, S. W. ; S. W., ; Lim, T. H. L. et al.(1996), *Journal of Membrane Science*, 111, 81.

Ishizumi, A. & Kanemitsu, Y. (2005): *Appl. Phys. Lett.* 86(253106).

Ishizumi, A.; Y. Taguchi, Yamamoto, A. & Kanemitsu, Y.(2005): *Thin Solid Films* 486(50).

Ibarra. L. & Alzorriz, M. J.(2002), *Appl. Polym. Sci.* 84(3), 605-615.

Ibarra, L. ; Macros-Fernandez, A. & Alzorriz., M.(2002), *Polymer*, 43(5), 1649-1655.

Ibarra, L. & Alzorriz, M. (2002), *J. Appl. Polym. Sci.*, 86(2), 335-340.

Ito, S., Zakeeruddin, S. M., Humphry-Baker, R., Liska, P., Charvet, R., Comte,P., Nazeeruddin, M., Péchy, P., Takata, M., Miura, H., Uchida, S. &Grätzel 2006, 'High-efficiency organicdye sensitized solar cells controlled by nanocrystalline-TiO2 electrodethickness,' *Adv. Mater.* 18, p. 1202.

Jain A, Kapoor A. Solar energy mater solar cells 2005; 86:197-205

Jingbin Han, Fengru Fan, chen xu, Shisheng Lin, Min wei, Xue duan, Zhong lin wang. Nanotechnology 2010, 21:405203(7p).

Janne Halme, thesis (2002), *Dye-sensitized nanostructured and organic photovoltaic cells: technical review and preliminary tests*, HELSINKI UNIVERSITY OF TECHNOLOGY.

Krebs, F. C. (2008), *Sol. Energy Mater. Sol. Cells*, 92,715.

Keis, K.; Bauer, C.; Boschloo, G.; Hagfeldt, A.; Westermark, K.; Rensmo, H. & Siegbahn, H.(2002), *J. Photochem. Photobiol., A*, 148.57 .

Keis, K.; Magnusson, E.; Lindström, H.; Lindquist, S. E. & Hagfeldt, A.(2002), *Sol. Energy Mater. Sol. Cells*, 73, 51.

Kakiuchi, K.; Hosono, E. & Fujihara, S.(2006). *J. Photochem. Photobiol. A* 179, 81.

Keis, K.; Lindgren, J.; S. E. Lindquist, S. E. & Hagfeldt, A.(2000), *Langmuir*, 16, 4688.

Kokoro et al. Appl. Phys. Express 1 (2008) 081202

Li, Z. Y.; Kobayashi, N.; Nishimura, A. & Hasatani.(2005), *M. Chem. Eng.Commun.* 12(7) ,18-932.

Liang, H. P.; Zhang, H. M.; Hu, J. S.; Guo, Y. G.; Wan, L. J.& Bai,C. L. Angew.(2004) *Chem., Int. Ed.* 43(12), 1540-1543.

Lee, K. T.; Jung, Y. S.; Oh, S. M.(2003), J. Am. Chem. Soc. 125(19), 5652-5653

Liu,X. M. ; Yin, W. D. ; Miao, S. B. & Ji, B. M.(2009). Mater. Chem. Phys. 113(2-3), 518-522

Li,B. X. ; Rong, G. X. ; Xie, Y. ; Huang, L. F. & Feng, C. Q.(2006), Inorg. Chem., 45(16), 6404-6410.

Liu, X. Y. ; Xi, G. C. ; Liu, Y. K. ; Xiong, S. L. ; Chai, L. L. & Qian, Y. T.(2007) *J. Nanosci. Nanotechnology.* 7(12), 4501-4507

Lou, X. W.; Archer, L. A. & Yang, Z. C.(2008), *Adv. Matter.* 20(21), 3987-4019.

Li, Z. Y.; Kobayashi, N.; Nishimura, A. & Hasatani, M.(2005), *Chem. Eng.Commun.* 12(7) ,18-932.

Lin, X. X. ; Zhu, Y. F. & Shen, W. Z.(2009), *J. Phys. Chem. C*, 113(5), 1812-1817 ,113(5), 1812-1817

Lira-Cantu, M. & Krebs, F. C.(2006), *Sol. Energy Mater. Sol. Cells*, 90,2076.

Lira-Cantu, M.; Norman, K.; Andreasen, J. W.; Casan-Pastor, N. & Krebs, F. C.(2007), *J. Electrichem. Soc.* 154(6), B508

Lira-Cantu, M.; Norrman, K.; Andreasen, J. W.& Krebs, F. C. (2006), Chem. Mater., 18, 5684.,

Molinari, R. ; Grande, C. ; Drioli, E. et al. (2001), *Catalysis Today*, 67,1.

Myo Than HTAY,; · Minori ITOH,; Yoshio HASHIMOTO,& Kentaro ITO(2008),J. *Appl. Phys.* (47)541

Negishi, N. & Takeuchi, K.(2001), *Thin Sold Films*, 392, 249.

Norman, V. j.(1978), *Australian J. Chem.*, 25(6),1189.

Nehaoua N, Chergui Y , Mekki D E. Vacuum 2010 , 84 : 326–329.

Otsuka, A.; K. Funabiki, Sugiyama,N.; Mase, H.; T. Yoshida,T.; Minoura, H. & Matsui, M.(2008), *Chem. Lett.* 37(2), 176.

Ottermann, C. R. ; Ottermann, R. ;Kischnereit,R. ; Anderson, O. et al.(1997), *Mat. Res. Soc. Symp. Proc.*, 436 ,251

Otsuka et al. Dalton Trans., 2008,5439

Pan, J.; Leygraf, C.; Thierry, D. et al.(1997), *Journal of Biomedical Materials Research*, 35, ,309.

Pierson, H. O.(1999), *Handbook of chemical vapour deposition(CVD): principles, technology and applications*, 2nd ed.(Park Ride, 1999).

Puhlfurss, P. ; Voigt, A. ; Weber, R. et al. (2000), *Journal of Membrane Science*, 174, 123.

Priyanka, Lal M , Singh S N. Solar energy material and solar cells 2007; 91:137-142.

Quintana,M.; Edvinsson,T.; Hagfeldt, A. & Boschloo, G.(2007), *J. Phys.Chem. C.* 111, 1035

Quintana, M.; Marinado, T.; Nonomura,K.; Boschloo, G. & Hagfeldt, A.(2009), J. *Photochem. Photobiol. A*, 202,159

Rabani, J.; Yamashita, K. ; Ushida K. et al. (1998). *J. Phys. Chem.*, 102, 1689.

K. Sakurai, T. Takagi, T. Kubo, D. Kajita, T. Tanabe, H. et al. J. Cryst. Growth 237-239(2002)514.

F. Verbakel, S. C. J. Meskers and R. A. J. Janssen.(2007), *J. Phys. C*, 111,10150

Verbakel,F.; Meskers, S. C. J. & Janssen, R. A. G..(2007), *J. Appl. Phys.* 102(8), 083701

Yamamoto, A. ; Atsuta S.; & Y. Kanemitsu, Y.(2005) ; *J. Lumin*, 112 (169).

Yamamoto,A., S. Atsuta, & Kanemitsue, Y.(2005): *Physica E* 26(96).

Wu, J. J. & Liu, S. C. (2002), *Adv. Mater.* 14(3), 215.

4

Chasing High Efficiency DSSC by Nano-Structural Surface Engineering at Low Processing Temperature for Titanium Dioxide Electrodes

Ying-Hung Chen, Chen-Hon Chen, Shu-Yuan Wu, Chiung-Hsun Chen,
Ming-Yi Hsu, Keh-Chang Chen and Ju-Liang He
Department of Materials Science and Engineering, Feng Chia University
Taichung, Taiwan,
R.O.C.

1. Introduction

The rapid shortage of petrochemical energy has led to the great demand in developing clean and renewable energy sources; such as solar cells in these years. The first commercially available photovoltaic cell (PV) by using solar energy is silicon-based solar cell however with high production cost and high energy payback time. This limited the usage and agitated vigorous studies on the next-generation solar cells in order to reduce cost and increase efficiency. It was until 1991, dye-sensitized solar cells (DSSCs) have attracted increasing interests by the pioneering work of O'Regan and Grätzel. They used a Ru-based dye to achieve higher conversion efficiency in a cell made of titania (TiO_2) as the active layer. Recent development of solar cells in dye-sensitized type devices is one great step forward in the field. The DSSCs take advantages in simple fabrication technique and low production costs in contrast to those conventional silicon-based solar cells.

The DSSC device (Fig. 1) is basically comprised of two facing electrodes: a transparent photoanode, consisting of a mesoporous large band gap semiconductor as an active layer, modified with a monolayer of dye molecules and a Pt counter electrode, both deposited on conductive glass substrates, for example: indium tin oxide (ITO) glass. An appropriate medium containing the redox couple (usually I^-/I_3^-) is placed between the two electrodes to transfer the charges. Among other semiconductors employed as the active layer of the DSSCs, titania known to have wide energy band gap, can absorb dye and is capable of generating electron-hole pairs via photovoltaic effect. DSSCs based on mesoporous titania, which exhibits very high specific surface area (and better dye-absorbing) has been drawn much attention over the past few years. A number of surface modification techniques have been reported to produce nanostructural TiO_2 layer. Moreover, researchers suggested that one dimensional nanostructural TiO_2 such as nano-rods, nano-wires or nano-tubes is an alternative approach for higher PV efficiency due to straightforward diffusion path of the free electron once being generated. For these reasons, we use several cost-effective manufacturing methods to develop the nanostructural TiO_2 electrode at near room temperature to form several types of DSSC device configuration and to investigate their PV efficiency. The aim is to develop feasible routes for commercializing DSSCs with high PV efficiency.

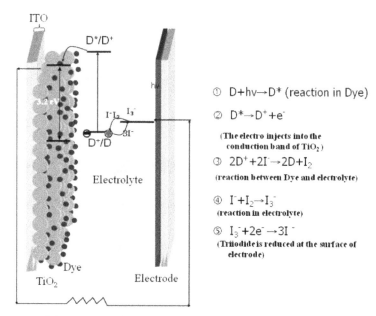

① $D + h\nu \rightarrow D^*$ (reaction in Dye)

② $D^* \rightarrow D^+ + e^-$

(The electro injects into the conduction band of TiO_2)

③ $2D^+ + 2I^- \rightarrow 2D + I_2$

(reaction between Dye and electrolyte)

④ $I^- + I_2 \rightarrow I_3^-$

(reaction in electrolyte)

⑤ $I_3^- + 2e^- \rightarrow 3I^-$

(Triiodide is reduced at the surface of electrode)

Fig. 1. Schematic of the principle for dye sensitized solar cell to indicate the electron energy level in different phases. (The electrode sensitizer, D; D*, electronically excited sensitizer; D+, oxidized sensitizer)

This chapter demonstrates four kinds of manufacturing methods to obtaion nanostructural photoanode for the purpose of achieving high efficiency DSSCs. These manufacturing methods were involved with each method chosen with good reason, but went out with different performance. These involves *liquid phase deposition (LPD) to grow TiO_2 nanoclusters layer, hydrothermal route (HR) to obtain TiO_2 nanowires, PVD titanium followed by anodic oxidation to grow TiO_2 nanotubes,* and *eventually microarc oxidation (MAO) /alkali etching to produce nanoflaky TiO_2.* The first three methods can directly grow TiO_2 layer on ITO glass and the specimens were assembled into ITO glass/$[TiO_2(N3 dye)]$/I_2+LiI/Pt/ITO glass device. The last method can only obtain TiO_2 layer on titanium and was assembled into Ti/$[TiO_2(N3 dye)]$/I_2+LiI/Pt/ITO glass inverted-type device. Microstructural characterization and observation work for the obtained nano featured TiO_2 were carried out using different material analyzing techniques such as field-emission scanning electron microscopy, high-resolution transmission electron microscopy and X-ray diffractometry. All the PV measurements were based on a large effective area of 1 cm x 1cm. The DSSC sample devices were then irradiated by using a xenon lamp with a light intensity of 6 mW/cm², which apparently is far lower than the standard solar simulator (100 mW/cm²). It would then be true for the photovoltaic data reported in this article for cross-reference within this article and not validated for inter-laboratory cross-reference. Photocurrent–voltage (*I–V*) characteristics were obtained using a potentiostat (EG&G 263A). Photovoltaic efficiency of each cell was calculated from *I-V* curves. The results for each study are reported and discussed with respect to their microstructure as below.

2. Nanocluster-TiO₂ layer prepared by liquid phase deposition

The LPD process, which was developed in recent years, is a designed wet chemical film process firstly by Nagayama in 1988. Than Herbig et al. used LPD to prepare TiO_2 thin film and studied its photocatalytic activity. Most vacuum-based technologies such as sputtering and evaporation are basically limited to the line-of-sight deposition of materials and cannot easily be applied to rather complex geometries. By contract, the easy production, no vacuum requirement, self-assembled and compliance to complicated geometry substrate has led many LPD applications for functional thin films. In order to directly grow nanocluster-TiO_2 on ITO glass, the simplest method - LPD process was firstly considered by using H_2TiF_6 and H_3BO_3 as precursors. The reaction steps involved to obtain nanocluster-TiO_2 are illustrated as followed. The H_3BO_3 pushes eq. (1) to form eventually $Ti(OH)_6^{2-}$ which transforms into TiO_2 after thermal annealing.

$$(TiF_6)^{2-} + nH_2O \leftrightarrow TiF_{6-n}(OH)_n^{2-} + nHF \tag{1}$$

$$H_3BO_3 + 4HF \leftrightarrow BF^{4+} + H_3O + 2H_2O \tag{2}$$

Here, the influence of deposition variables including deposition time and post-heat treatment on the microstructure of TiO_2 layer and the photovoltaic property was studied. The LPD system to deposit titania film is schematically shown in Fig. 2.

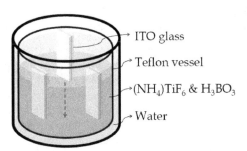

ITO glass

Teflon vessel

$(NH_4)TiF_6$ & H_3BO_3

Water

Fig. 2. Schematic diagram of LPD-TiO_2 deposition system.

Figure. 3 shows the *I-V* characteristics of the DSSCs assembled by using TiO_2 films deposited for different time, with their corresponding surface and cross sectional film morphology also shown. It was indeed capable of producing nanocluster featured TiO_2 films shown in the surface morphology, regardless of the deposition time. It can also be found that the *I-V* characteristics are sensitive to the TiO_2 film deposition time, but unfortunately non-linearly responded to the deposition time. By careful examination on the surface morphology of these TiO_2 films deposited at different deposition time, the film obtained at longer period of deposition time, say 60 h presents no longer nanocluster feature, but cracked-chips feature instead. This significantly reduces the open circuit voltage (V_{oc}) as well as the short circuit current density (J_{sc}). It shall be a consequence of the cracks that leads to the direct electrolyte contact to the front window layer (to reduce V_{oc}) and the reduced specific surface area (to reduce J_{sc}). Further exam cross sectional morphology of the TiO$_2$ films as a function of deposition time, it was found that the film thickness does not linearly respond to the deposition time. This shall be the gradual loss of reactivity of the electrolyte liquid. Therefore, it is not practical to increase the film thickness by an extended deposition time. Still, we believed that by constant precursor supplement into the electrolyte liquid, it would refresh the liquid and certainly the increased film growth rate, of course with the price of process monitoring automation.

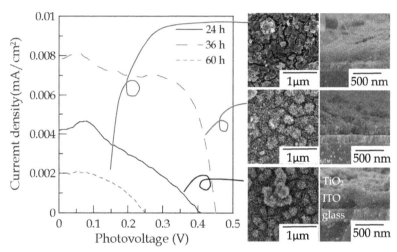

Fig. 3. I-V characteristic of the cell assembled by LPD-TiO$_2$ under different deposition time, with their corresponding surface and cross sectional film morphology.

Fig. 4 shows the XRD patterns of the TiO$_2$ film with different annealing temperature. The results indicate that the as-deposited film was amorphous due to the low LPD growth temperature. Annealing provides thermal energy as a driving force to overcome activation energy that required for crystal nucleation and growth. The exact TiO$_2$ phase to be effective for DSSC has been known to be anatase, which can found that the peak ascribed to anatase phase A(101) can only appear over 400 °C and become stronger over 600 °C, ie. better crystallinity of the film annealed at higher temperature. Over an annealing temperature of 600 °C leads to the ITO glass distortion.

The I-V characteristics of the DSSCs assembled by using TiO$_2$ films with different annealing temperatures, with their corresponding surface and cross sectional film morphology are shown in Fig. 5. The TiO$_2$ film surface forms numerous tiny nanocracks and needle-like structures with increasing annealing temperature. It can be found that the I-V characteristics are sensitive to the TiO$_2$ film annealing temperatures and the J_{sc} increases straight up to a maximum when annealed at 600 °C. Apparently, the increase of J_{sc} shall be associated with the reformation of the TiO$_2$ film morphology and the increased film crystallinity. By reforming numerous tiny nanocracks and needle-like structures, the TiO$_2$ film has more specific surface area after post-annealing and achieves higher efficiency dye adsorbing. However, the negative effect of annealing occurred to the significant increase of the ITO electrical resistance that causes the V_{oc} drop off as can be seen in Fig. 5. Anyhow, the overall increased photovoltaic efficiency as a function of annealing temperature is an encouraging

Chasing High Efficiency DSSC by Nano-Structural Surface Engineering at Low Processing Temperature for
Titanium Dioxide Electrodes

69

result of this study using PLD to obtain TiO$_2$ film and post-annealing for DSSC photoanode preparation.

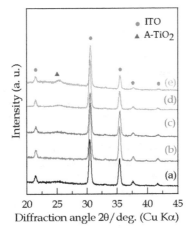

Fig. 4. XRD patterns of (a) ITO glass substrate, (b) TiO$_2$ as-deposited specimen, and the post annealed specimens obtained at (c) 200, (d) 400 and (e) 600 °C for 30 min.

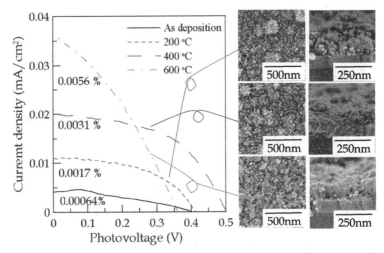

Fig. 5. I-V characteristic of the cell assembled by LPD-TiO$_2$ under different annealing temperature, with their corresponding surface and cross sectional film morphology.

2.1 Summary

In this paragraph, a LPD system is used to prepare the TiO$_2$ layer on ITO glass at the room temperature followed by post-annealing as the photoanode in DSSC. The result is closely connected to the variation of microstructure including both the specific surface area and crystal structure. This demonstration work confirms the truth that the LPD method is capable of obtaining nanocluster TiO$_2$ and with crystallinic anatase structure through suitable annealing treatment. Unfortunately, the unacceptable LPD-TiO$_2$ film growth has led some other attempts to obtain nano-structural TiO$_2$ layer. These methods are sketched as below.

3. TiO$_2$ nanowires growth on TiO$_2$ template via hydrothermal route

As being well acknowledged that pressurized hydrothermal route is able to synthesize 1D nanomaterials without using catalysts. Due to 1D nanomaterials (such as nanowires) having a relatively higher interfacial charge transfer rate and specific surface area compared with the spherical TiO$_2$ particles and nanocluster TiO$_2$, the simple operation, fast formation and low cost process interested us using this method to produce TiO$_2$ nanowires. The idea was that via the hydrothermal (HR) growth of TiO$_2$ nanowires on an arc ion plated (AIP) TiO$_2$ layer (as a template during HR and a barrier layer during service that pre-deposited on ITO glass), the obtained film would be able to exhibit the desired photoanode properties. AIP is known to be capable of producing high growth rate, high density and strong adhesion films without additional substrate heating, the pre-deposited AIP-TiO$_2$ template might also be able to get rid of the autoclave while at least well-aligned or randomly-oriented TiO$_2$ nanowire can be grown. In this study, anatase Degussa TG-P25 powder was used as starting material. Eventually, the experimental result showed the randomly-orientated TiO$_2$ nanowires were formed on AIP-TiO$_2$ template. TiO$_2$ powder content in the HR bath (g/l) and post-annealing temperature were evaluated their microstructure and photovoltaic efficiency of the assembled DSSC devices. The HR system and preparation method to obtain TiO$_2$ nanowires is illustrated in Fig. 6.

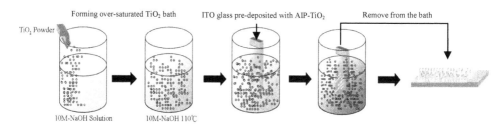

Fig. 6. The HR system and preparation method to obtain TiO$_2$ nanowires.

Figure. 7 shows the *I-V* characteristics of the DSSCs assembled by using HR-TiO$_2$ as the photoanode deposited at different TiO$_2$ powder content, with their corresponding surface and cross sectional film morphology also shown. The dense columnar AIP-TiO$_2$ bottom layer can partially be seen in cross sectional view for each specimen. The result of *I-V* curve for the DSSC assembled directly from Degussa TG-P25 as the photoanode is also shown. The HR was indeed capable of generating randomly-stacked TiO$_2$ nanowires on template, regardless of the TiO$_2$ content. It can also be found that the *I-V* characteristics are sensitive to the TiO$_2$ powder content, but unfortunately non-linearly responded. The HR-TiO$_2$ obtained at high TiO$_2$ content, say 75 g/l, presents no longer nanowires, but agglomerated powdery feature instead. This corresponds to a less specific surface area for dye adsorption and a decreased overall photovoltaic efficiency (mainly cause a reduction of the J_{sc}). From cross sectional image, the as-grown HR-TiO$_2$ thickness is insusceptible to the TiO$_2$ powder

Chasing High Efficiency DSSC by Nano-Structural Surface Engineering at Low Processing Temperature for
Titanium Dioxide Electrodes

71

content. Ultimately, the highest photovoltaic efficiency of 3.63 % is achieved for the HR-TiO$_2$
obtained at a 50 g/l TiO$_2$ powder content. Interestingly, some of the photovoltaic efficiency
of DSSCs assembled from HR-TiO$_2$ nanowires surpassing that of the DSSC assembled from
Degussa P-25 powder, proves that using the one-dimensional structure to enhance DSSC
efficiency is conceptually correct.

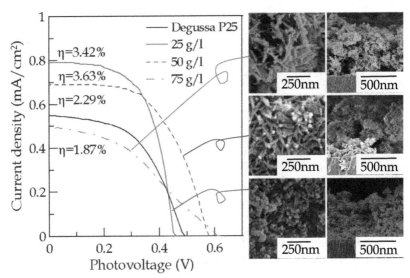

Fig. 7. *I-V* characteristic of the DSSCs assembled by HR-TiO$_2$ as photoanode prepared under
different amount of TiO$_2$ powder, with their corresponding surface and cross sectional film
morphology.

The *I-V* characteristics of the DSSCs assembled with HR-TiO$_2$ nanowires on template and
annealed at different temperatures, with their corresponding XRD pattern and surface
morphology are also shown in Fig. 8. Basically, the as-grown HR-TiO$_2$ nanowires are
amorphous and account for the lowest J_{sc} of the assembled DSSCs. However, the high
crystallinity of the AIP-TiO$_2$ bottom layer facilitates the diffraction peaks shown in the XRD
patterns, even though amorphous HR-TiO$_2$ nanowires cover all over the top. By knowing
this, the specimens with the HR-TiO$_2$ nanowires on template shown in XRD patterns give a
gradual increase in peak intensity when annealing temperature is increased. Apparently,
this shall be due to the improved crystallinity of the HR-TiO$_2$ nanowires by the annealing
process. This helps for the increased J_{sc} of the assembled DSSCs as can be observed in Fig. 8.
The annealing crystallized HR-TiO$_2$ nanowires provides more surface area for dye absorbing
and thus the increased J_{sc} of the assembled DSSCs. The side effect accompanied with
annealing to the TiO$_2$ nanowires is the decrease in V_{oc} of the assembled DSSCs as can be seen
again in Fig. 8. This can be ascribed to the volume change of the re-grown HR-TiO$_2$ that pays
for the open channel for the I$_2$+LiI liquid electrolyte to be in direct contact with the AIP-TiO$_2$
bottom layer. The ultimate PV efficiency of 3.63% can be achieved in this study. By using
this method, annealing temperature shall however be carefully selected to trade-off the J_{sc}
and V_{oc} of the assembled DSSCs.

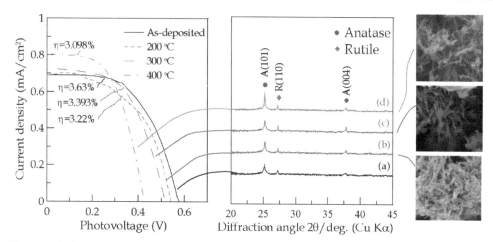

Fig. 8. (a) *I-V* characteristics of the DSSCs assembled with HR grown TiO_2 nanowires on template and annealed at (b) 200, (c) 300 and (d) 400 °C for 30 min, with their corresponding XRD pattern and surface morphology

3.1 Summary

In this study, hydrothermal method was demostrated to successfully prepare the randomly-orientated TiO_2 nanowires on AIP-TiO_2 template. A cell of ITO glass/AIP-TiO_2/[nanowire-TiO_2 (N3 dye)]/I_2+LiI electrolyte/Pt/ITO glass was constructed. Although TiO_2 nanowires randomly-orientated, it possesses remarkable PV efficiency. By optimizing hydrothermal process condition and annealing treatment, an ultimate PV efficiency of 3.63% can be achieved. The AIP-TiO_2 accidentally acts as a block layer for the I_2+LiI electrolyte in the assembled PV device. A hydrothermal treatment time so long as 24 hours shall be required for achieving this, which however has shorter treatment time than the LPD process and a fair PV efficiency without post-thermal annealing. This study also implicates a new possibility for 1-D nanomaterial, such as nanotubes, that can rapidly transfer of the charge carriers along the length of TiO_2 nanotubes. The method to grow the TiO_2 nanotubes is sketched as below.

4. PVD titanium followed by anodic oxidation to grow TiO_2 nanotubes

Anodization is one promising route to prepare long and highly ordered TiO_2 nanotubes array. This has been demonstrated by Shankar et al. who synthesized TiO_2 nanotube array on titanium foil with a tube length up to 220 μm. Very short anodic oxidation treatment time is required as compared to LPD and HR and might bring this technique a step further toward industrial practice. However, this tube-on-foil design may potentially only be applied as a back-side illuminated DSSCs which are predestined to deplete certain quantity of incident light while traveling through the I_2+LiI electrolyte. Direct growth of TiO_2 nanotubes array on transparent conducting oxide (TCO) glass substrate via anodizing a sputtering-deposited or evaporation-deposited titanium layer on TCO for constructing front-side illuminated DSSCs has been attempted, but suffering with a problem of easy detachment of the TiO_2 nanotubes array. By considering this, a two-step method involving

Chasing High Efficiency DSSC by Nano-Structural Surface Engineering at Low Processing Temperature for
Titanium Dioxide Electrodes

73

AIP metal titanium film on ITO glass followed by anodic oxidation was proposed. A tenaciously and dense AIP titanium layer was obtained and was bearable for subsequent anodic oxidation. In this approach, TiO₂ nanotubes array was successfully formed by anodizing the pre-deposited AIP metal Ti on ITO glass. A 5 μm-thick metal Ti layer can be used to convert into a 10 μm-thick amorphous TiO₂ nanotubes array by anodic oxidation for 2 h. NH₄F and H₂O addition in the ethylene glycol (EG) bath and post-heat treatment on the microstructure of TiO₂ nanotube array in responding to the photovoltaic property of the assembled DSSCs were investigated.

For better morphological control of TiO₂ nanotubes before further evaluation on the microstructure and photovoltaic property, the TiO₂ nanotubes growth mechanism was revealed during anodic oxidation, anodic current occurring to the specimen was recorded and the accompanied surface morphology was observed through the whole stages as shown in Fig. 9. It is seen that a rapid decrease of current density is caused when a thin passivated oxide layer was developed on the Ti surface in the beginning stage as can be seen in Fig. 9(a). Then, localized dissolution of the oxide layer begins to form pits over the entire oxide layer surface. This causes a small turbulent current density as presented in Fig. 9(b). At the bottom of each pit, the relatively thinner oxide layer (than that around the periphery) facilitates a localized electric field intensity across the oxide layer and drives the pit growth inward further. The continuing growth of the pit pushes oxide/metal interface inward while charge exchange occurs to the inner wall of the pit to form nanotube. At the same time, a steady-state current density is observed as can be seen in Fig. 9(c). An extended anodizing time can completely consumes the pre-deposited titanium metal layer and rapid decrease in current density is observed as shown in Fig. 9(d).

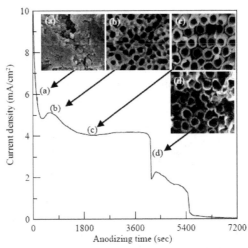

Fig. 9. Current density and surface morphology variation during anodic oxidation.

For exploring the effect of anodizing bath composition on the microstructural evolution of the grown TiO₂ nanotubes, five different types of bath composition were evaluated and their composition were listed in Table 1, where bath A, B and C are different in content of H₂O addition and bath B, D and E are different in content of NH₄F addition. Fig. 10 shows the SEM observation result of the TiO₂ nanotubes arrays anodized in the bath A, B, C, D and E.

It was indeed capable of producing nanotube featured TiO_2 films shown in the SEM morphology, regardless of the electrolyte composition. It can be noted that the entire grown TiO_2 nanotubes array (using whatever electrolyte bath) are strongly adhered on the ITO glass. Some TiO_2 nanotubes grow slower at specific electrolyte composition (bath D and E for example) and leave a remnant titanium layer (identified separately as α-titanium) beneath the nanotubes.

Electrolyte bath	Composition
A	1 L EG + 3 g NH_4F + 0.1 g H_2O
B	1 L EG + 3 g NH_4F + 20 g H_2O
C	1 L EG + 3 g NH_4F + 40 g H_2O
D	1 L EG + 1.5 g NH_4F + 20 g H_2O
E	1 L EG + 2 g NH_4F + 20 g H_2O

Table 1. Electrolyte composition used in this study for anodization to obtain TiO_2 nanotubes.

More quantitative comparison of the SEM observations, the tube length, inner diameter and outer diameter of the TiO_2 nanotubes anodized in electrolytes A, B, C, D and E are measured and drawn in Fig. 10. For electrolyte A, B and C (in sequence of increasing water content of the electrolyte), the tube length and tube diameter (inner and outer) are bar chart illustrated in Fig. 10 (upper right). The water content is found to not only influence the tube diameter but also the tube length. With increasing water content, the tube length decreases, but the tube diameters increase. One explanation for this is that the H_2O not only inhibits nanotubes growth but also dilutes the reactivity of NH_4F of the electrolyte. On the other hand for electrolyte D, B and E (in sequence of increasing NH_4F content of the electrolyte), the tube length and tube diameter (inner and outer) are also bar chart illustrated in Fig. 10 (bottom right). It can be found that increasing NH_4F content of the electrolyte prompts the tube growth rate to obtain longer tubes while at the same time with a decreased diameter. In this regard, NH_4F behaves as the active regent for the formation of nanotubes and restricts lateral growth of the nanotubes. As a whole of anodizing variables study here, it demonstrates a feasible way to convert AIP metal titanium layer into TiO_2 nanotubes array on the ITO glass by anodic oxidization procedure. The firmly adhered AIP metal titanium layer guarantees the successful growth of TiO_2 nanotubes. By knowing this and compromising tube length and diameters, the following I-V characteristics study for the DSSCs assembled by using TiO_2 nanowirs are based on the electrolyte B.

The I-V characteristics of the DSSCs assembled by using TiO_2 nanowirss with different annealing temperatures, with their corresponding XRD pattern was also shown in Fig. 11. As opposed to those Ti layer obtained by using sputter deposition, the AIP-deposited Ti layer exhibits mainly crystallinic α-Ti phase and account for the strong film adhesion. The as-anodized TiO_2-nanotube array presents an X-ray amorphous structure with trace amount of remnant α-Ti. The diffraction peaks corresponding to anatase phase TiO_2 can be found to appear in the specimens annealed over 250 °C indicating that the crystallization occurs to the amorphous TiO_2 nanotubes after post-annealing. The intensity increase of the diffraction peaks corresponding to the anatase phase TiO_2 shows that crystallinity of the nanotubes increases with the annealing temperature. However, the disappearing of the diffraction peak corresponding to remnant α-Ti can only be observed for the specimen annealed at 450 °C. This suggests that complete thermal oxidation of remnant α-Ti layer took place at a temperature over 450 °C. Furthermore, it can also be found that the I-V characteristics are

Chasing High Efficiency DSSC by Nano-Structural Surface Engineering at Low Processing Temperature for
Titanium Dioxide Electrodes

75

sensitive to the annealing temperature, the J_{sc} in particular (due to the enhanced crystallinity of nanotubes). The V_{oc} unfortunately on the contrary decreases with increasing post-annealing temperature because of the negative effect devastated by the increased sheet resistance of the ITO film (measured but not shown). An overview of the photovoltaic efficiency of the cell assembled from the as-anodized and post-annealed TiO_2-nanotube array, when an annealing temperature is over 350 °C, a maximum efficiency of 1.88% can be obtained and subsequent a decrease in J_{sc} occurs leading to a decreased efficiency of 0.53%. Apparently, this result from two opposite competitive factors, i.e. the sheet resistance of the ITO film and the profitable crystallinity of TiO_2 nanotubes, which can be affected by the post-annealing. When increasing post-annealing temperature, the improved crystallinity of the anatase TiO_2 nanotubes array facilitates a more ideal electron migration path from dye to ITO front electrode, therefore an increased J_{sc}, but the abrupt increase in sheet resistance of the ITO film (over 450 °C) seriously hinders electron current flowing through it.

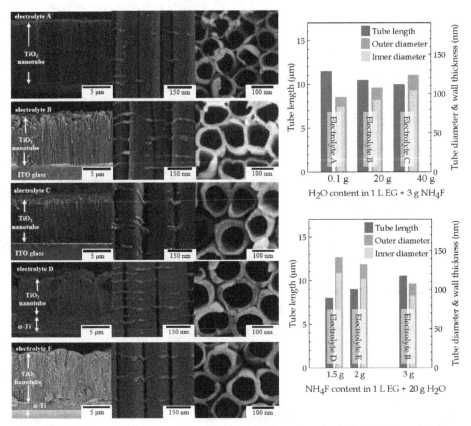

Fig. 10. SEM images of TiO_2 nanotubes anodized in electrolyte bath A, B, C, D and E. The left pictures are through-thickness cross sectional view of the TiO_2 nanotubes at low magnification, the middle pictures are magnified image of the tubes, the right pictures are top-view of the tubes. The tube length, inner diameter and outer diameter of the TiO_2 nanotubes anodized in electrolytes A, B, C, D and E are also measured and compared.

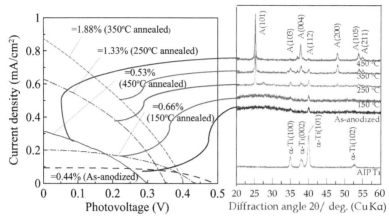

Fig. 11. *I-V* characteristic of the cell assembled from the as-anodized and post-annealed TiO₂-nanotube array which was produced by anodic oxidation with their corresponding XRD patterns of AIP-deposited Ti, as anodized TiO₂ nanotubes array and post-annealed TiO₂ nanotubes array.

4.1 Summary

Successful demonstration to prepare TiO₂ nanotubes array by arc ion plating pre-deposit metal Ti layer on ITO glass followed by anodic oxidation has been carried out in this study to reveal the influence of anodization electrolyte variables and post-heat treatment on the microstructure of TiO₂ nanotubes array and the photovoltaic behavior of the assembled DSSCs device. The key to successfully develop 10 micrometer long TiO₂ nanotubes array lies in the strongly adhered Ti-layer which tolorates the electrolyte attack during anodic oxidation. Ultimate photovoltaic efficiency of 1.88% appears on the DSSC assembled from TiO₂ nanotubes array which was annealed at 350 °C. However, the annealing temperature that requires to form anatase phase through post-annealing would be detrimetal to the ITO front electrode and limits further increase in photovoltaic efficiency.

5. Micro-arc oxidation and alkali etching to produce nanoflaky TiO₂

Micro-arc oxidation (MAO) technique is a relatively convenient and effective technique for producing micrometer scale porous crystalline anatase TiO₂ over a metal titanium surface. This technique involves the anodically charging of a metal (similar to conventional anodic oxidation but with a higher level of discharge voltage) in a specific electrolyte to reach a critical value at which dielectric breakdown takes place, and initiates micro-arc discharge over the entire metal surface. The micro-arc discharge enables the rapid oxidation of the metal due to the effect of impact or tunneling ionization over the metal surface. The schematic MAO system to obtain TiO₂ films is shown in Fig. 12. First attempt using MAO technique to grow microporous TiO₂ over a Ti surface for applying as DSSC electrode has also demonstrated, with however limited photovoltaic efficiency due to unsatisfactory specific surface area. In responding to the demanding in high efficiency PV device, we have developed another two-step method for the Ti foil to grow nanoflaky TiO₂. An idea is proposed in this study simply by using alkali etching to develop nanoflaky morphology

Chasing High Efficiency DSSC by Nano-Structural Surface Engineering at Low Processing Temperature for
Titanium Dioxide Electrodes

77

over the pre-micro-arc oxidized Ti (i.e. MAO-TiO₂) as the ideal electron emitter (or TiO₂ electrode). Such a nano featured TiO₂ layer shall be able to exhibit very large specific surface area and capable of efficient dye absorbing and eventually high photovoltaic efficiency. The alkali etching began with the immersion the MAO treated titanium foil into a NaOH solution and soaking for 12 h to develop nano-featured TiO₂. Later on, an alkali etching treatment followed by MAO was proposed to develop 3D-network nanostructural anatase TiO₂ without annealing, with the accompanied photovoltaic efficiency substantially improved. In this work, a further detailed observation on the microstructural development of the nanostructural anatase TiO₂ is carried out as a function of alkali bath concentration and post-heat treatment effect to the associated photovoltaic efficiency is correlated.

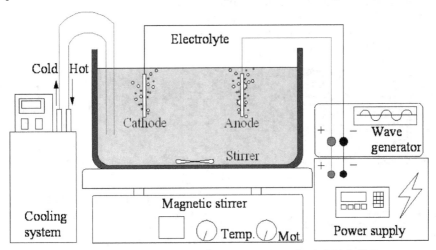

Fig. 12. Schematic diagram of micro-arc oxidation system to obtain MAO-TiO₂.

Figure. 13 shows the surface and cross sectional morphology of the MAO formed titania layer as well as the alkali etched TiO₂ layers obtained at different bath concentration. After MAO treatment, the titanium forms porous crystallinic anatase TiO₂ layer (as identified and described elsewhere) with numerous micrometer scale holes as observed in Fig. 13(a). These holes are discharge channels induced by the electrical breakdown of the oxide layer during the MAO treatment. It is worth noting that the surface is roughened, which is based on the fact that an intensive microdischarge occurs at a high voltage; as a result (Fig. 13(a)), the coating itself appears to be a microscopically splashed surface under the strong discharge effect. The morphology of the specimens alkali etched at different NaOH concentration shown in Fig. 13(b)~(d) reveal that nanoflaky TiO₂ can be developed through the alkali etching. The nano featured layer was developed over the MAO-TiO₂ scaffold surface with free interspace and nanoflakes of about 50~100 nm in size. As can be seen from the figure, these nanoflakes uniformly distribute over the entire surface of the treated specimen. The results revealed that alkali solution concentration appear to be an important variable in nanostructural control. Moreover, the higher NaOH concentration leads to much bigger free interspace and deeper nanoflaky TiO₂ layer as well as bigger nanoflake size. It is therefore out of question that the TiO₂ layer reformed by the alkali etching can have higher specific surface area than the MAO-TiO₂. Through the evaluation of a series of alkali-etched specimen at different NaOH concentrations, the size of the developed nanoflakes is found to be determined by the NaOH concentration. The morphological development of the nanoflakes is thought to be associated with the complicated dissolution and re-precipitation mechanism that involves the attack by hydroxyl groups and negatively charged $HTiO_3^-$ ions formed on the surface. The $HTiO_3^-$ ions are thought to be consequently attracted and dissolved by the positively charged ions in the NaOH solution. In our case, it is hypothetically proposed that the low-concentration NaOH solution gives rise to the diffusion control mode enabling charged ion exchange between the MAO specimen surface and the alkali solution, where a limited ion flux yields a low reaction rate that favors fine structure formation. Contrarily, the high NaOH bath concentration enables fast exchange of the charged ion species and fast structure formation (accompanied by the flakes grown in larger dimension and larger interspace). In addition, cracks occur to the nanoflaky TiO_2 layer when NaOH bath concentration is increased. The results reflected in Fig. 13(c) and (d) indicate that the cracks began to form on the MAO specimen surface and grow with the increasing NaOH concentration.

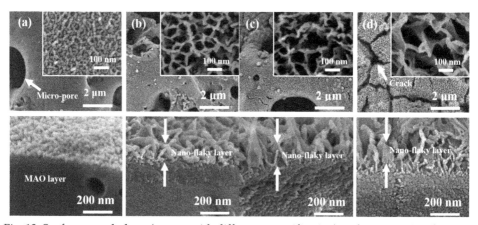

Fig. 13. Surface morphology (upper, with different magnification) and cross sectional morphology (lower) of the (a) MAO treated specimen, and alkali etched specimen in NaOH bath concentration of (b) 0.50 M (c) 1.25M and (d) 2.50 M, respectively.

Further exam of the detailed microstructure of nanoflakes by using transmission electron microscope (TEM) in high magnification bright field images taken from specimen with alkali etching at 40 °C for 12 h are shown in Fig. 14. It can be seen that the hair-like structure (corresponds to the nanoflaky structure as been observed in Fig. 13) exists over the TiO_2 surface as shown in Fig. 14(a). Here, it clearly presents a 3D network fine structure. In addition, the hair-like structure grown from the inner wall of the pore as also observed in the Fig. 14(b) is again seen as a 3D network feature. These 3D nanoflakes led to a significant increase in specific surface area and presumably photovoltaic efficiency. It should also be noted that these pores and voids are opened to the alkali etched and their surfaces are also involved with the reforming process via dissolution and re-deposition. This means that the nanoflakes grow not only on the TiO_2 surface but also grow deep into the inner surfaces, thereby significantly increase specific surface area, even though these nanoflakes unfortunately appear to be amorphous as identified by TEM selected area diffraction technique and described elsewhere.

Chasing High Efficiency DSSC by Nano-Structural Surface Engineering at Low Processing Temperature for
Titanium Dioxide Electrodes

79

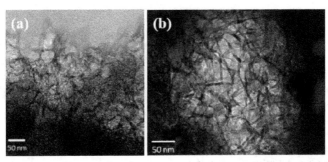

Fig. 14. Bright field image of nanoflaky TiO$_2$ grown from (a) the MAO-TiO$_2$ surface and (b) the inner pore of the MAO-TiO$_2$.

The I-V curves of DSSCs assembled with the MAO-TiO$_2$ and alkali etched TiO$_2$ obtained at different concentrations are shown in Fig. 15.

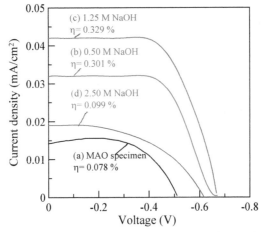

Fig. 15. I-V characteristic of the DSSC device assembled using (a) MAO treated specimen and alkali etched specimens at different NaOH bath concentration.

Photovoltaic efficiency of the assembled DSSC is substantially increased by alkali etching. Apparently, the remarkable increase in the J_{sc} and V_{oc} of the cell assembled from alkali etched specimens appear to be contributed to by the nanoflaky surface structure, which possesses a markedly higher specific surface area than the MAO layer. Note that the J_{sc} is significantly dropped for the DSSC using alkali etched TiO$_2$ specimen prepared at 2.5 M NaOH. This is due to the cracks formed and distributed over the entire oxide layer leaving the I$_2$+LiI electrolyte to directly contact with fresh metallic titanium plate. A close look at Fig. 13(b), (c) and (d), the DSSC assembled by the alkali etched specimen at 1.25 M NaOH solution performs the highest J_{sc} and V_{oc} among the three alkali etched specimens. Good explanation is that this is a compromising of the effect of the enlarged specific surface area and the effect of crack formation caused by the alkali etching, i.e. the increased NaOH bath concentration not only results in the increased specific surface area but also the increased free interspace and even worse the crack formation. As revealed in Fig. 13(d), the cracks causing the discontinuity of path for charge carrier shall be the main reason for the significant decrease in photovoltaic efficiency of the assembled DSSC which employs specimen alkali etched at 0.50 M NaOH. By comparison, the DSSC assembled by using MAO scaffold presents a photovoltaic efficiency of 0.078%, while it present a highest photovoltaic efficiency of 0.329% (over four times increment) for the DSSC assembled by the alkali etched specimen. Through this simple and low cost alkali etching route, it is able to produce nano structural TiO$_2$ electrode for photovoltaic DSSC.

The I-V characteristics of the DSSCs assembled by using specimens with MAO-TiO$_2$ and nanoflaky TiO$_2$ (as-etched and annealed at 400 °C), with their corresponding XRD patterns are shown in Fig. 16. The annealing work significantly improves the crystallinity of the nanoflakes and consequently photovoltaic efficiency can be dramatically increased for the device assembled with the specimen with nanoflaky TiO$_2$. It was 0.329% for the specimen with the as-etched nanoflaky TiO$_2$ and 2.194% for the specimen with annealed nanoflaky TiO$_2$. Both are however greater than that of the MAO-TiO$_2$ specimen only with 0.061%. By contrast, the J_{sc} and V_{oc} of the solar cell assembled by alkali etched specimen are substantially higher. Apparently, the dramatic increase in J_{sc} and V_{oc} of the cell assembled by alkali treated specimens is contributed by the nanoflaky surface structure, which possesses far higher specific surface area than MAO layer does. With this simple and low cost post-alkali etching demonstration, the photovoltaic efficiency of the DSSC using the MAO treated Ti foil as the back electrode can be significantly increased. The increased crystallinity provides higher dye-absorption for generating more electron-hole pairs and suppresses the electron loss due to the recombination of electron-hole pairs. Therefore, the DSSC assembled with Ti electrode which synthesized by MAO, treated by alkali etching and annealing, presents highest photovoltaic efficiency.

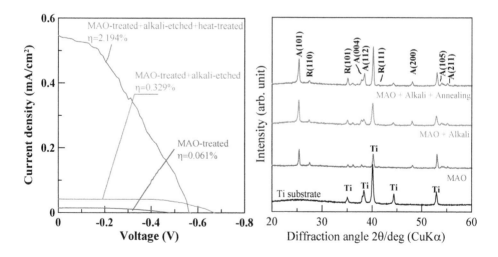

Fig. 16. The I-V characteristics of the DSSCs assembled by using specimens with MAO-TiO$_2$ and nanoflaky TiO$_2$ (as-etched and annealed at 400 °C), with their corresponding XRD patterns.

5.1 Summary

Here at last, a method to combine with micro-arc oxidation and post-alkali etching has succeeded in forming an 3D network nanoflaky anatase TiO_2 layer on the surface of a Ti substrate. The nanoflaky TiO_2 completely cover the upmost surface of the MAO TiO_2 layer as well as the inner pores and voids, therefore provides very large surface area for dye absorption to increase the efficiency of the assembled DSSCs. Without post-annealing, maximum photovoltaic efficiency of 0.329% for the DSSC is achieved with the amorphous nanoflaky TiO_2 layer alkali etched at 1.25 M NaOH. Post-annealing at 400 °C significantly enhances crystallinity of the nanoflaky TiO_2 layer and ultimately photovoltaic efficiency 2.194% for the DSSC is achieved. The above results have shown that the method to combine with micro-arc oxidation and post-alkali etching was a potential and low-cost process for developing the nano featured TiO_2 photoanode for obtaining high efficiency DSSC. However, post-annealing shall not be abandoned for additional windfall of the photovoltaic efficiency.

6. Conclusion

Here, we begin our conclusion by reviewing the results in previously described. We started developing nano featured TiO_2 layer by using LPD method and found that it is capable of obtaining nanocluster TiO_2 with unacceptable growth rate. Crystallinic anatase TiO_2 layer can be obtained through suitable annealing treatment to achieve only 0.0056% photovoltaic efficiency. In the work of TiO_2 nanowires growth on AIP-TiO_2 template via hydrothermal route, an ultimate PV efficiency of 3.63% can be achieved by optimizing hydrothermal process condition and annealing treatment. A hydrothermal treatment time so long as 24 hours shall be required for achieving this, which however has shorter treatment time than the LPD process and a fair PV efficiency. In the work of preparing TiO_2 nanotubes array by arc ion plating pre-deposit metal Ti layer on ITO glass followed by anodic oxidation, the key to successfully develop 10 micrometer long TiO_2 nanotubes array lies in the strongly adhered Ti-layer which tolerates the electrolyte attack during anodic oxidation. Ultimate photovoltaic efficiency of 1.88% appears on the DSSC assembled from TiO_2 nanotubes array which was annealed at 350 °C. Although the tube length and diameter is controllable, it is expected to exhibit higher photovoltaic efficiency by further reducing tube diameter for more specific surface area of the photoanode. At last, a DSSC assembled from the nanoflaky TiO_2 prepared by using micro-arc oxidation and alkali etching was demonstrated. Ultimate photovoltaic efficiency 2.194% for the DSSC is achieved.

Results in these studies are remarkably consistent with what we expected. Cost saving and easy operation processes for obtaining TiO_2 photoanode has been achieved. Despite the encouraging result of this study as the positive effect of nanostructural surface engineering, future research is required in a number of directions about chasing high efficiency DSSCs. However, a step further has been taken in the improved photovoltaic efficiency by nanostructural surface engineering and an opportunity for commercializing DSSC using low-cost process.

7. References

O'Regan, B. & Grätzel, M. (1991). *Letters to nature*, Vol.353, pp. 733-740

Nagayama, H., Honda, H. & Kawahara., H. (1988). *Journal of the Electrochemical Society*, Vol.135, pp. 2013-2014

Herbig, B. & Löbmann, P. (2004). *Journal of Photochemistry and Photobiology A: Chemistry*, Vol.163, pp. 359-365

He, J. L., Chen, C. H. & Hsu, M. Y. (2006). *The Chinese Journal of Process Engineering*, Vol.6, pp. 224-227

Rao, C. N. R., Satishkumar, B. C., Govindaraj, A., Vogl, E. M. & Basumallick, L. (1997). *Journal of Materials Research*, Vol.12, pp. 604-606

Kasuga, T., Hiramatsu, M., Hoson, A., Sekino, T. & Niihara, K. (1999). *Advanced Materials*, Vol.11, pp. 15-18

He, J. L., Hsu, M. Y., Li, H. F. & Chen, C. H. (2006). *The Chinese Journal of Process Engineering*, Vol.2, pp. 228-234

Shankar, K., Mor, G. K., Prakasam, H. E., Yoriya, S., Paulose, M., Varghese, O. K. & Grimes, C. A. (2007). *Nanotechnology*, Vol.18, pp.065707

Chen, C. H., Chen, K. C. & He, J. L. (2010). *Current Applied Physics*, Vol.10, pp. S176

Yerokhin, A. L., Nie, X., Leyland, A., Matthews, A. & Dowey, S.J. (1999). Surface and Coatings Technology, Vol.122, pp. 73-93

Song, W., Xiaohong, W., Wei, Q. & Zhaohua, J. (2007). *Electrochimica Acta*, Vol.53, pp.1883-1889

Wei, D., Zhou, Y., Jia, D. & Wang, Y. (2008). *Acta Biomaterialia*, Vol.254, pp.1775-1782

Wu, S. Y., Lo, W. C., Chen, K. C. & He, J. L. (2010). *Current Applied Physics*, Vol.10, pp.S180-S183.

Wu, S. Y., Chen, Y. H., Chen, K. C. & He, J. L. (2010). *Japanese Journal of Applied Physics*, Vol.10, pp.180-183.

Chu, P. J, Wu, S. Y., Yerokhin, A., Matthews, A. & He, J. L. (2009). *Program and Abstract Book of TACT 2009 International Thin Films Conference*, (December 2009.) pp.115-116

5

The Application of Inorganic Nanomaterials in Dye-Sensitized Solar Cells

Zhigang Chen, Qiwei Tian, Minghua Tang and Junqing Hu
State Key Laboratory for Modification of Chemical Fibers and Polymer Materials,
College of Materials Science and Engineering, Donghua University
China

1. Introduction

Energy crisis and environment pollution cause a great quest and need for environmentally sustainable energy technologies. Among all the renewable energy technologies, photovoltaic technology utilizing solar cell has been considered as the most promising one (Chen et al., 2007a; Gratzel, 2001). As early as in 1954, researchers demonstrated the first practical conversion from solar radiation to electricity by a *p-n* junction type solar cell with 6% efficiency (Chapin et al., 1954). Up to now, the common solar power conversion efficiencies of this type solar cell are beyond 15% (Tributsch, 2004). Unfortunately, the relatively high cost of manufacturing and the use of toxic chemicals have prevented their widespread use, which prompts the search for high efficient, low cost and environmentally friendly solar cells.

Semiconductor with a very large bandgap, such as TiO_2, ZnO and SnO_2, can be employed to construct solar cells. But these materials can only be excited by ultraviolet or near-ultraviolet radiation that occupies only about 4% of the solar light. Dye molecules as light absorbers for energy conversion have shaped evolution via the process of photosynthesis and photo-sensoric mechanisms (Tributsch, 2004). Dye sensitization of semiconductor with a wide band-gap has provided a successful solution to extending the absorption range of the cells to long wavelength region. This approach also presents advantages over the direct band-to-band excitation in conventional solar cells, since attached dyes, rather than the semiconductor itself, are the absorbing species (Garcia et al., 2000). Importantly, light absorption and charge carrier transport are separated, and the charge separation takes place at the interface between semiconductor and sensitizer, preventing electron–hole recombination. Since the discovery of the photocurrents resulting from dye sensitization of semiconductor electrodes in 1968, dyes have been widely used in electrochemical energy converting cells (Tributsch, 1972).

During the first years of the sensitized solar cell research, most studies were made with single crystal oxide samples, because by eliminating grain boundaries and high concentrations of surface states the interpretation of the results became more transparent (Tributsch, 2004). However, at that time, the power conversion efficiencies were very low (< 1%). Tsubomura *et al.* reported a breakthrough in the conversion efficiency in 1976 (Tsubomura et al., 1976). They used the powdered high porosity multi-crystalline ZnO instead of single crystal semiconductor, resulting in a significant increase of the surface area of the electrode. When the dye (Rose Bengal) was used as the sensitizer, an energy efficiency of 1.5% was obtained for light incident within the absorption spectrum of the sensitizer.

Later, in 1980, Tsubomura group increased the surface roughness of ZnO samples and demonstrated an energy efficiency of 2.5%, also for light incident within the absorption spectrum of the sensitizer (Matsumura et al., 1980).

A significant advance in the field of dye-sensitized solar cells (DSCs, also considered as Grätzel cell) was made through the efforts of Grätzel and coworkers in 1991 (Fig. 1) (Oregan & Gratzel, 1991). They greatly improved the power energy conversion efficiency from lower than 2.5 to 7%, and the main reasons for the improvement were as follows: (1) the preparation of nanostructured TiO$_2$ film (Desilvestro et al., 1985), (2) the use of ruthenium complex that was adequately bonded to TiO$_2$ nanoparticles and (3) the selected organic liquid electrolyte based on iodide/triiodide. The jump in solar energy conversion efficiency has attracted considerable attentions and motivated significant optimism with respect to the feasibility of DSCs as a cost-effective alternative to conventional solar cells.

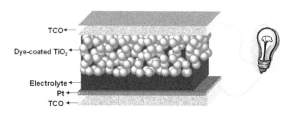

Fig. 1. Typical sandwich-type dye-sensitized TiO$_2$ nanocrystal solar cell.

The typical DSCs are composed of a transparent nanoporous semiconductor electrode on transparent conducting optically (TCO) glass, a very thin layer of light-absorbing material (dyes) — sensitizer on the entire surface of semiconductor (such as TiO$_2$) electrode, a counter electrode (such as Pt), and a transparent hole conductor (electrolyte) filling the pores, in contact with the light-absorbing layer at all points. The typical dye-sensitized TiO$_2$ nanocrystal solar cell is shown in Fig. 1. The operational principle of DSCs is described as follows. Photons enter DSCs and can be absorbed by dye molecules (D) at various depths in the film. The dye molecule after absorbing photon will then be promoted into its excited state (D*) from where it is now energetically able to inject an electron into the conduction band of semiconductor, leaving an oxidized dye (D$^+$) on the semiconductor surface. The injected electrons percolate via the interconnected nanoparticles to the substrate and are fed into an electrical circuit, where it can deliver work. Subsequently, the electrons, which now carry less energy, enter the cell again via the counter electrode and transport to hole-conductor materials (commonly the organic electrolyte containing iodide/triiodide, this process can be described as: I$_3^-$ +2e$^-$ →3I$^-$). The oxidized dye (D$^+$) is then reduced back to its original state by hole-conductor materials (such as, this process can be typically described as: 2D$^+$ + 3 I$^-$ → 2 D + I$_3^-$). Then the cycle is completed.

The main issues for the development of DSCs are to improve their photoelectric conversion efficiency, thermostability and long-term stability, which are strongly dependent on the advances in nanotechnology and nanomaterials. It is well known that nanotechnology opens a door to tailing materials and creating various nanostructures for the use in dye-sensitized solar cells. A predominant feature of these nanostructures is that the size of their basic units is on nanometer scale (10^{-9} m), and inorganic nanomaterials therefore present an internal surface area significantly larger than that of bulk materials. Currently, inorganic nanomaterials have been widely used in all components in DSCs, including transparent semiconductor electrode, counter electrode, electrolyte and light-absorbing material. In this chapter, the recent progress on the selection and utilization of inorganic nanomaterials in dye-sensitized solar cells are mainly introduced and discussed.

2. The application of semiconductor nanomaterials in photoanodes

Compared to bulk materials, semiconductor nanomaterials as photoanodes can offer a larger surface area for dye adsorption, contributing to optical absorption and leading to an improvement in the solar cell conversion efficiency. As photoelectrode materials in DSCs, the semiconductor nanostructures are usually classified into two types: (1) nanoparticles, which offer large surface area to photoanodes for dye-adsorption, however have recombination problem due to the existence of considerable grain boundaries in the film. To settle this issue, core-shell structure derived from the nanoparticles by forming a coating layer has been developed and applied to DSCs with a consideration of suppressing the interfacial charge recombination, while this kind of structure has been proved to be less effective and lack of consistency and reproducibility; (2) one-dimensional nanostructures such as nanowires and nanotubes, which are advantageous in providing direct pathways for electron transport much faster than in the nanoparticle film, however face drawback of insufficient internal surface area of the photoelectrode film, leading to relatively low conversion efficiency (Zhang & Cao, 2011). This section aims to demonstrate semiconductor nanomaterials as photoanodes, including nanoparticles, nanowires and nanotubes.

2.1 Semiconductor nanoparticles

Among different nanostructures, nanoparticles have been most widely studied for the use in DSCs to form photoelectrode film (Oregan & Gratzel, 1991). This is because, to a large extent, the photoelectrode films comprised of nanoparticles can give a high specific surface area, resulting in an efficient extinction of incident light within film which is a few microns thick. And the anodes of DSCs are typically constructed with the nanoparticles film (thickness: ~10 μm) of wide bandgap semiconductor including SnO_2, ZnO, and TiO_2.

Among these wide bandgap semiconductors, titania (TiO_2), an n-type semiconductor with a wide bandgap (3.2 eV for anatase), has been well known and widely used in the photoanode of DSCs. As early as 1985, Desilvestro et al (Desilvestro et al., 1985) have showed that if TiO_2 is used in a nanoparticle form, the power conversion efficiency of DSC can be drastically enhanced. The improvement of conversion efficiency lies in the superiority of nanoparticles to create large surface, which was demonstrated by a comparison between a flat film and a 10-μm-thick film that consisted of nanoparticles with an average size of 15 nm (Oregan & Gratzel, 1991). The latter, nanoparticle film, showed a porosity of 50-65% and gave rise to almost 2000-fold increase in the surface area. Fig. 2 shows typical cross-section morphology of TiO_2 nanoparticle film with thickness of about 4 μm.

For state-of-the-art DSCs, the employed architecture of the mesoporous TiO_2 electrode is as follows (Hagfeldt et al., 2010): (a) a TiO_2 blocking layer (thickness ~50 nm), coating TCO glass to prevent contact between the redox mediator in the electrolyte and TCO glass ; (b) a light absorption layer consisting of a ~10 μm thick film of mesoporous TiO_2 with ~20 nm particle size that provides a large surface area for sensitizer adsorption and good electron transport to the substrate; (c) a light scattering layer on the top of the mesoporous film, consisting of a ~3 μm porous layer containing ~400 nm sized TiO_2 particles; (d) an ultrathin overcoating of TiO_2 on the whole structure, deposited by means of chemical bath deposition(using aqueous $TiCl_4$ solution), followed by heat treatment.

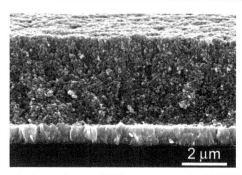

Fig. 2. Cross-section morphology of typical TiO$_2$ nanoparticle film.

Furthermore, it has mentioned that the performance of solar cell is intimately linked to the structure and morphology of the nanoporous oxide layer. For efficient dye distribution, the surface area of the membrane film must be large and porous. So the preparation procedure of TiO$_2$ nanoparticle film must be optimized so as to provide an optimal particle size and porosity features. Up to now, the main synthesis methods of TiO$_2$ nanoparticle film for DSCs include sol-gel/hydrothermal synthesis (Muniz et al., 2011), electrochemical deposition (Murakami et al., 2004), magnetron sputtering (Sung & Kim, 2007), chemical vapor deposition (Murakami et al., 2004) and so on.

The most common technique for the preparation of TiO$_2$ nanoparticles is sol-gel/hydrothermal synthesis, which involves the hydrolysis of a titanium precursor such as titanium (IV) alkoxide with excess water catalyzed by acid or base, followed by hydrothermal growth and crystallization. Acid or basic hydrolysis gives materials of different shapes and properties; while the rate of hydrolysis, temperature, and water content can be tuned to produce particles with different sizes. Transmission electron microscopy measurements revealed that for TiO$_2$ nanoparticles prepared under acidic conditions, crystalline anatase particles were formed exposing mainly the <101> surface (Zaban et al., 2000). Compared with nitric acid, employing acetic acid increases the proportion of the <101> face about 3-fold (Zaban et al., 2000). The differences can be explained by different growth rates: in acetic acid crystal growth is enhanced in the <001> direction compared with the growth in the presence of nitric acid (Neale & Frank, 2007). Hore et al. (Hore et al., 2005) found that base-catalyzed conditions led to mesoporous TiO$_2$ that gave slower recombination in DSCs and higher Voc but a reduced dye adsorption compared with the acid-catalyzed TiO$_2$. The produced TiO$_2$ nanoparticles are formulated in a paste with polymer additives and deposited onto TCO glass using screen printing techniques. Finally, the film is sintered at about 450 °C in air to remove organic components and to make electrical connection between the nanoparticles.

It has been found that pure TiO$_2$ nanoparticles are not perfect in terms of solar cell efficiency, since the charge recombination between the injected electrons in conductor band of TiO$_2$ and electron acceptors in the electrolyte is unavoidable to diminish both photovoltage and photocurrent, thus limiting the device efficiency (Gregg et al., 2001). To improve device efficiency, one effective approach is to grow a thin coating layer of another oxide on the surface of TiO$_2$ particles. Thus, coating TiO$_2$ nanoparticles with a different metal oxide to build core-shell structure has received much attention (Kay & Gratzel, 2002).

Now, two approaches have been developed to create such core-shell structure (Zhang & Cao, 2011). One involves a first synthesis of nanoparticles and then fabricating a shell layer on the surface of nanoparticles. This leads to the formation of core-shell structured nanoparticles, with which the film photoanode is prepared then. Such an approach builds up a photoelectrode structure as shown in Fig. 3a, and an energy barrier is formed at the interfaces not only between nanoparticle/electrolyte but also between the individual core nanoparticles. In another approach, the nanoparticle film photoanode is prepared prior to the deposition of shell layer, receiving a structure as show in Fig. 3b. The latter approach is obviously advantageous in electron transport that happens within single material, but there is usually a challenge in the fabrication of shell layer regarding a complete penetration and ideal coating of the shell material. Metal oxides such as ZnO, $CaCO_3$, Nb_2O_3, $SrTiO_3$, MgO and Al_2O_3 have been usually used as the coating layer for TiO_2. Table 1 shows photoelectric conversion efficiencies of DSCs based on TiO_2 and metal oxides-coated TiO_2 electrodes, which demonstrates the improved efficiency of DSCs employing a core-shell structured TiO_2 electrode. This improvement should be attributed to the following two factors (Jung et al., 2005): First, the wide bandgap coating layer retards the back transfer of electrons to the electrolyte solution and minimizes electron-hole recombination. Second, the coating layer enhances the dye adsorption and increases the volume of the optically active component, leading to the improved cell performance. If pH of the coating oxides is more basic than that of TiO_2, the carboxyl groups in a dye molecule are more easily adsorbed to their surface.

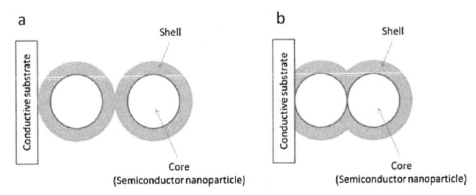

Fig. 3. Core-shell structures used in DSCs. (a) The shell layer is formed prior to the film deposition, (b) the shell layer is coated after the film deposition (Zhang & Cao, 2011).

metal oxides	ZnO	$CaCO_3$	Nb_2O_3	$SrTiO_3$	MgO	Al_2O_3
$\eta^b(\%)$	7.7	6.9	3.6	3.81	3.1	3.93
$\eta^c(\%)$	9.8	7.9	5.0	4.39	4.5	5.91
Ref.	(Wang et al., 2001)	(Wang et al., 2006)	(Chen et al., 2001)	(Diamant et al., 2003)	(Jung et al., 2005)	(Wu et al., 2008)

η^b : Conversion efficiency of DSC based on the bare TiO_2 film
η^c : Conversion efficiency of DSC based on the metal oxides-coated TiO_2 film

Table 1. Photoelectric conversion efficiency of DSCs based on TiO_2 and metal oxides-coated TiO_2 electrodes.

2.2 Semiconductor nanowires

Nanoparticle films have been regarded as a paradigm of porous photoanodes. However, the nanoparticle films are not thought to be ideal in structure with regard to electron transport. While one-dimensional nanostructures, such as nanowires, are advantageous in providing direct pathways for electron transport much faster and therefore giving electron diffusion length larger than in the nanoparticle films. Hence, semiconductor nanowires become a kind of promising candidate for making up the drawbacks of nanoparticle films.

2.2.1 TiO$_2$ nanowires

Oriented single-crystalline TiO$_2$ nanowires are an important one-dimensional nanostructure that has also attracted a lot of interests regarding an application in DSCs (Liu & Aydil, 2009). Hydrothermal growth has been reported to be a novel method for the synthesis of TiO$_2$ nanowire array on TCO glass substrate. In this method, tetrabutyl titanate and/or titanium tetrachloride are used as the precursor, to which an HCl solution is added to stabilize and control the pH of the reaction solution.

Liu et al. (Liu & Aydil, 2009) developed a facile hydrothermal method to grow oriented, single-crystalline rutile TiO$_2$ nanorod films on TCO glass substrate, as shown in Fig. 4a. The diameter, length, and density of the nanorods could be varied by changing the growth parameters, such as growth time, temperature, initial reactant concentration, acidity, and additives. The epitaxial relation between TCO galss and rutile TiO$_2$ with a small lattice mismatch played a key role in driving the nucleation and growth of the rutile TiO$_2$ nanorods. With TiCl$_4$-treatment, DSC based on 4 µm-long TiO$_2$ nanorod film exhibited a power conversion efficiency of 3%. Furthermore, Feng et al. (Feng et al., 2008) presented a straightforward hydrothermal method to prepare single crystal rutile TiO$_2$ nanowire arrays up to 5 µm long on TCO glass via a non-polar solvent/hydrophilic substrate interfacial reaction (Fig. 4b). The as-prepared densely packed nanowires grew vertically oriented from TCO glass along the (110) crystal plane with a preferred (001) orientation (Fig. 4c). The Cl$^-$ ions were explained to play an important role in the growth of TiO$_2$ nanowires by attaching on the (110) plane of TiO$_2$ nanocrystal and thus suppressing the growth of this plane. DSCs based on 2-3 µm long nanowire array demonstrated a very encouraging power conversion efficiency of 5.02%.

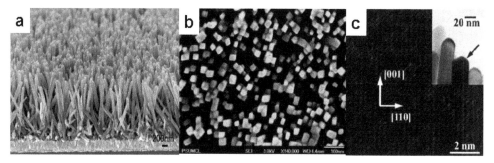

Fig. 4. DSCs with TiO$_2$ nanowires: (a) Cross sectional view of oriented rutile TiO$_2$ nanorod film grown on TCO substrate (Liu & Aydil, 2009); (b) top-view images of vertically oriented self-organized TiO$_2$ nanowire array grown on FTO coated glass (Feng et al., 2008); (c) TEM image illustrating the [001] orientation of TiO$_2$ nanowires (Feng et al., 2008).

In addition, Park et al. (Park et al., 2000) compared the performance between TiO_2 nanowires and nanoparticles which were both rutile phase. The efficiency achieved by TiO_2 nanowire film in 2-3 μm thick was higher than that obtained for the nanoparticle film in 5 μm thick. This is good evidence that the one-dimensional nanostructures may offer better charge transport than the nanoparticles. A further increase in the conversion efficiency most likely relies on the development of new fabrication techniques that can achieve longer TiO_2 nanowires.

2.2.2 ZnO nanowires

Zinc oxide (ZnO) has remarkable optical properties with a wide bandgap (3.2 eV) and a large exciton binding energy (60 meV). In 2005, Law et al. (Law et al., 2005) reported the preparation of ZnO nanowire array on TCO glass by seed mediated liquid phase synthesis method. The experiment was purposely designed to grow ZnO nanowires with a high aspect ratio so as to attain a nanowire film with high density and sufficient surface area (Fig. 5a and b). A ~25-μm-thick film consisting of ZnO nanowires in diameter of ~130 nm was mentioned to be able to achieve a surface area up to one-fifth as large as a nanoparticle film used in the conventional DSCs. The superiority of ZnO nanowires for DSC application was firstly demonstrated by their high electron diffusion coefficient, 0.05-0.5 cm^2 s^{-1}, which is several hundred times larger than that of nanoparticle films. Larger diffusion coefficient means longer diffusion length. In other words, the photoanode made of nanowires allows for thickness larger than that in the case of nanoparticles. This can compensate for the insufficiency of surface area of the nanowire-based photoanode. DSC based on ZnO nanowire gave a power conversion efficiency of 1.5%. In the same year, Prof. Aydil group also successfully prepared ZnO nanowire array on TCO glass and then fabricated a DSC with a conversion efficiency of 0.5% (Baxter & Aydil, 2005).

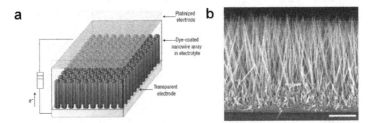

Fig. 5. Nanowire dye-sensitized cell based on ZnO wire array. (a) Schematic diagram of the cell; (b) Typical scanning electron microscopy cross-section of a cleaved nanowire array on TCO Scale bar=5 μm. (Law et al., 2005).

2.3 Semiconductor nanotubes

Nanotubes are a class of very important one-dimensional nanostructure since their hollow structure may usually give surface area larger than that of nanowires or nanorods. The nanotube arrays are ordered and strongly interconnected, which eliminates randomization of the grain network and increases contact points for good electrical connection.

2.3.1 TiO_2 nanotubes

TiO_2 nanotubes have been prepared by several methods including anodization, sol-gel, and hydrothermal synthesis and so on. Among these methods, anodization of titanium metal has been an intensively employed method for the fabrication of the oriented TiO$_2$ nanotube array which could combine high surface area with well-defined pore geometry (Zhu et al., 2007). The vertical pore geometry of the nanotubes appears to be more suitable than the conventional random pore network for the fabrication of DSCs, especially DSCs with quasi-solid/solid-state electrolytes. It has been also reported that nanotube arrays give enhanced light scattering and improved collection efficiencies compared to conventional sol-gel-derived TiO$_2$ films with the same thickness (Zhu et al., 2007).

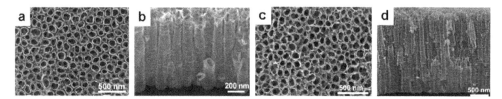

Fig. 6. SEM top-views (a, c) and cross-sections (b, d) for the "short" and "long" tubes (Macak et al., 2005).

By anodization of Ti metal in fluoride-based electrolytes, TiO$_2$ nanotubes can be prepared from Ti foil, but nanotube growth can also be obtained from Ti thin film deposited on TCO glass. The length of the nanotubes, wall thickness, pore diameter, and tube-to-tube spacing can be controlled by the preparation conditions, such as the anodization potential, time, and temperature, and the electrolyte composition (water content, cation size, conductivity, and viscosity) (Hagfeldt et al., 2010). In 2005, the first report on dye-sensitized TiO$_2$ nanotubes appeared (Macak et al., 2005) and the tubes were grown by Ti anodization in two different forms as "long" tubes (tube lengths ~2.5 μm) and "short" tubes (tube lengths ~500 nm), as shown in Fig. 6. Clearly sub-bandgap sensitization with Ru-dye (N3) was successful and led to considerable incident photon-to-current conversion efficiency (IPCE: up to 3.3%). They suggested that main factors that affect IPCE in the visible range were the structure of the tube (anatase better than amorphous), the dye concentration and the tube length. The increase in dye concentration and/or tube length leads to an increase in IPCE, which can be ascribed to a higher packing density of the dye on TiO$_2$ surface with a higher concentration and a higher light absorption length. So it turns out that for the longer tubes, a significantly lower dye concentration is needed to achieve maximum IPCE values.

TiO$_2$ nanotubes can also be obtained by anodization of aluminum films on TCO and subsequent immersion in a titanium precursor solution, followed by sintering in a furnace at 400 °C. The alumina template is then removed by immersing the samples in 6 M NaOH solution. Kang (Kang et al., 2009) fabricated highly ordered TiO$_2$ nanotubes using such nanoporous alumina template method. Such nanotubes with 15 μm lengths were heat treated at 500 °C for 30 min and soaked in N3 dye for 24 h. DSC based on the nanotubes showed a conversion efficiency of as high as 3.5% and a maximum IPCE of 20% at 520 nm.

In order to improve the overall solar cell efficiency, the amount of dye adsorbed by a unit solar cell volume needs to be higher enough. TiCl$_4$ treatment is an effective way to achieve a higher surface area for nanotube-based photoanodes. Roy et al. (Roy et al., 2009) investigated the effect of TiCl$_4$ treatments on the conversion efficiency of TiO$_2$ nanotube arrays. Typical morphology characterizations obtained for nanotube layers before and after this treatment are shown in Fig. 7. The results clearly show that by an appropriate treatment, the inner as well as the outer wall of TiO_2 nanotubes are covered with a ~25 nm thick nanoparticle coating of TiO_2 nanoparticles with diameter of about 3 nm. This leads to a significant increase in surface area and therefore more dye can be adsorbed to the nanotube walls. Thus, the power conversion efficiency improves from 1.9% for untreated nanotutes to 3.8% for treated nanotubes.

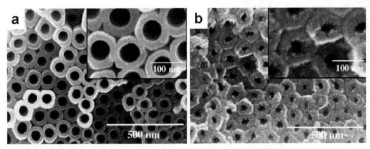

Fig. 7. Influence of $TiCl_4$ treatment: SEM images of TiO_2 nanotubes (a) before and (b) after $TiCl_4$ treatment (Roy et al., 2009).

2.3.2 ZnO nanotubes

ZnO nanotube arrays have been prepared by different methods for DSCs. There are chiefly three kinds of methods. Firstly, atomic layer deposition is an interesting technique to prepare well-defined and ordered ZnO nanotubes. For example, Martinson et al. (Martinson et al., 2007) reported ZnO nanotube photoanodes through atomic layer deposition in the pores of anodic aluminum oxide (AAO) membrane. However, the surface area of ZnO nanotubes is also very low due to a limit of the available size and pore density of the AAO membranes. Secondly, chemical etching is also effective for fabricating ZnO nanotubes, which involves two steps of process: a first growth of ZnO nanorods and a consequent treatment in alkaline solution; the latter is to convert the nanorods into nanotube structure through a chemical etching. Han et al (Han et al., 2010) fabricated high-density vertically aligned ZnO nanotube arrays on TCO substrates by such simple and facile chemical etching process from electrodeposited ZnO nanorods. The nanotube formation was rationalized in terms of selective dissolution of the (001) polar face. And the morphology of the nanotubes can be readily controlled by electrodeposition parameters for the nanorod precursor. DSC based on 5.1 μm-length ZnO nanotubes exhibited a power conversion efficiency of 1.18%. The conversion efficiency is generally low, which probably results from a fact that the length of ZnO nanotubes is limited by the fabrication method based on an etching mechanism. During the etching treatment, an accompanying dissolution of the ZnO occurs simultaneously and thus leads to a shortening of the nanorods (Zhang & Cao, 2011). At last, electrochemical deposition can also be used to directly prepare ZnO nanotube on TCO glass. Prof. Tang group reported for the first time the electrochemical deposition of large-scale single-crystalline ZnO nanotube arrays on TCO glass substrate from an aqueous solution (Fig. 8a) (Tang et al., 2007). The nanotubes had a preferential orientation along the [0001] direction and hexagon-shaped cross sections. The growth mechanism of ZnO nanotubes was investigated (Fig. 8b). They believed that the key growth step is the formation of oriented nanowires and their self-assembly to hexagonal circle shapes. The nanowires initiated subsequent growth of nanotubular structure. But not all the nanowire circle planes are parallel to the horizontal plane because of the roughness of F–SnO$_2$ surface. Some nanotubes have certain angles with the substrate. This nanotube array has great potential for the application in DSCs.

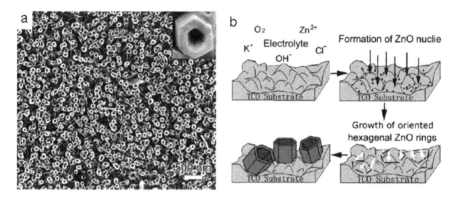

Fig. 8. Surface morphology of ZnO nanotube film (a) and illustration of the growth mechanism of ZnO nanotube array (b). (Tang et al., 2007)

3. The application of inorganic nanomaterials in photocathodes

The photocathode, namely counter electrode (CE) where the regeneration of the charge mediator (typical reaction: $I_3^- + 2e^-$ (catalyst) $\rightarrow 3I^-$) takes place, is one of the most important components in DSCs. The task of CE is twofold: firstly, it transfers electrons arriving from the external circuit back to the redox system (Fig. 9a), and secondly, it catalyzes the reduction of the redox species (Fig. 9b). In order to obtain an effective CE, main requirements for a material to be used as CE are good catalytic activity for the reaction (I_3^-/I^-), a low charge transfer resistance, chemical/electrochemical stability in the electrolyte system used in the cell, mechanical stability and robustness.

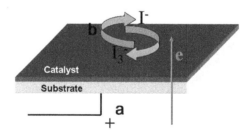

Fig. 9. Schematic representation of the counter electrode based on a I^-/I_3^- redox couple.

At present, many kinds of CEs have been introduced, for example, metal material CEs, carbon material CEs, conducting polymer CEs, hybrid material CEs, and metallic compound nanomaterial CEs. Since each kind of CEs has their own unique advantages and disadvantages, the CE is chosen according to the particular application of DSCs. For example, the noble metal Pt CE is extremely expensive for large-scale production and may be corroded by the iodide solution, but for 'champion cells' one must choose this highly catalytic CE with the lowest possible sheet resistance and a high rate of reduction of the redox electrolyte to speed up the reaction ($I_3^- + 2e^- \rightarrow 3I^-$). For power-producing windows or metal-foil-supported DSCs, one must employ a transparent counter electrode, e.g. a small amount of platinum deposited on TCO glass. On the other hand, large solar conversion systems producing electric power on the terawatt scale will prefer materials that are abundantly available. Carbonaceous material CE will be a good choice due to their advantages including good catalytic properties, electronic conductivity, corrosion resistance towards iodine, high reactivity, and abundance. In this section, we briefly introduce three kinds of inorganic nanomaterials as CEs with respect to their application in DSCs, including noble metal materials, carbon materials and metallic compounds nanomaterials.

3.1 Metal materials CEs
Up to now, noble metal-loaded substrates have already been widely used as the standard for the CE of DSCs, due to their unique properties, including (1) high electrochemical activity that can reduce the voltage loss due to charge-transfer overpotential of CE; (2) a low charge transfer resistance which can lead to minimum energy loss. A thin layer of noble metals, e.g. Pt, Au, is well established as the catalyst on CE substrate, such as TCO glass and metal foil. One of the important roles of noble metal in CE is to catalyze the reduction of triiodide (I_3^-) ions. This means that the available catalytic surface in the electrode plays a crucial role in determining the overall device current. So the rough/porous electrodes, which are characterized by a higher surface, are expected to assure a higher number density of catalytic sites. It is obvious that the use of metal nanoparticle film results in CE with high surface area. Therefore, the preparation methods of noble metal films will influence the final film structure/properties.

Generally all the synthetic methods of the noble metal film are roughly divided into two categories: the physical and chemical approaches. The difference between the two approaches arises from the starting point in the synthetic route to prepare the films. For the physical approaches, the film is prepared by the macroscopic precursors through subsequent subdivision in ever smaller particles by strong milling of solids or through lithographic processes including sputtering, laser ablation, vapor phase deposition, lithography, etc. While the chemical approaches start from their atomic and molecular precursors, through chemical reactions and modulating their self-assembling in order. The physical approaches (Terauchi et al., 1995) are generally quite expensive and resource-consuming, while chemical methods are generally cheaper and better fit for large scale applications. Moreover, since chemical methods allow in principle the control at a molecular level, they allow a fine size and polydispersity control and can be implemented to prepare 2D and/or 3D nanoparticle arrays to enhance the available catalytic site.

The catalytic activity is expressed in terms of the exchange current density (J_0), which is calculated from the charge-transfer resistance (R_{ct}) using the equation $R_{ct} = RT/nFJ_0$, in which R, T, n, and F are the gas constant, temperature, number of electrons transferred in the elementary electrode reaction ($n = 2$) and Faraday constant, respectively. Yoon et al (Yoon et al., 2008) used Nyquist plots to investigate R_{ct} on platinized TCO CEs prepared by either the electrochemical deposition (ED) method, the sputter-deposited (SD) method, or the thermal deposited (TD) method. The Nyquist plots suggest qualitatively that R_{ct} of a cell increases in the order of R_{ct} of ED-Pt film < R_{ct} of SD-Pt film < R_{ct} of TD-Pt film. The DSC fabricated with the ED-Pt CE rendered the highest power conversion efficiency of 7.6%, compared with approximately 6.4% of the cells fabricated with the SD-Pt or most commonly-employed TD-Pt CEs. The improved performance of DSC with the ED-Pt CE is attributed to the improved catalytic activity of the reduction reaction ($I_3^- +2e^- \rightarrow 3I^-$) and the decreased charge transfer resistance at CE/electrolyte interface. Hauch et al (Hauch & Georg, 2001) also used impedance spectra to investigate R_{ct} on platinized TCO CEs prepared by either the electron-beam evaporation (EB), SD or TD method. The 450 nm thick platinum electrode prepared by SD method gave the lowest R_{ct} of of 0.05 ohm/cm^2. The TD of a Pt film (<10 nm thick), using H_2PtCl_6 as a precursor on a TCO substrate, produced a low R_{ct} of 1.3 ohm/cm^2 comparable to the R_{ct} of the 40 nm thick sputtered Pt, confirming the superiority of the TD method.

3.2 Carbon material CEs

DSC is well known as potentially low-cost photovoltaic devices (Gratzel, 2004); from this perspective, the application of low-cost materials should be important. Low-cost carbon is the second most widely studied material for CEs after metal materials. Carbonaceous materials feature good catalytic properties, electronic conductivity, corrosion resistance towards iodine, high reactivity and abundance (Wroblowa & Saunders, 1973). Since the fact that Kay and Grätzel found good electrocatalytic activity of graphite/carbon black mixture in 1996 (Kay & Gratzel, 1996), various kinds of carbon are studied, such as hard carbon spherules (Huang et al., 2007), activated carbon (Imoto et al., 2003), mesoporous carbon (Wang et al., 2009a), nanocarbon (Ramasamy et al., 2007), single-walled carbon nanotubes (Suzuki et al., 2003), multiwalled carbon nanotubes (MWNTs) (Seo et al., 2010), carbon fiber (Joshi et al., 2010) and graphene nanoplates (Kavan et al., 2011). Nevertheless carbonaceous electrodes which would be superior to Pt were reported only rarely for certain kinds of activated carbon (Imoto et al., 2003). This result is mainly attributed to the poor catalytic activity for I_3^-/I^- redox reaction. In addressing this issue, several optimizing methods have been developed, such as increasing the surface area, functionalizing the carbon materials to get more active sites for I_3^-/I^- redox reaction.

To achieve a comparable activity to platinum, carbon-based CEs must have sufficiently high surface area. Although carbonaceous electrodes have poor catalytic activity for I_3^-/I^- redox reaction, its exceptional surface area and conductivity, such as mesoporous carbon and graphene, have been shown to be quite effective and in some cases even exceeded the performance of platinum. Imoto et al (Imoto et al., 2003) compared several types of activated carbon with different surface areas ranging from 1000 m^2 g^{-1} to 2000 m^2 g^{-1} as the CE catalyst, assessing in addition the activity of several different types of activated carbon, glassy carbon, and graphite. The surface area of the glassy carbon and graphite was three orders of magnitude lower than those of the activated carbon catalysts. In the preparation of the latter electrodes, a certain amount of carbon black was included. In their results, the electrodes consisting of the lower sheet resistance materials, graphite and glassy carbon, gave lower J_{sc} values and fill factors, indicating the importance of the roughness of the carbon materials in achieving a better performance. They demonstrated an improvement in the J_{sc} and FF with increasing thickness (>30 μm) of the carbon material. Robert et al (Sayer et al., 2010) used the dense, vertical, undoped MWCNT arrays grown directly on the electrode substrate as CEs and got a greater short-circuit current density and higher efficiency than DSCs with Pt CE. The improved performance is attributed to increased surface area at the electrolyte/counter electrode interface that provides more pathways for charge transport. Prakash et al (Joshi et al., 2010) investigated the electrospun carbon nanofibers as CEs. The results of electrochemical impedance spectroscopy (EIS) and cyclic voltammetry measurements indicated that the carbon nanofiber based CEs exhibited low charge-transfer resistance, large capacitance, and fast reaction rates for triiodide reduction. Joseph et al (Roy-Mayhew et al., 2010) found that functionalized grapheme sheets with oxygen-containing sites perform comparably to platinum. Using cyclic voltammetry, they demonstrated that tuning the grapheme sheets by increasing the amount of oxygen-containing functional groups can improve its apparent catalytic activity.

3.3 Metal compounds CEs

A possible approach is to utilize transition-metal compounds with similar properties to those of noble metals (Shi et al., 2004). Compared with metallic materials, transition-metal compounds have a great potential to be used as cost-effective CEs in DSCs due to their unique properties including a broad variety of low cost materials, a good plasticity, a simple fabrication, high catalytic activity, selectivity, and good thermal stability under rigorous conditions (Wu et al., 2011a). Recently, several kinds of transition-metal compounds CEs, e.g., nitrided Ni particle film, TiN nanotube arrays, MoC have been reported in DSCs which have a conversion efficiency superior to Pt.

Jiang et al (Jiang et al., 2009) used TiN nanotube arrays as CE in DSC for the first time, and the resulting DSC had photovoltaic performances comparable to those using the conventional TCO/Pt counter electrodes, which should be attributed to the obviously lower charge-transfer resistances at the CE/electrolyte interfaces and ohmic internal resistances. The exciting photovoltaic performances comparable to Pt CE inspire the researches on the transition-metal compounds used as the new kind of CEs. They also investigated the surface-nitrided nickel film as a low cost CE material, and the resulting DSCs presented an excellent photovoltaic performance competing with that with the conventional Pt CE (Jiang et al., 2010). Molybdenum and tungsten carbides embedded in ordered mesoporous carbon materials (MoC-OMC, WC-OMC) as well as Mo_2C and WC were prepared respectively by Wu et al (Wu et al., 2011a). They demonstrated that DSCs equipped with optimized MoC-OMC, WC-OMC, Mo_2C, and WC showed higher power conversion efficiency than those devices with a Pt CE. Very recently, they have developed another kind of CE, tungsten oxides, based on their excellent catalytic activity (Wu et al., 2011b). They found that WO_2 nanorods showed excellent catalytic activity for triiodide reduction, and the DSC based on a WO_2 CE reached a high energy conversion efficiency of 7.25%, close to that of the DSC using Pt CE (7.57%). Their results exhibit that tungsten oxides are promising alternative catalysts to replace the expensive Pt in DSCs system. Wang et al (Wang et al., 2009b) have demonstrated, for the first time, that CoS is very effective in catalyzing the reduction of triiodide to iodide in a DSC, superseding the performance of Pt as an electrocatalyst. They deposited the CoS layer on a flexible ITO/polyethylene naphthalate films. CoS based flexible and transparent CEs not only matched the performance of Pt as a triiodide reduction catalyst in DSCs, but also showed excellent stability in ionic liquids-based DSCs under prolonged light soaking at 60 °C. Clearly, the CoS has an advantage for large scale application as being a much more abundant, transparent and cheaper CE.

4. The application of inorganic nanomaterials in quasi-solid/solid-state electrolytes

The power conversion efficiency of DSCs with organic solvent-based electrolyte was reported to exceed 11% (Gratzel, 2004). However, the presence of organic liquid electrolytes in cells causes problems, such as leakage, evaporation of solvent, high-temperature instability, and flammability, and therefore results in practical limitations to sealing and long-term operation. At present, many attempts have been made to substitute liquid electrolytes with solid state electrolytes (e.g. p-type semiconductors, organic hole-transport materials, solid polymer electrolytes, and plastic crystal electrolytes) or quasi-solid-state electrolytes (e.g. polymer gel, low-molecular-weight gel). Here, we focus our attention on the application of inorganic nanomaterials in quasi-solid/solid-state electrolytes, including p-type semiconductor nanoparticles as solid-state electrolytes and solidification of liquid electrolyte by inorganic nanoparticles.

4.1 p-type semiconductor nanoparticles as solid-state electrolytes

P-type semiconductors are the most common hole-transporting materials to fabricate solid-state DSCs. Several aspects are essential for any p-type semiconductor in a DSC: (a) It must be able to transfer holes from the oxidized dye; (b) It must be able to be deposited within the porous TiO$_2$ nanocrystal layer; (c) A method must be available for depositing the p-type semiconductors without dissolving or degrading the dye on TiO$_2$ nanocrystal; (d) It must be transparent in the visible spectrum, otherwise, it must be as efficient in electron injection as the dye. Copper-based materials, especially CuI and CuSCN (Tennakone et al., 1995), are found to meet all these requirements. CuI and CuSCN share good conductivity in excess of 10^{-2} Scm^{-1}, which facilitates their hole conducting ability (Smestad et al., 2003), hence, they have been widely used as complete hole-transporting layer for fabricating solid-state DSCs. Tennakone et al (Tennakone et al., 1995) first reported a nano-porous solid-state DSC based on CuI in 1995. DSCs were fabricated by sandwiching a monolayer of the pigment cyanidin adsorbed on nano-porous n-TiO$_2$ film within a transparent polycrystalline film of p-CuI, filling the intercrystallite pores of the porous n-TiO$_2$ film. The short-circuit current density reached about 1.5–2.0 mAcm^{-2} in sunlight (about 800 Wm^{-2}), however, they found that the polarization arising from mobile Cu$^+$ ions tends to decrease the open-circuit voltage of the cell. By replacing cyanidin with a Ru-bipyridyl complex dye, they (Tennakone et al., 1998) then fabricated a solid-state DSC with the structure of TiO$_2$/Ru(II)(dcbpy)$_2$(SCN)$_2$/CuI in 1998. The efficiency and the stability of the solid-state DSC was less than those of the typical DSCs. They attributed it to the loosening of the contact between the p-type semiconductor and the dye monolayer, and to short-circuiting across the voids in the nanoporous TiO$_2$ film, which allows direct contact between CuI and the tin oxide surface. To solve this issue, they (Kumara et al., 2002) incorporated a small quantity ($\sim 10^{-3}$ M) of 1-methyl-3-ethylimidazolium thiocyanate (MEISCN) into the coating solution (i.e., CuI in acetonitrile) and the stability of the CuI-based DSC was greatly improved. In this case, MEISCN acted as a CuI crystal growth inhibitor, which enables filling of the pores of the porous matrix, resulting in the formation of more complete and secure contacts between the hole collector and the dyed surface. Meng et al (Meng et al., 2003) also utilized MEISCN to control the CuI crystal growth and act as a protective coating for CuI nanocrystals, improving the cells' efficiency to 3.8% with improved stability under continuous illumination for about 2 weeks. Another alternative for solid-state DSCs is CuSCN. O'Regan et al (Oregan & Schwartz, 1995) used CuSCN as a hole transport layer, and demonstrated that CuSCN is a promising candidate material as the hole-conducting layer. They found the UV illumination created an interfacial layer of (SCN)$_3^-$, and/or its polymerization product (SCN)$_x$, between the TiO$_2$ and the CuSCN, causing a dramatic improvement in the efficiency of DSCs (O'Regan & Schwartz, 1998). Later, they (O'Regan et al., 2000) substituted TiO$_2$ for ZnO, fabricating

ZnO/dye/CuSCN solar cell with a conversion efficiency of 1.5%. In 2005, they (O'Regan et al., 2005) made another solid-state DSC with the structure of TiO$_2$/dye/CuSCN with thin Al$_2$O$_3$ barriers between the TiO$_2$ and the dye. It is noteworthy that the Al$_2$O$_3$-treated cells showed improved voltages and fill factors but lower short-circuit currents. Nevertheless, the performance of DSCs with CuSCN is still lower than that of cells utilizing CuI, probably due to the relatively lower hole conductance.

In summary, compared to a liquid electrolyte DSC, the solid-state counterpart presents a relatively low conversion efficiency, which is probably due to three reasons (Kron et al., 2003): (a) the less favorable equilibrium Fermi-level position in the TiO$_2$; (b) poor conductivity of hole-transporting materials; (c) the much larger recombination probability of photogenerated electrons from the TiO$_2$ with holes as compared to recombination with the I$^-$/I$_3^-$ redox couple. So it is necessary to make further efforts to design new and more efficient inorganic nanomaterial electrolytes for DSCs.

4.2 Solidification of liquid electrolyte by inorganic nanoparticles

Room-temperature ionic liquids (RTILs) such as imidazolium iodide have been widely used in DSCs as a solvent and a source of I$^-$ or other ions, because of their favorable properties such as thermal stability, nonflammability, high ionic conductivity, negligible vapor pressure, and a possible wide electrochemical window. However, the fluidity of RTIL-based electrolytes, resulting in difficulty in seal, is still an obstacle for long-term operation. To reduce its fluidity, combination of RTILs with a framework material including small-molecular organogels, inorganic nanoparticles, and polymer, has been attempted by many groups.

Among these framework materials, inorganic nanoparticles have drawn more attention. In 2003, silica nanoparticles were used for the first time to solidify ionic liquids by Prof. Grätzel group (Wang et al., 2003a). The presence of silica nanoparticles has no adverse effect on the conversion efficiency, and the ionic liquid-based quasi-solid-state electrolytes are successfully employed for fabricating DSC with a conversion efficiency of 7%. This means that quasi-solid-state electrolytes offer specific benefits over the ionic liquids and will enable the fabrication of flexible, compact, laminated quasi-solid-state devices free of leakage and available in varied geometries. In addition, for their pore structures (2-50 nm) and large surface area, mesoporous materials may solidify liquid electrolytes and provide favourable channels for the triiodide/iodide diffusion. By using the mesoporous SiO$_2$ material (SBA-15) as the framework material, Yang et al (Yang et al., 2005) fabricated a quasi-solid-state electrolyte and then fabricated DSC with a energy conversion efficiency of 4.34%. ZnO nanoparticle also can be used to solidify the liquid electrolyte. For example, Huang group (Xia et al., 2007) used ZnO nanoparticles as a framework to form a quasi-solid-state electrolyte for DSC. The quasi-solid-state DSC with the quasi-solid-state electrolyte showed higher stability in comparison with that of the liquid device, and gave a comparable overall efficiency of 6.8% under AM 1.5 illumination.

Thermal stability is an urgent concern for quasi-solid DSCs based on RTIL gel electrolytes. Some room-temperature quasi-solid-state electrolytes usually become liquid at high temperature (40–80 °C), for example, 3-methoxypropionitrile (MPN)-based polymer gel electrolyte (viscosity: 4.34 MPa s at 80 °C) (Wang et al., 2003b) and plastic crystal electrolytes (m.p. 40–45 °C) (Wang et al., 2004). Since the working temperature of DSCs may reach 60 °C under full sunlight, it is necessary that at high temperature (60–80 °C), the electrolytes are still in the quasi-solid or solid state and that the DSCs maintain high overall energy-conversion efficiency (>4%). We (Chen et al., 2007b) have developed a succinonitrile-based gel electrolyte by introducing a hydrogen bond (O–H...F) network upon addition of silica nanoparticles and BMI·BF$_4$ (1-Butyl-3-methylimidazolium tetrafluoroborate) to succinonitrile. When the content of fumed silica nanoparticles was over 5 wt%, the succinonitrile–BMI·BF$_4$–silica system became a gel, and the succinonitrile–BMI·BF$_4$–silica (7 wt%) system still remained in the gel state even at 80 °C, as shown in the inset of Fig. 10, which confirms that the addition of silica nanoparticles and BMI·BF$_4$ is critical for the gelation and thermostability of succinonitrile-based electrolytes. The appropriate addition of BMI·BF$_4$ and silica nanoparticles in this gel electrolyte can greatly improve the thermostability but has no adverse effects on the conductivity, ionic diffusion coefficients and the cell performance. Moreover, the relatively high succinonitrile content in the electrolyte is also very important because the electrolyte without succinonitrile has very low conductivity and results in poor cell performance. Herein, the obtained succinonitrile-based gel electrolyte satisfies the need for both thermostability and high conductivity in electrolytes. DSCs with this gel electrolyte showed power conversion efficiencies of 5.0–5.3% over a wide temperature range (20–80 °C). Furthermore, the aging test revealed that the cell still maintained 93% of its initial value for the conversion efficiency after being stored at 60 °C for 1000 h, indicating an excellent long-time durability.

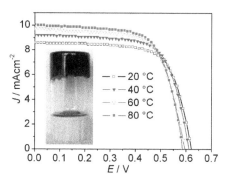

Fig. 10. Photos of succinontrile–BMI·BF$_4$–silica (silica: 7 wt% of succinontrile–BMI·BF$_4$) at 80 °C (Chen et al., 2007b).

5. Inorganic nanomaterials as light-absorbing materials

5.1 Semiconductor QDs

The emergence of semiconductor quantum dots (QDs) has opened up new ways to utilize them as a possible replacement for ruthenium complexe dyes (sensitizers) in DSCs. QDs have many advantages including good thermal stability under rigorous conditions, tunable bandgaps, sharp absorption onset, large absorption coefficients and low cost etc. Hence, the QD-sensitized solar cells (QD-SCs) have attracted more attention. In the past few years, a rapid increase of the conversion efficiency of QD-SCs has been reported, reaching values of around 4-5% at 1 sun (Chang et al., 2010). The efficiency of QD-SC still lags behind those of DSC; however, a further performance improvement for QD-SCs can be anticipated.

Recently, various QDs (CdSe (Levy-Clement et al., 2005), PbSe (Leschkies et al., 2009), CuInS$_2$ (Kaiser et al., 2001), etc.) as the sensitizer have been proposed, and various strategies have been developed for maximizing photoinduced charge separation and electron transfer processes to improve the power conversion efficiency. This section aims to introduce the recent developments in the synthesis methods of high quality QDs and the performance optimization strategies of QD-SC.

5.1.1 Synthesis of QDs

QDs used as sensitizer in DSC have been fabricated by using two fundamentally different approaches. The first and most common route employs the *in situ* preparation of QDs onto the nanostructured semiconductor metal oxide film. The second approach is *ex situ* growth approach, which can take advantage of the tremendous developments in controlling the growth of monodisperse, highly crystalline and diverse QDs. This method is to synthesis the QDs independently, and subsequently to attach the pre-synthesized QDs to the photoanode by a bifunctional linker molecule or direct adsorption. Both of these preparation methods have their limitation respectively, hence it is still necessary to develop a new method in the future.

5.1.1.1 *In situ* preparation of QDs

The *in situ* preparation method, where the QDs are directly generated on the surface of metal oxide film electrode, mainly includes chemical bath deposition (CBD) (Diguna et al., 2007) and successive ionic layer adsorption and reaction (SILAR) (Lee et al., 2009a). There are many of advantages of the *in situ* deposition approaches. Firstly, the *in situ* deposition approaches are easy to process since it does not need any expensive equipment and multiple complicated steps. Secondly, the QDs are in direct electronic contact with metal oxide, which not only makes the QDs good anchoring to the electrodes but also shortens the electron diffusion length. Thirdly, they can easily produce metal oxide films with high surface coverage of the sensitizing QDs, which increases the absorption of light. However, this method still has several intrinsic limitations. For example, it is difficult to control chemical composition, crystallinity, size distribution and surface properties of QDs, which may hamper the effective exploitation of QDs advantages. .

For CBD method, QDs are deposited *in situ* by immersing the wide-bandgap nanostructured electrode (usually metal oxide) into a solution that contains the cationic and anionic precursors, which react slowly in one bath under different temperature. Most of the sulfides and selenides can be prepared by this method. Lee et al developed a method coupling self-assembled monolayer and CBD, as well as a modified CBD process performed in an alcohol system to assemble CdS into a TiO$_2$ film (Lin et al., 2007). These modified processes have been proved to be efficient for CdS QD-SC, and CdS QD-SCs exhibit a power conversion efficiencies of 1.84% and 1.15% respectively for iodide/triiodide and polysulfide electrolytes (Lin et al., 2007).

In the SILAR approach, the cationic and anionic precursor solution is placed in two vessels respectively. Firstly, the nanostructured electrode is immersed into the solution containing the metal cation, and then the nanostructured electrode absorbing the metal cation on the surface is taken out from the metal cation solution and dip into the second precursor solution containing the anion. After the second rinsing step, the deposition cycle completes. The average QD size can be controlled by the number of deposition cycles. This method has been used in particular to prepare metal sulfides, but recently it has been expanded to metal selenides and tellurides. For example, Grätzel et al deposited the CdSe and CdTe QDs *in situ* onto mesoporous TiO$_2$ films using SILAR approach (Lee et al., 2009b). After some optimization of these QD-sensitized TiO$_2$ films in solar cells, over 4% overall efficiency was achieved at 100 W/m^2 with about 50% IPCE at its maximum (Lee et al., 2009b).

5.1.1.2 *Ex situ* growth approach

The *ex situ* growth approach is based on a two-step process, whereby QDs are first independently synthesized using established colloidal synthesis method and then QDs are subsequently linked on to the metal oxide film electrodes to achieve effective QD-electrode junctions that would promote charge separation while minimizing surface trapping and hence losses (Kamat, 2008). This approach can take advantage of the tremendous developments in controlling over the chemical, structural, and electronic properties of QDs compared to the in situ approaches. However, since QDs, unlike dyes, do not possess an anchoring functional group for coupling to the metal oxide film surface, this approach provides limited control over the metal oxide film sensitization process and degree of QD-metal oxide film electronic coupling. In addition, the as prepared QDs are typically passivated with a layer of organic ligands, such as tri-n-octylphosphine oxide (TOPO), aliphatic amines, or acids, which serve as an impediment to effective metal oxide film sensitization and as a barrier to efficient charge transfer across the QD/metal oxide film and QD/electrolyte interfaces.

The first step of the *ex situ* growth approach is the synthesis of the monodisperse QDs (Fig. 11a). The most common synthesis approach is to control the nucleation and growth process of particles in a solution of chemical precursors containing the metal and the anion sources (Overbeek, 1982). In a typical process, the solvent containing molecules (e.g. trioctylphosphine and trioctylphosphine oxide) is heated to 150-350 °C under the vigorously stirred with protective atmosphere, and then the organometallic precursor and related species are injected. Consequently, a large number of nucleation centers are initially formed, and the coordinating ligands in the hot solvent prevent or limit particle growth via Ostwald ripening. Typical reactions used for the synthesis of II-VI (CdSe, CdTe, CdS) (Peng & Peng, 2002), III-V (InP, InAs) (Talapin et al., 2002), and IV-VI (PbSe (Talapin & Murray, 2005)) QDs are outlined by reactions 1-3 (Fig. 11a).

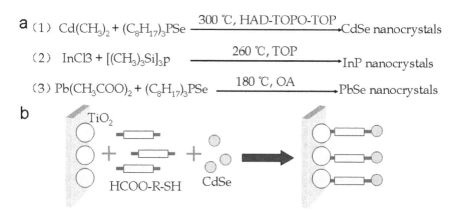

Fig. 11. (a) Typical reactions used for the synthesis of QDs. (b) Scheme of attaching the pre-synthesized QDs to the electrode material by a bifunctional linker molecule (Robel et al., 2006).

The second step of the *ex situ* growth approach is to attach the pre-synthesized QDs to the electrode material by a bifunctional linker molecule (usually HOOC-R-SH, where R is the organic core of the linker) or direct adsorption (Fig. 11b). The carboxyl group attaches to the nanostructured metal oxide film, while the thiol remains free for connecting QDs. Then the modified film is immersed in the QD solution (for several hours or days) for the adsorption of QDs, which typically involves fractional ligand exchange. For instance, Kamat et al (Robel et al., 2006) immersed the TiO$_2$ films into a solution of acetonitrile containing carboxy alkane thiols (Mercaptopropionic acid, thiolacetic acid and mercaptohexadecanoic acid) for ~4 h as shown in the Fig. 11b. Resulting TiO$_2$ films functionalized with these bifunctional surface modifiers, were then washed with both acetonitrile and toluene, and transferred to a glass vial containing a suspension of CdSe QDs in toluene. The electrodes were kept immersed in the CdSe solution for approximately 12 h. QDs self-assembly had also been used on dispersed TiO$_2$ crystals, where the photoelectrode was fabricated after QD sensitization by a pressing route (Ardalan et al., 2011). Direct adsorption was recently proposed for deposition of monodisperse QDs without molecular linkers to the surface of the metal oxide nanostructure. This procedure has already been employed to sensitize TiO$_2$ with QDs, such as CdSe QDs (Shen et al., 2009), although the obtained photocurrents at 1 sun intensity are low. One possible explanation is that directly adsorbed colloidal QDs provides a low surface coverage of about 14% (Gimenez et al., 2009). It is obvious that higher surface coverage of colloidal QDs on TiO$_2$ substrate will result in QD-SCs with higher conversion efficiency.

5.1.2 Performance optimization of QDs in the QD-SCs

Theoretically, there are many advantages of the QDs for sensitizing the nanoporous oxide film electrode, such as higher absorption of QD coating, greater stability of the semiconductor, and tailoring of optical absorption over a wider wavelength range (Hodes, 2008). Moreover, the demonstration of multiple exciton generation (MEG) by impact ionization in colloidal QDs could push the thermodynamic efficiency limit of these devices up to 44% (Klimov, 2006) instead of the current 31% of the Schockley-Queisser detailed balance limit. Although, up to now, the efficiencies of QD-SCs are far behind those of DSCs (DSC currently exceeds 11% at 1 sun illumination (Cao et al., 2009)), QD-sensitized nanostructured solar cells are attracting increasing attention among researchers and are progressing rapidly to values around 4-5% from quite low conversion efficiencies (Gonzalez-Pedro et al., 2010). Several methods have been developed to optimize the performance of the QD-SCs including (1) tuning of effective bandgaps from visible down to the IR range by changing their sizes and compositions, (2) utilizing the phenomena of MEG, Förster resonance energy transfer (FRET)-based charge collection, and direct charge transfer schemes, and (3) surface treatment.

5.1.2.1 Tuning of bandgaps of QDs

The most striking property of QDs is the massive changes in electronic structure as a function of size. As the size decreases, the electronic excitations shift to higher energy, and the oscillator strength is concentrated into just a few transitions (Murray et al., 1993). Therefore, controlling quantum size confinement in monodisperse QDs is the most obvious method not only to extend the range of the QDs absorbance from the visible to near infrared range but also to align the energy levels with respect to the wide-bandgap nanostructure. Herein, the CdSe QDs and TiO$_2$ system as a model is introduced for the direction of the optimization. The driving force for the electron separation and transfer is dictated by the energy difference between the conduction band energies. The conduction band of TiO_2 is at -0.5 V vs NHE. If we assume the larger CdSe particles have band energy close to the reported value of -0.8 V vs NHE, we can use the increase in bandgap as the increase in driving force for the electron transfer. Since the shift in the conduction band energy is significantly greater than the shift in valence band energy for quantized particles (Norris & Bawendi, 1996), we can expect the conduction band of CdSe QDs to become more negative (on NHE scale) with decreasing particle size. As the particle size decreases from 7.5 to 2.4 nm, the first excitonic peak shifts from 645 nm to 509 nm and the conduction band shifts from -0.8 V vs NHE to -1.31 V, the electron transfer rate improve by nearly 3 orders of magnitude (Robel et al., 2007). PbS nanocrystals have similar properties with the light absorption range extending from visible to near infrared (Hyun et al., 2008).

Another method that can broaden the spectral absorption range is the use of nanocomposite absorbers (Lee & Lo, 2009). Semiconductor QDs are excellent building blocks for more sophisticated nanocomposite absorbers, which the QDs combined with each other with different size or type. The well-known example of the nanocomposite is the combination of CdS and CdSe QDs. The combination can be used as co-sensitizers to provide enhanced performance compared to the use of each individual semiconductor QDs. When CdSe QDs are assembled on a TiO_2/CdS electrode, the co-sensitized electrode (TiO_2/CdS/CdSe) has an absorption edge close to that of TiO_2/CdSe electrode but its absorbance is higher than those of TiO_2/CdS and TiO_2/CdSe electrodes both in the short wavelength region (<550 nm) where both CdS and CdSe are photoactive and long wavelength region (ca. 550–700 nm) which belong to the CdSe due to the complementary effect of the composite sensitizers. When CdS is located between CdSe and TiO_2, both the conduction and valence bands edges of the three materials increase in the order: TiO_2<CdS<CdSe, which is advantageous to the electron injection and hole recovery of CdS and CdSe. This clearly shows that nanocomposite absorbers can improve these systems through two different beneficial effects. On the one hand, the spectral absorption range can be broadened. On the other, the re-organization of energy levels between CdS and CdSe forms a stepwise structure of band-edge levels (Lee & Lo, 2009).

5.1.2.2 Utilizing the phenomena of MEG and energy/charge transfer

MEG can occur when absorption of a high-energy photon leads to production of an excited electron or a hole with an excess energy at least equal to or greater than the QD bandgap (Eg). These hot carriers can transfer the entire excess energy, or part of it, to one or more valence electrons, and excite them across the bandgap. In this way, absorption of a single photon leads to generation of two or more electron-hole pairs. The quantum yield for exciton generation is defined as the average number of electron-hole pairs produced by absorption of a single photon. The analogous phenomenon of multiple charge carrier generation per photon in bulk semiconductors is termed impact ionization and is considered an inverse of Auger recombination. MEG in QDs has mostly been described as a coherent process in which single and multi-exciton states are coupled via the Coulomb interaction (Beard et al., 2007). In 1982 it was recognized that it could be possible to increase solar cell efficiency for a single junction by utilizing MEG (Ross & Nozik, 1982). Under 1 sun AM1.5 spectrum the theoretical efficiency of a MEG-enhanced cell is over 44% (Klimov, 2006). Among the various QDs materials, PbS and PbSe are good candidates for solar cells, because they not only can be made to overlap the solar spectrum optimally, which the absorption wavelength of the first exciton peak can easily be extended into the infrared by controlling their sizes (Vogel et al., 1994), but also show that two or more excitons can be generated with a single photon of energy greater than the bandgap (Ellingson et al., 2005). PbSe and PbS QDs have been shown to have quantum yields above 300% (Ellingson et al., 2005) and even above 700% (Schaller et al., 2006).

As a new approach, long range FRET, also known as electronic energy transfer, has been also utilized in the DSCs recently. The donor dye molecules, which are added to the redox solution, upon absorption of light, transfer the excitation energy to an acceptor dye adsorbed on the electrode followed by the standard charge separation process. However, the donor molecules are heavily quenched in the liquid electrolytes (Shankar et al., 2009). Similar geometries using other sensitizers, such as inorganic semiconductor nanocrystals, have been proposed (Chen et al., 2008). QDs have the advantage of a broad absorption spectrum, stretching from the band edge to higher energies, in contrast with the narrow absorption spectra typically exhibited by molecular dyes (Buhbut et al., 2010). The use of FRET has been contemplated as an alternative mechanism for charge separation and a way to improve exciton harvesting by placing the exciton close to the heterojunction interface (Liu et al., 2006). In inorganic QD-DSCs, the use of FRET to transfer the exciton generated in the QD to a high mobility conducting channel, such as a nanowire or a quantum well, has been proposed as a way to bypass the traditional limitations of charge separation and transport (Lu & Madhukar, 2007). Very recently, Sophia et al (Buhbut et al., 2010) presented a design which combines the benefits of QDs in terms of their broad absorption spectrum with the evolved charge transfer mechanism of DSC. They demonstrated that QDs serving as "antennas" could enhance light absorption, broaden the absorption spectrum, increase the number of photons harvested by solar cell effectively and funnel absorbed energy to nearby dye molecules via FRET successfully. Their design introduced new degrees of freedom in the utilization of QDs sensitizers for solar cells. In particular, it opens the way toward the utilization of new materials whose band offsets do not allow direct charge injection.

5.1.2.3 Surface treatment

Regarding the surface treatment, it is important to remark a significant difference between molecular dyes and semiconductors used as sensitizers. As suggested by Hodes (Hodes, 2008), the probable existence of surface states in the sensitizing semiconductor is a major difference between DSCs and QD-SCs. These surface states have been widely studied in colloidal QDs and can be detected by different techniques such as photoluminescence or scanning tunnelling microscopy. The observed behaviors strongly depended on the surface treatment or the type of capping ligand (Frederick & Weiss, 2010). Surface modification by dipoles provides a simple and efficient way to shift the QD energy level. For monodisperse QDs, fractional exchange of the capping ligands by molecular dipoles can be used to shift the electronic QD states with respect to their environment in a systematic fashion, as shown in the Fig. 12 (Ruhle et al., 2010).

5.2 Rare-earth up-converting nanophosphors

Light-absorbing materials developed for solar cells, such as silicon and dyes, can not efficiently absorb near infrared light. In contrast, some rare-earth phosphors, especially those co-doped with Yb^{3+} and Er^{3+}, can efficiently absorb near infrared light, such as 980-nm laser light, after which they exhibit up-converted luminescence in the visible range (Heer et al., 2004). Doping of rare-earth materials into solar cells has been demonstrated as a better

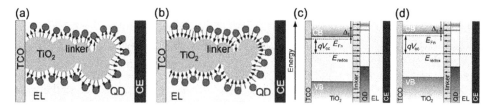

Fig. 12. (a) Schematic drawing of a mesoporous TiO_2 film, covered with linker molecules with a molecular dipole pointing towards the QD monolayer. (b) As in (a) but for linker molecules with a dipole pointing in the opposite direction. (c) Energy-band diagram showing a downward shift of the energy levels of the wide-bandgap nanostructure due to the dipole moment. (d) As in (c) but with linker molecules with a dipole moment pointing away from the QDs (Ruhle et al., 2010).

absorption of the long wavelength light in the solar spectrum, thus enhance the performance of solar cells (Shalav et al., 2005).

Recently, our group has developed 980-nm laser-driven photovoltaic cells (980LD-PVCs) firstly by introducing of a film of rare earth up-converting nanophosphors in conventional dye-sensitized solar cells. The novel photovoltaic cells were constructed by coating $Na(Y_{1.5}Na_{0.5})F_6:Yb,Er$ nanorod films on N3 dye-sensitized TiO_2 films to create direct-current electricity under the irradiation of a 980-nm laser. Under the irradiation of a 980-nm laser with a power of 1 W, the visible up-converting luminescence of rare-earth nanophosphors could be efficiently absorbed by the dyes in 980LD-PVCs, and they exhibited a maximal output power of 0.47 mW. Because of the rather high transparency of biological tissue to 980-nm light, 980LD-PVCs efficiently provided electricity output even when covered by intestinal layers, which promises a new strategy to supply the electrical power for in-vivo nanorobots and other nanobiodevices (Chen et al., 2009).

6. Conclusion

DSCs have currently been attracting widespread scientific and technological interest as an alternative to conventional solar cells. The development and future of DSCs are seriously dependent on the optimal combination of high efficiency, good stability, and low cost, which lie on the design, preparation and modification of materials, especially inorganic materials with excellent properties. Nanotechnology opens a door to tailing inorganic materials and creating various nanostructures for the use in DSCs. Currently, inorganic nanomaterials have been widely used in all components. In this chapter, we describe four topics about the application of inorganic nanomaterials in DSCs. Firstly, the preparation and development of semiconductor nanomaterials, including nanoparticles, nanowires and nanotubes for film photoanodes have been demonstrated. Secondly, different nanomaterials, including metal materials, carbon materials and transition-metal compound, have been introduced for the use in photocathodes (counter electrode). Thirdly, the applications of inorganic nanomaterials in quasi-solid/solid-state electrolytes, including p-type semiconductor nanoparticles as solid-state electrolytes and solidification of liquid electrolyte by inorganic nanoparticles, have been discussed. At last, the development of semiconductor quantum dots and rare-earth up-converting nanophosphors as light-absorbing material has been presented.

However, these are initial results, and further research is still needed to improve the conversion efficiency and stability of DSCs. In order to further improve the efficiency and facilitate the practical application of DSCs, the following challenges should be addressed: (1) design and preparation transparent nanocrystalline film electrodes with higher surface area and faster electron transport property; (2) development of inorganic nanomaterials with low cost and high catalytic activity as counter electrode materials; (3) preparation of novel quasi-solid/solid-state electrolytes with high thermostability and excellent hole-transporting property; (4) design and construction of semiconductor quantum dots as effective light-absorbing material, especially quantum dots with excellent multiple exciton generation ability; (5) development of new techniques to better control the kinetic electron transfer processes at all interfaces.

7. Acknowledgment

This work was financially supported by the National Natural Science Foundation of China (Grant No. 50872020 and 50902021), the Science and Technology Commission of Shanghai-based "Innovation Action Plan" Project (Grant No. 10JC1400100), Shanghai Rising-Star Program (Grant No. 11QA1400100), Shanghai "Chen Guang" project (Grant No. 09CG27), the Fundamental Research Funds for the Central Universities.

8. References

Ardalan, P.; Brennan, T. P.; Lee, H. B. R.; Bakke, J. R.; Ding, I. K.; Mcgehee, M. D. & Bent, S. F. (2011). Effects of self-assembled monolayers on solid-state cds quantum dot sensitized solar cells. *ACS Nano*, VOL.5, NO.2, pp. 1495-1504.

Baxter, J. B. & Aydil, E. S. (2005). Nanowire-based dye-sensitized solar cells. *Appl. Phys. Lett.* , VOL.86, NO.5, pp. 053114.

Beard, M. C.; Knutsen, K. P.; Yu, P. R.; Luther, J. M.; Song, Q.; Metzger, W. K.; Ellingson, R. J. & Nozik, A. J. (2007). Multiple exciton generation in colloidal silicon nanocrystals. *Nano Lett.* , VOL.7, NO.8, pp. 2506-2512.

Buhbut, S.; Itzhakov, S.; Tauber, E.; Shalom, M.; Hod, I.; Geiger, T.; Garini, Y.; Oron, D. & Zaban, A. (2010). Built-in quantum dot antennas in dye-sensitized solar cells. *ACS Nano*, VOL.4, NO.3, pp. 1293-1298.

Cao, Y. M.; Bai, Y.; Yu, Q. J.; Cheng, Y. M.; Liu, S.; Shi, D.; Gao, F. F. & Wang, P. (2009). Dye-sensitized solar cells with a high absorptivity ruthenium sensitizer featuring a 2-(hexylthio)thiophene conjugated bipyridine. *J. Phys. Chem. C* VOL.113, NO.15, pp. 6290-6297.

Chang, J. A.; Rhee, J. H.; Im, S. H.; Lee, Y. H.; Kim, H. J.; Seok, S. I.; Nazeeruddin, M. K. & Gratzel, M. (2010). High-performance nanostructured inorganic-organic heterojunction solar cells. *Nano Lett.* , VOL.10, NO.7, pp. 2609-2612.

Chapin, D. M.; Fuller, C. S. & Pearson, G. L. (1954). A new silicon p-n junction photocell for converting solar radiation into electrical power. *J. Appl. Phys.* , VOL.25, NO.5, pp. 676-677.

Chen, S. G.; Chappel, S.; Diamant, Y. & Zaban, A. (2001). Preparation of Nb_2O_5 coated TiO_2 nanoporous electrodes and their application in dye-sensitized solar cells. *Chem. Mater.* , VOL.13, NO.12, pp. 4629-4634.

Chen, Z. G.; Chen, H. L.; Hu, H.; Yu, M. X.; Li, F. Y.; Zhang, Q.; Zhou, Z. G.; Yi, T. & Huang, C. H. (2008). Versatile synthesis strategy for carboxylic acid-functionalized upconverting nanophosphors as biological labels. *J. Am. Chem. Soc.* , VOL.130, NO.10, pp. 3023-3029.

Chen, Z. G.; Li, F. Y. & Huang, C. H. (2007a). Organic d-Π-a dyes for dye-sensitized solar cell. *Curr. Org. Chem.* , VOL.11, pp. 1241-1258.

Chen, Z. G.; Yang, H.; Li, X. H.; Li, F. Y.; Yi, T. & Huang, C. H. (2007b). Thermostable succinonitrile-based gel electrolyte for efficient, long-life dye-sensitized solar cells. *J. Mater. Chem.* , VOL.17, NO.16, pp. 1602-1607.

Chen, Z. G.; Zhang, L. S.; Sun, Y. G.; Hu, J. Q. & Wang, D. Y. (2009). 980-nm laser-driven photovoltaic cells based on rare-earth up-converting phosphors for biomedical applications. *Adv. Funct. Mater.* , VOL.19, NO.23, pp. 3815-3820.

Desilvestro, J.; Gratzel, M.; Kavan, L.; Moser, J. & Augustynski, J. (1985). Highly efficient sensitization of titanium-dioxide. *J. Am. Chem. Soc.* , VOL.107, NO.10, pp. 2988-2990.

Diamant, Y.; Chen, S. G.; Melamed, O. & Zaban, A. (2003). Core-shell nanoporous electrode for dye sensitized solar cells: the effect of the $SrTiO_3$ shell on the electronic properties of the TiO_2 core. *J. Phys. Chem. B* VOL.107, NO.9, pp. 1977-1981.

Diguna, L. J.; Shen, Q.; Kobayashi, J. & Toyoda, T. (2007). High efficiency of CdSe quantum-dot-sensitized TiO_2 inverse opal solar cells. *Appl. Phys. Lett.* , VOL.91, NO.2, pp.

Ellingson, R. J.; Beard, M. C.; Johnson, J. C.; Yu, P. R.; Micic, O. I.; Nozik, A. J.; Shabaev, A. & Efros, A. L. (2005). Highly efficient multiple exciton generation in colloidal PbSe and PbS quantum dots. *Nano Lett.* , VOL.5, NO.5, pp. 865-871.

Feng, X. J.; Shankar, K.; Varghese, O. K.; Paulose, M.; Latempa, T. J. & Grimes, C. A. (2008). Vertically aligned single crystal TiO_2 nanowire arrays grown directly on transparent conducting oxide coated glass: synthesis details and applications. *Nano Lett.* , VOL.8, NO.11, pp. 3781-3786.

Frederick, M. T. & Weiss, E. A. (2010). Relaxation of exciton confinement in CdSe quantum dots by modification with a conjugated dithiocarbamate ligand. *ACS Nano*, VOL.4, NO.6, pp. 3195-3200.

Garcia, C. G.; de Lima, J. F. & Iha, N. Y. M. (2000). Energy conversion: from the ligand field photochemistry to solar cells. *Coord. Chem. Rev.* , VOL.196, pp. 219-247.

Gimenez, S.; Mora-Sero, I.; Macor, L.; Guijarro, N.; Lana-Villarreal, T.; Gomez, R.; Diguna, L. J.; Shen, Q.; Toyoda, T. & Bisquert, J. (2009). Improving the performance of colloidal quantum-dot-sensitized solar cells. *Nanotechnology*, VOL.20, NO.29, pp. 295204.

Gonzalez-Pedro, V.; Xu, X. Q.; Mora-Sero, I. & Bisquert, J. (2010). Modeling high-efficiency quantum dot sensitized solar cells. *ACS Nano*, VOL.4, NO.10, pp. 5783-5790.

Gratzel, M. (2004). Conversion of sunlight to electric power by nanocrystalline dye-sensitized solar cells. *Journal of Photochemistry and Photobiology a-Chemistry*, VOL.164, NO.1-3, pp. 3-14.

Gratzel, M. (2001). Photoelectrochemical cells. *Nature*, VOL.414, NO.6861, pp. 338-344.

Gregg, B. A.; Pichot, F.; Ferrere, S. & Fields, C. L. (2001). Interfacial recombination processes in dye-sensitized solar cells and methods to passivate the interfaces. *J. Phys. Chem. B* VOL.105, NO.7, pp. 1422-1429.

Hagfeldt, A.; Boschloo, G.; Sun, L. C.; Kloo, L. & Pettersson, H. (2010). Dye-sensitized solar cells. *Chem. Rev.* , VOL.110, NO.11, pp. 6595-6663.

Han, J. B.; Fan, F. R.; Xu, C.; Lin, S. S.; Wei, M.; Duan, X. & Wang, Z. L. (2010). ZnO nanotube-based dye-sensitized solar cell and its application in self-powered devices. *Nanotechnology*, VOL.21, NO.40, pp. 405203.

Hauch, A. & Georg, A. (2001). Diffusion in the electrolyte and charge-transfer reaction at the platinum electrode in dye-sensitized solar cells. *Electrochim. Acta* VOL.46, NO.22, pp. 3457-3466.

Heer, S.; Kompe, K.; Gudel, H. U. & Haase, M. (2004). Highly efficient multicolour upconversion emission in transparent colloids of lanthanide-doped NaYF$_4$ nanocrystals. *Adv. Mater.* , VOL.16, NO.23-24, pp. 2102-2105.

Hodes, G. (2008). Comparison of dye- and semiconductor-sensitized porous nanocrystalline liquid junction solar cells. *J. Phys. Chem. C* VOL.112, NO.46, pp. 17778-17787.

Hore, S.; Palomares, E.; Smit, H.; Bakker, N. J.; Comte, P.; Liska, P.; Thampi, K. R.; Kroon, J. M.; Hinsch, A. & Durrant, J. R. (2005). Acid versus base peptization of mesoporous nanocrystalline TiO$_2$ films: functional studies in dye sensitized solar cellst. *J. Mater. Chem.* , VOL.15, NO.3, pp. 412-418.

Huang, Z.; Liu, X. H.; Li, K. X.; Li, D. M.; Luo, Y. H.; Li, H.; Song, W. B.; Chen, L. Q. & Meng, Q. B. (2007). Application of carbon materials as counter electrodes of dye-sensitized solar cells. *Electrochem. Commun.* , VOL.9, NO.4, pp. 596-598.

Hyun, B. R.; Zhong, Y. W.; Bartnik, A. C.; Sun, L. F.; Abruna, H. D.; Wise, F. W.; Goodreau, J. D.; Matthews, J. R.; Leslie, T. M. & Borrelli, N. F. (2008). Electron injection from colloidal PbS quantum dots into titanium dioxide nanoparticles. *ACS Nano*, VOL.2, NO.11, pp. 2206-2212.

Imoto, K.; Takahashi, K.; Yamaguchi, T.; Komura, T.; Nakamura, J. & Murata, K. (2003). High-performance carbon counter electrode for dye-sensitized solar cells. *Sol. Energy Mater. Sol. Cells* VOL.79, NO.4, pp. 459-469.

Jiang, Q. W.; Li, G. R. & Gao, X. P. (2009). Highly ordered TiN nanotube arrays as counter electrodes for dye-sensitized solar cells. *Chem. Commun.* , NO.44, pp. 6720-6722.

Jiang, Q. W.; Li, G. R.; Liu, S. & Gao, X. P. (2010). Surface-nitrided nickel with bifunctional structure as low-cost counter electrode for dye-sensitized solar cells. *J. Phys. Chem. C* VOL.114, NO.31, pp. 13397-13401.

Joshi, P.; Zhang, L. F.; Chen, Q. L.; Galipeau, D.; Fong, H. & Qiao, Q. Q. (2010). Electrospun carbon nanofibers as low-cost counter electrode for dye-sensitized solar cells. *ACS Appl. Mater. Interfaces* VOL.2, NO.12, pp. 3572-3577.

Jung, H. S.; Lee, J. K.; Nastasi, M.; Lee, S. W.; Kim, J. Y.; Park, J. S.; Hong, K. S. & Shin, H. (2005). Preparation of nanoporous MgO-Coated TiO$_2$ nanoparticles and their application to the electrode of dye-sensitized solar cells. *Langmuir*, VOL.21, NO.23, pp. 10332-10335.

Kaiser, I.; Ernst, K.; Fischer, C. H.; Konenkamp, R.; Rost, C.; Sieber, I. & Lux-Steiner, M. C. (2001). The eta-solar cell with CuInS$_2$: A photovoltaic cell concept using an extremely thin absorber (eta). *Sol. Energy Mater. Sol. Cells* VOL.67, NO.1-4, pp. 89-96.

Kamat, P. V. (2008). Quantum dot solar cells. Semiconductor nanocrystals as light harvesters. *J. Phys. Chem. C* VOL.112, NO.48, pp. 18737-18753.

Kang, T. S.; Smith, A. P.; Taylor, B. E. & Durstock, M. F. (2009). Fabrication of highly-ordered TiO$_2$ nanotube arrays and their use in dye-sensitized solar cells. *Nano Lett.* , VOL.9, NO.2, pp. 601-606.

Kavan, L.; Yum, J. H. & Gratzel, M. (2011). Optically transparent cathode for dye-sensitized solar cells based on graphene nanoplatelets. *ACS Nano,* VOL.5, NO.1, pp. 165-172.

Kay, A. & Gratzel, M. (1996). Low cost photovoltaic modules based on dye sensitized nanocrystalline titanium dioxide and carbon powder. *Sol. Energy Mater. Sol. Cells* VOL.44, NO.1, pp. 99-117.

Kay, A. & Gratzel, M. (2002). Dye-sensitized core-shell nanocrystals: Improved efficiency of mesoporous tin oxide electrodes coated with a thin layer of an insulating oxide. *Chem. Mater.* , VOL.14, NO.7, pp. 2930-2935.

Klimov, V. I. (2006). Mechanisms for photogeneration and recombination of multiexcitons in semiconductor nanocrystals: Implications for lasing and solar energy conversion. *J. Phys. Chem. B* VOL.110, NO.34, pp. 16827-16845.

Kron, G.; Egerter, T.; Werner, J. H. & Rau, U. (2003). Electronic transport in dye-sensitized nanoporous TiO_2 solar cells-comparison of electrolyte and solid-state devices. *J. Phys. Chem. B* VOL.107, NO.15, pp. 3556-3564.

Kumara, G. R. A.; Kaneko, S.; Okuya, M. & Tennakone, K. (2002). Fabrication of dye-sensitized solar cells using triethylamine hydrothiocyanate as a CuI crystal growth inhibitor. *Langmuir,* VOL.18, NO.26, pp. 10493-10495.

Law, M.; Greene, L. E.; Johnson, J. C.; Saykally, R. & Yang, P. D. (2005). Nanowire dye-sensitized solar cells. *Nat. Mater.* , VOL.4, NO.6, pp. 455-459.

Lee, H.; Leventis, H. C.; Moon, S. J.; Chen, P.; Ito, S.; Haque, S. A.; Torres, T.; Nuesch, F.; Geiger, T.; Zakeeruddin, S. M.; Gratzel, M. & Nazeeruddin, M. K. (2009a). PbS and US quantum dot-sensitized solid-state solar cells: "old concepts, new results". *Adv. Funct. Mater.* , VOL.19, NO.17, pp. 2735-2742.

Lee, H.; Wang, M. K.; Chen, P.; Gamelin, D. R.; Zakeeruddin, S. M.; Gratzel, M. & Nazeeruddin, M. K. (2009b). Efficient CdSe quantum dot-sensitized solar cells prepared by an improved successive ionic layer adsorption and reaction process. *Nano Lett.* , VOL.9, NO.12, pp. 4221-4227.

Lee, Y. L. & Lo, Y. S. (2009). Highly efficient quantum-dot-sensitized solar cell based on co-sensitization of CdS/CdSe. *Adv. Funct. Mater.* , VOL.19, NO.4, pp. 604-609.

Leschkies, K. S.; Jacobs, A. G.; Norris, D. J. & Aydil, E. S. (2009). Nanowire-quantum-dot solar cells and the influence of nanowire length on the charge collection efficiency. *Appl. Phys. Lett.* , VOL.95, NO.19, pp. 193103.

Levy-Clement, C.; Tena-Zaera, R.; Ryan, M. A.; Katty, A. & Hodes, G. (2005). CdSe-Sensitized p-CuSCN/nanowire n-ZnO heterojunctions. *Adv. Mater.* , VOL.17, NO.12, pp. 1512-1515.

Lin, S.-C.; Lee, Y.-L.; Chang, C.-H.; Shen, Y.-J. & Yang, Y.-M. (2007). Quantum-dot-sensitized solar cells: Assembly of CdS-quantum-dots coupling techniques of self-assembled monolayer and chemical bath deposition. *Appl. Phys. Lett.* , VOL.90, NO.14, pp. 143517.

Liu, B. & Aydil, E. S. (2009). Growth of oriented single-crystalline rutile TiO_2 nanorods on transparent conducting substrates for dye-sensitized solar cells. *J. Am. Chem. Soc.* , VOL.131, NO.11, pp. 3985-3990.

Liu, Y.-X.; Summers, M. A.; Scully, S. R. & McGehee, M. D. (2006). Resonance energy transfer from organic chromophores to fullerene molecules. *J. Appl. Phys.* , VOL.99, NO.9, pp. 093521.

Lu, S. & Madhukar, A. (2007). Nonradiative resonant excitation transfer from nanocrystal quantum dots to adjacent quantum channels. *Nano Lett.* , VOL.7, NO.11, pp. 3443-3451.

Macak, J. M.; Tsuchiya, H.; Ghicov, A. & Schmuki, P. (2005). Dye-sensitized anodic TiO$_2$ nanotubes. *Electrochem. Commun.* , VOL.7, NO.11, pp. 1133-1137.

Martinson, A. B. F.; Elam, J. W.; Hupp, J. T. & Pellin, M. J. (2007). ZnO nanotube based dye-sensitized solar cells. *Nano Lett.* , VOL.7, NO.8, pp. 2183-2187.

Matsumura, M.; Matsudaira, S.; Tsubomura, H.; Takata, M. & Yanagida, H. (1980). Dye sensitization and surface structures of semiconductor electrodes. *Ind. Eng. Chem. Prod. Res. Dev.* , VOL.19, NO.3, pp. 415-421.

Meng, Q. B.; Takahashi, K.; Zhang, X. T.; Sutanto, I.; Rao, T. N.; Sato, O.; Fujishima, A.; Watanabe, H.; Nakamori, T. & Uragami, M. (2003). Fabrication of an efficient solid-state dye-sensitized solar cell. *Langmuir*, VOL.19, NO.9, pp. 3572-3574.

Muniz, E. C.; Goes, M. S.; Silva, J. J.; Varela, J. A.; Joanni, E.; Parra, R. & Bueno, P. R. (2011). Synthesis and characterization of mesoporous TiO$_2$ nanostructured films prepared by a modified sol-gel method for application in dye solar cells. *Ceram. Int.* , VOL.37, NO.3, pp. 1017-1024.

Murakami, T. N.; Kijitori, Y.; Kawashima, N. & Miyasaka, T. (2004). Low temperature preparation of mesoporous TiO$_2$ films for efficient dye-sensitized photoelectrode by chemical vapor deposition combined with UV light irradiation. *Journal of Photochemistry and Photobiology a-Chemistry*, VOL.164, NO.1-3, pp. 187-191.

Murray, C. B.; Norris, D. J. & Bawendi, M. G. (1993). Synthesis and characterization of nearly monodisperse CdE (E = S, Se, Te) semiconductor nanocrystallites. *J. Am. Chem. Soc.* , VOL.115, NO.19, pp. 8706-8715.

Neale, N. R. & Frank, A. J. (2007). Size and shape control of nanocrystallites in mesoporous TiO$_2$ films. *J. Mater. Chem.* , VOL.17, NO.30, pp. 3216-3221.

Norris, D. J. & Bawendi, M. G. (1996). Measurement and assignment of the size-dependent optical spectrum in CdSe quantum dots. *Physical Review B*, VOL.53, NO.24, pp. 16338-16346.

O'Regan, B. & Schwartz, D. T. (1998). Large enhancement in photocurrent efficiency caused by UV illumination of the dye-sensitized heterojunction TiO$_2$/RuLL ' NCS/CuSCN: Initiation and potential mechanisms. *Chem. Mater.* , VOL.10, NO.6, pp. 1501-1509.

O'Regan, B.; Schwartz, D. T.; Zakeeruddin, S. M. & Gratzel, M. (2000). Electrodeposited nanocomposite n-p heterojunctions for solid-state dye-sensitized photovoltaics. *Adv. Mater.* , VOL.12, NO.17, pp. 1263-1267.

O'Regan, B.; Scully, S.; Mayer, A. C.; Palomares, E. & Durrant, J. (2005). The effect of Al$_2$O$_3$ barrier layers in TiO$_2$/Dye/CuSCN photovoltaic cells explored by recombination and DOS characterization using transient photovoltage measurements. *J. Phys. Chem. B* VOL.109, NO.10, pp. 4616-4623.

O'Regan, B. & Gratzel, M. (1991). A low-cost, high-efficiency solar-cell based on dye-sensitized colloidal TiO$_2$ films. *Nature*, VOL.353, NO.6346, pp. 737-740.

O'Regan, B. & Schwartz, D. T. (1995). Efficient photo-hole injection from adsorbed cyanine dyes into electrodeposited copper(I) thiocyanate thin-films. *Chem. Mater.* , VOL.7, NO.7, pp. 1349-1354.

Overbeek, J. T. G. (1982). Monodisperse colloidal systems, fascinating and useful. *Adv. Colloid Interface Sci.* , VOL.15, NO.3-4, pp. 251-277.

Park, N. G.; van de Lagemaat, J. & Frank, A. J. (2000). Comparison of dye-sensitized rutile- and anatase-based TiO₂ solar cells. *J. Phys. Chem. B* VOL.104, NO.38, pp. 8989-8994.

Peng, Z. A. & Peng, X. G. (2002). Nearly monodisperse and shape-controlled CdSe nanocrystals via alternative routes: Nucleation and growth. *J. Am. Chem. Soc.* , VOL.124, NO.13, pp. 3343-3353.

Ramasamy, E.; Lee, W. J.; Lee, D. Y. & Song, J. S. (2007). Nanocarbon counterelectrode for dye sensitized solar cells. *Appl. Phys. Lett.* , VOL.90, NO.17, pp.

Robel, I.; Kuno, M. & Kamat, P. V. (2007). Size-dependent electron injection from excited CdSe quantum dots into TiO₂ nanoparticles. *J. Am. Chem. Soc.* , VOL.129, NO.14, pp. 4136-4137.

Robel, I.; Subramanian, V.; Kuno, M. & Kamat, P. V. (2006). Quantum dot solar cells. Harvesting light energy with CdSe nanocrystals molecularly linked to mesoscopic TiO₂ films. *J. Am. Chem. Soc.* , VOL.128, NO.7, pp. 2385-2393.

Ross, R. T. & Nozik, A. J. (1982). Efficiency of hot-carrier solar energy converters. *J. Appl. Phys.* , VOL.53, NO.5, pp. 3813-3818.

Roy-Mayhew, J. D.; Bozym, D. J.; Punckt, C. & Aksay, I. A. (2010). Functionalized graphene as a catalytic counter electrode in dye-sensitized solar cells. *ACS Nano*, VOL.4, NO.10, pp. 6203-6211.

Roy, P.; Kim, D.; Paramasivam, I. & Schmuki, P. (2009). Improved efficiency of TiO₂ nanotubes in dye sensitized solar cells by decoration with TiO₂ nanoparticles. *Electrochem. Commun.* , VOL.11, NO.5, pp. 1001-1004.

Ruhle, S.; Shalom, M. & Zaban, A. (2010). Quantum-dot-sensitized solar cells. *ChemPhysChem* VOL.11, NO.11, pp. 2290-2304.

Sayer, R. A.; Hodson, S. L. & Fisher, T. S. (2010). Improved efficiency of dye-sensitized solar cells using a vertically aligned carbon nanotube counter electrode. *Journal of Solar Energy Engineering-Transactions of the Asme*, VOL.132, NO.2, pp. 021007.

Schaller, R. D.; Sykora, M.; Pietryga, J. M. & Klimov, V. I. (2006). Seven excitons at a cost of one: Redefining the limits for conversion efficiency of photons into charge carriers. *Nano Lett.* , VOL.6, pp. 424-429.

Seo, S. H.; Kim, S. Y.; Koo, B. K.; Cha, S. I. & Lee, D. Y. (2010). Influence of electrolyte composition on the photovoltaic performance and stability of dye-sensitized solar cells with multiwalled carbon nanotube catalysts. *Langmuir*, VOL.26, NO.12, pp. 10341-10346.

Shalav, A.; Richards, B. S.; Trupke, T.; Kramer, K. W. & Gudel, H. U. (2005). Application of NaYF₄:Er³⁺ up-converting phosphors for enhanced near-infrared silicon solar cell response. *Appl. Phys. Lett.* , VOL.86, NO.1, pp. 013505.

Shankar, K.; Feng, X. & Grimes, C. A. (2009). Enhanced harvesting of red photons in nanowire solar cells: Evidence of resonance energy transfer. *ACS Nano*, VOL.3, NO.4, pp. 788-794.

Shen, Y.; Bao, J.; Dai, N.; Wu, J.; Gu, F.; Tao, J. C. & Zhang, J. C. (2009). Speedy photoelectric exchange of CdSe quantum dots/mesoporous titania composite system. *Appl. Surf. Sci.* , VOL.255, NO.6, pp. 3908-3911.

Shi, C.; Zhu, A. M.; Yang, X. F. & Au, C. T. (2004). On the catalytic nature of VN, Mo₂N, and W₂N nitrides for NO reduction with hydrogen. *Applied Catalysis a-General*, VOL.276, NO.1-2, pp. 223-230.

Smestad, G. P.; Spiekermann, S.; Kowalik, J.; Grant, C. D.; Schwartzberg, A. M.; Zhang, J.; Tolbert, L. M. & Moons, E. (2003). A technique to compare polythiophene solid-state dye sensitized TiO_2 solar cells to liquid junction devices. Sol. Energy Mater. Sol. Cells VOL.76, NO.1, pp. 85-105.

Sung, Y. M. & Kim, H. J. (2007). Sputter deposition and surface treatment of TiO_2 films for dye-sensitized solar cells using reactive RF plasma. Thin Solid Films VOL.515, NO.12, pp. 4996-4999.

Suzuki, K.; Yamaguchi, M.; Kumagai, M. & Yanagida, S. (2003). Application of carbon nanotubes to counter electrodes of dye-sensitized solar cells. Chem. Lett. , VOL.32, NO.1, pp. 28-29.

Talapin, D. V. & Murray, C. B. (2005). PbSe nanocrystal solids for n- and p-channel thin film field-effect transistors. Science, VOL.310, NO.5745, pp. 86-89.

Talapin, D. V.; Rogach, A. L.; Shevchenko, E. V.; Kornowski, A.; Haase, M. & Weller, H. (2002). Dynamic distribution of growth rates within the ensembles of colloidal II-VI and III-V semiconductor nanocrystals as a factor governing their photoluminescence efficiency. J. Am. Chem. Soc. , VOL.124, NO.20, pp. 5782-5790.

Tang, Y. W.; Luo, L. J.; Chen, Z. G.; Jiang, Y.; Li, B. H.; Jia, Z. Y. & Xu, L. (2007). Electrodeposition of ZnO nanotube arrays on TCO glass substrates. Electrochem. Commun. , VOL.9, NO.2, pp. 289-292.

Tennakone, K.; Kumara, G.; Kottegoda, I. R. M.; Wijayantha, K. G. U. & Perera, V. P. S. (1998). A solid-state photovoltaic cell sensitized with a ruthenium bipyridyl complex. Journal of Physics D-Applied Physics, VOL.31, NO.12, pp. 1492-1496.

Tennakone, K.; Kumara, G.; Kumarasinghe, A. R.; Wijayantha, K. G. U. & Sirimanne, P. M. (1995). A dye-sensitized nano-porous solid-state photovoltaic cell. Semicond. Sci. Technol. , VOL.10, NO.12, pp. 1689-1693.

Terauchi, S.; Koshizaki, N. & Umehara, H. (1995). Fabrication of Au nanoparticles by radio-frequency magnetron sputtering. Nanostruct. Mater. , VOL.5, NO.1, pp. 71-78.

Tributsch, H. (1972). Reaction of excited chlorophyll molecules at electrodes and in photosynthesis. Photochem. Photobiol. , VOL.16, NO.4, pp. 261-269.

Tributsch, H. (2004). Dye sensitization solar cells: a critical assessment of the learning curve. Coord. Chem. Rev. , VOL.248, NO.13-14, pp. 1511-1530.

Tsubomura, H.; Matsumura, M.; Nomura, Y. & Amamiya, T. (1976). Dye sensitised zinc oxide: aqueous electrolyte: platinum photocell. Nature, VOL.261, NO.5559, pp. 402-403.

Vogel, R.; Hoyer, P. & Weller, H. (1994). Quantum-sized PbS, CdS, Ag_2S, Sb_2S_3, and Bi_2S_3 particles as sensitizers for various nanoporous wide-bandgap semiconductors. J. Phys. Chem. , VOL.98, NO.12, pp. 3183-3188.

Wang, G. Q.; Xing, W. & Zhuo, S. P. (2009a). Application of mesoporous carbon to counter electrode for dye-sensitized solar cells. J. Power Sources VOL.194, NO.1, pp. 568-573.

Wang, M. K.; Anghel, A. M.; Marsan, B.; Ha, N. L. C.; Pootrakulchote, N.; Zakeeruddin, S. M. & Gratzel, M. (2009b). CoS supersedes Pt as efficient electrocatalyst for triiodide reduction in dye-sensitized solar cells. J. Am. Chem. Soc. , VOL.131, NO.44, pp. 15976-15977.

Wang, P.; Dai, Q.; Zakeeruddin, S. M.; Forsyth, M.; MacFarlane, D. R. & Gratzel, M. (2004). Ambient temperature plastic crystal electrolyte for efficient, all-solid-state dye-sensitized solar CeN. J. Am. Chem. Soc. , VOL.126, NO.42, pp. 13590-13591.

Wang, P.; Zakeeruddin, S. M.; Comte, P.; Exnar, I. & Gratzel, M. (2003a). Gelation of ionic liquid-based electrolytes with silica nanoparticles for quasi-solid-state dye-sensitized solar cells. *J. Am. Chem. Soc.* , VOL.125, NO.5, pp. 1166-1167.

Wang, P.; Zakeeruddin, S. M.; Moser, J. E.; Nazeeruddin, M. K.; Sekiguchi, T. & Gratzel, M. (2003b). A stable quasi-solid-state dye-sensitized solar cell with an amphiphilic ruthenium sensitizer and polymer gel electrolyte. *Nat. Mater.* , VOL.2, NO.6, pp. 402-407.

Wang, Z.-S.; Yanagida, M.; Sayama, K. & Sugihara, H. (2006). Electronic-insulating coating of $CaCO_3$ on TiO_2 electrode in dye-sensitized solar cells: Improvement of electron lifetime and efficiency. *Chem. Mater.* , VOL.18, NO.12, pp. 2912-2916.

Wang, Z. S.; Huang, C. H.; Huang, Y. Y.; Hou, Y. J.; Xie, P. H.; Zhang, B. W. & Cheng, H. M. (2001). A highly efficient solar cell made from a dye-modified ZnO-covered TiO_2 nanoporous electrode. *Chem. Mater.* , VOL.13, NO.2, pp. 678-682.

Wroblowa, H. S. & Saunders, A. (1973). Flow-through electrodes: II. The I_3^-/I^- redox couple. *J. Electroanal. Chem. Interfacial Electrochem.* , VOL.42, NO.3, pp. 329-346.

Wu, M.; Lin, X.; Hagfeldt, A. & Ma, T. (2011a). Low-cost molybdenum carbide and tungsten carbide counter electrodes for dye-sensitized solar cells. *Angew. Chem. Int. Ed.*, VOL.50, NO.15, pp. 3520-3524.

Wu, M. X.; Lin, X. A.; Hagfeldt, A. & Ma, T. L. (2011b). A novel catalyst of WO_2 nanorod for the counter electrode of dye-sensitized solar cells. *Chem. Commun.* , VOL.47, NO.15, pp. 4535-4537.

Wu, S. J.; Han, H. W.; Tai, Q. D.; Zhang, J.; Chen, B. L.; Xu, S.; Zhou, C. H.; Yang, Y.; Hu, H. & Zhao, X. Z. (2008). Improvement in dye-sensitized solar cells with a ZnO-coated TiO_2 electrode by rf magnetron sputtering. *Appl. Phys. Lett.* , VOL.92, NO.12, pp. 122106.

Xia, J. B.; Li, F. Y.; Yang, H.; Li, X. H. & Huang, C. H. (2007). A novel quasi-solid-state dye-sensitized solar cell based on monolayer capped nanoparticles framework materials. *Journal of Materials Science*, VOL.42, NO.15, pp. 6412-6416.

Yang, H.; Cheng, Y. F.; Li, F. Y.; Zhou, Z. G.; Yi, T.; Huang, C. H. & Jia, N. Q. (2005). Quasi-solid-state dye-sensitized solar cells based on mesoporous silica SBA-15 framework materials. *Chin. Phys. Lett.* , VOL.22, NO.8, pp. 2116-2118.

Yoon, C. H.; Vittal, R.; Lee, J.; Chae, W. S. & Kim, K. J. (2008). Enhanced performance of a dye-sensitized solar cell with an electrodeposited-platinum counter electrode. *Electrochim. Acta* VOL.53, NO.6, pp. 2890-2896.

Zaban, A.; Aruna, S. T.; Tirosh, S.; Gregg, B. A. & Mastai, Y. (2000). The effect of the preparation condition of TiO_2 colloids on their surface structures. *J. Phys. Chem. B* VOL.104, NO.17, pp. 4130-4133.

Zhang, Q. F. & Cao, G. Z. (2011). Nanostructured photoelectrodes for dye-sensitized solar cells. *Nano Today*, VOL.6, NO.1, pp. 91-109.

Zhu, K.; Neale, N. R.; Miedaner, A. & Frank, A. J. (2007). Enhanced charge-collection efficiencies and light scattering in dye-sensitized solar cells using oriented TiO_2 nanotubes arrays. *Nano Lett.* , VOL.7, NO.1, pp. 69-74.

6

Fabrication, Doping and Characterization of Polyaniline and Metal Oxides: Dye Sensitized Solar Cells

Sadia Ameen[1], M. Shaheer Akhtar[2],
Young Soon Kim[1] and Hyung-Shik Shin[1]
*[1]Energy Materials & Surface Science Laboratory, Solar Energy Research Center,
School of Chemical Engineering, Chonbuk National University, Jeonju,
[2]New & Renewable Energy Material Development Center (NewREC),
Chonbuk National University, Jeonbuk,
Republic of Korea*

1. Introduction

The photoelectrochemical devices like dye sensitized solar cell (DSSCs) are the promising photovoltaic device for the utilization of solar into electricity energy by converting solar radiation through the generation of photogenerated carriers. DSSCs possess the benign properties of low cost, high conversion efficiency and ease of fabrication (Gratzel, 2005, 2004 & Frank, et al., 2004). Several significant advantages are associated to DSSCs such as the semiconductor-electrolyte interface (SEI) is easy to manufacture and it is cost effective for production, non sensitive to the defects, the two functions of light harvesting and charge-carrier transport are remain separated and favours the direct energy transfer from photons to chemical energy. Although, DSSCs are the promising photovoltaic technology for achieving reasonably high conversion efficiency but the improvements are still demanded to develop a high potential technology.

A typical DSSC is comprised of a dye-coated mesoporous nanocrystalline metal oxide semiconductor, machinated between two the conductive transparent electrodes as shown in Fig. 1. A liquid iodide/tri-iodide redox couple as an electrolyte is introduced to fill the pores of the film and to improve the contact between the nanoparticles (Gratzel, 2000). Upon photo excitation of the dye, the electrons are injected into the conduction band of the nanocrystalline metal oxide semiconductor and the original state of the dye is restored by the electron donation from the hole conductor. The regeneration of the sensitizer by the hole conductor intercepts the recapture of the conduction band electron by the oxidized dye and the hole conductor is regenerated at the counter-electrode. The circuit is completed via electron migration through the external load.

Nanocrystalline metal oxide semiconductor like TiO_2, ZnO and SnO_2 etc have been accepted as the effective photoanode materials for DSSCs due to their good optical and electronic properties. Till date, DSSCs constructed from TiO_2 nanocrystalline metal oxide electrodes has presented the highest solar to electricity conversion efficiency of 11.4%. These nanocrystalline metal oxide semiconductors possess high surface to volume ratio required

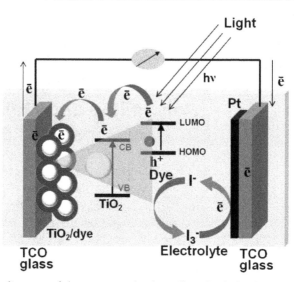

Fig. 1. Mechanistic diagram of dye sensitized solar cell under light illumination.

for the high absorption of dye molecules for the improved light harvesting efficiency but the challenges like enhancement in the electron transport rate, reduction of recombination rate and the improvement in the energy conversion efficiency are still associated with metal oxide semiconductor based DSSCs. It has been reported that the photoelectrodes made of metal oxides nanostructures like nanoparticles, nanorods, nanowires and nanotubes improve the photovoltaic properties of DSSCs such as J_{SC}, V_{OC}, FF and conversion efficiency by the effective electron transfer during the performance of DSSCs. The conducting polymers are the semiconducting polymers which exhibit good electrical properties similar to metals and possess unique properties like light weight, resistance to corrosion, flexibility and low cost. The nanostructures of conducting polymers are excessively exploited in the electronic devices due to the large surface area of the nanomaterials and their unique electronic properties (Huang et al., 2006, Zhnag et al., 2006, Wan, 2008, 2009 & Li et al., 2009). These perspectives have prompted the researchers for the rapid development of conducting polymers and the nanocrystalline metal oxide semiconductors for improving the processability, functionality and the conversion efficiency of DSSCs.

In this chapter, we have briefly surveyed the conducting polymers, its doping and the different morphologies of nanocrystalline metal oxide semiconductors for the effective processing, functionality and the fabrication of DSSCs. The chapter focuses the recent developments in the heterostructures diodes, Ohmic devices and DSSCs fabricated with polyaniline (PANI) and the nanostructured metal oxides. The morphology, structure and physiochemical properties of TiO_2 and ZnO nanostructure semiconductors have been reviewed as photoelectrode for the application of DSSCs. It has been reviewed that the performances of DSSCs are considerably affected by the preparation, morphology and the electrical properties of conducting polymers and metal oxide semiconductors. TiO_2 and ZnO nanostructures based photoelectrodes have shown comparable photovoltaic performance compared to the conventional TiO_2 nanoparticles based DSSCs. On the other hand, the conducting polymers as a hole transport materials are found as the potential candidates for replacing the liquid electrolyte in DSSCs and the dopants have been discussed for enhancing the conducting properties of a particular polymer i.e. PANI.

2. Aspects of conducting polymers

The polymers are insulators due to the covalent bonded atoms in the polymer chain and thus, there is no scope for the delocalization of the valence electrons and consequently neither charge carriers nor the path for their movement are available (Stejskal et al., 2002). However, the important classes of polymers called as conducting polymers are semiconductors (Friend et al., 1999) and their band gaps could be tuned by alternating the chemical nature of either the polymer backbone or the side groups present in the polymeric chain. The overlapping of the molecular orbital for the formation of the delocalized molecular wave functions and secondly the partially filled molecular orbital for a free movement of electrons throughout the polymeric structure are the basic requirements for the polymers to become conducting in nature. These conducting polymers exhibit the unusual electronic properties such as electrical conductivity, low energy optical transitions, low ionization potential and high electron affinity (Park et al., 2010) due to the presence of extended π-conjugated systems along the polymeric chain (Gerard et al., 2002). The electrical properties of the conducting polymers could be easily tuned from insulating to metallic range by the reversible chemical, electrochemical and physical properties controlled by a doping/de-doping procedure. The electrical properties are changed by the partial oxidation or reduction of the polymer chain during the doping procedure. The conducting polymers possess the appended advantage of tailor-made property for accomplishing the requirements of the particular application through the modifications in the polymer structure. It is reported that the nanostructures of conducting polymer not only retain these unique characteristics, but also exhibit the properties like large surface area, size, and quantum effect etc which are important for designing and making the novel photovoltaic devices (Xia et al., 2010). Out of several conducting polymers, polyaniline (PANI) is the centre of scientific interest and has been considered as one of the most potential conducting polymers. The interest in PANI could possibly be linked to the numerous applications for various electrochemical, electro rheological and in the electronic fields such as batteries, sensors, controlling systems and organic displays (Smith, 1998, Wessling, 2001, Cho et al., 2002 & Choi et al. 2001) because of its facile synthetic process, good environmental stability, easy conductivity control and cheap production in large quantities. The high yield of PANI demands several essential conditions such as highly pured monomers, chemicals and solvents for obtaining the high quality polymer, the strict control on the polymerization conditions, inert and dry environment for the polymerization since the small variation in the polymerization conditions might alter the nature of the product, and as PANI might undergo isomerization reaction under light and heat therefore, the synthesized PANI products should be kept in a dark, cool and dry place (Ameen, et al., 2010)

3. General approaches for the synthesis of PANI

3.1 Interfacial polymerization

In a typical reaction (Huang, et al., 2003, Virji, et al., 2004, King, et al., 2005 & Huang et al., 2004) aniline monomer is dissolved into the organic solvents which have either low or high density than water like hexane, benzene, toluene, carbon tetrachloride, chloroform etc. A separate aqueous solution of ammonium peroxydisulphate (APS) as an oxidant is prepared with the acids such as HCl, sulfuric acid, CSA, toluene sulfonic acid etc. The two solutions are then gently transferred to a glass vial for the generating the interface between the two layers. As the reaction proceeds, the green colored PANI is formed at the interface and gradually migrates into the aqueous phase and slowly the entire aqueous phase get filled homogeneously with dark-green PANI. The aqueous phase is collected through filtration after 24 h. The non-conducting PANI could be obtained by dialyzing the synthesized PANI with ammonia solution

3.2 Rapid mixing of reactants
This is the simplest method for synthesizing PANI and in place of conventionally slow addition of the aqueous APS solution, the aniline monomer is mixed rapidly with APS solution. Here, the initiator molecules are consumed rapidly after the initialization of the polymerization reaction due to the even distribution of aniline monomer and APS molecules in the solution.

3.3 Sonochemical synthesis
Jing et al (Jing, et al., 2006, 2007) has reported the synthesis of PANI similar to the synthesis procedure of conventional PANI. The acidic APS solution is added dropwise to an acidic aniline solution, and subjected to the ultrasonic irradiation which results to the high yield of PANI. It has been found that the excessive addition of aniline monomer or APS could contribute to the continuous formation of the primary PANI products or could lead the agglomeration of PANI. Unlike conventional procedure where irregular morphology of PANI is obtained, the sonochemical synthesis strictly prevents the further growth and agglomeration and thus, generates uniform morphology of PANI.

3.4 Electrophoretic synthesis
The first step of a reaction requires a preparation of conducting PANI through the chemical procedure using APS as oxidant. In the second step, a stock solution of PANI (1 mg/ml) is prepared in the formic acid and the colloidal suspension is prepared by adding 100 ml of the stock solution into acetonitrile for the preparation of an electrolyte. In the colloidal suspension, PANI dissociate into ions due to high dielectric medium which is offered by acetonitrile and thus, results to the formation of the positively charged PANI. The last step involves the electrophoretic film deposition where under the influence of applied voltage for a required time duration, positively charged colloid spheres of PANI in acidic colloidal suspension starts moving towards the negatively biased fluorinated tin oxide glass (FTO) electrode. This stepwise growth synthesis produces the uniform nanostructures of PANI on the surfaces of the FTO glass substrates.

3.5 Plasma polymerization
The plasma polymerization, as shown in Fig. 2 requires four parts (1) a reactor chamber quartz tube (2 cm), (2) Cu coil (4 in.), (3) plasma system (R.F. generator: 0-600 W, matching network frequency of 13.56 MHz) and (4) mechanical vacuum pump (speed 600 l/min). The glow discharges are introduced through RF amplifier with a resistive coupling mechanism at 13.5 MHz and power of 120 W. These discharges are set without carrier gas or any other additional chemical elements to prevent the contamination. The cleaned FTO glass

substrates are placed just below the RF coil, placed inside the quartz tube. Initially, the chamber is evacuated to a base pressure of 10^{-3} Torr through a rotator vacuum pump. After attaining the base pressure, aniline monomer is injected, using a hypodermic syringe and the reaction is promoted by the collisions of the aniline monomer molecules with the ions/particles present in the plasma.

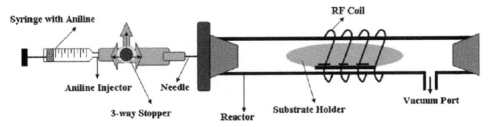

Fig. 2. Reactor setup for plasma synthesis of PANI. Reprinted with permission from [Ameen S. et al 2009], Superlatt. Microstruc. 46 (2009) 745. © 2009, Elsevier Ltd.

4. Concept of doping in PANI

Doping is the unique, central, underlying and unifying concept, which distinguishes PANI as conducting polymer from other insulating polymers. The controlled addition of known chemical species results in the dramatic changes in the electronic, electrical, magnetic, optical and structural properties of PANI. The doping procedures of inorganic semiconductors and for conducting polymers including PANI are fundamentally different from each other. In PANI, the dopant concentrations are exceptionally high compared to commonly doped inorganic semiconductors (typically in parts per million ranges) and the doping of PANI involves random dispersion or aggregation of dopants in molar concentrations in the disordered structure of entangled chains and fibrils. However, for inorganic semiconductors, the doping generates either holes in the valence band or electrons in the conduction band. The doping of PANI generates high conductivity primarily by increasing the carrier concentrations and leads to the conjugational defect like solitons, polarons, bipolarons in the polymer chain. Doping is a reversible produce which could be carried out chemically or electrochemically with the oxidation or reduction by accepting or donating the electrons respectively and thus, results to the positive or negative charges. Ameen et al have used various acids and binary mixtures as dopants to improve the electrical conductivity of PANI.

4.1 Doping with sulfamic acid

Sulfamic acid is one of the promising dopant which displays the properties of handling ease, solubility, and low corrosiveness. It is a dry, non-volatile stable solid acid, which is moderately soluble in water at room temperature. Wadia et al prepared the thin film of sulfamic acid doped PANI as electrochemical sensor for the determination of nitrite (Dhaoui, et al., 2007, 2008). Ameen et al reported the electrical conductivity of PANI nanostructures in the range of 1.1×10^{-4} – 4.50×10^{-3} ohm^{-1}cm^{-1}, doped with different wt% of sulfamic acid (Ameen, et al., 2007). Moreover, PANI nanofibers doped with sulfamic acid thin film was applied as counter electrodes for the fabrication of DSSCs and obtained the conductivity of 2.06×10^{-4} ohm^{-1}cm^{-1} by Ameen et al (Ameen, et al., 2010).

4.2 Doping with sodium thiosulphate

Sodium thiosulphate is an odorless translucent crystals or white crystalline powder which is soluble in water and exhibit the stability. It is incompatible with strong acid and is not hazardous. Ameen et al reported the electrical conductivity of PANI nanostructures in the range of $8.55\times10^{-9} - 3.67\times10^{-7}$ ohm^{-1}cm^{-1} after doping with sodium thiosulphate (Ameen, et al., 2009).

4.3 Doping with samarium (III) chloride

Samarium (III) chloride is a light yellow colored strong Lewis acid. This dopant is non hazardous and stable in nature and is reported to increase the conductivity of PANI nanostructure from 3.70×10^{-11} to 1.30×10^{-9} ohm^{-1}cm^{-1} (Ameen, et al., 2009).

4.4 Doping with praseodymium (III) chloride

Praseodymium (III) Chloride is a green colored powder which does not show thermal decomposition and is stable in nature. It is an excellent water-soluble crystalline strong Lewis acid and after doping is reported to show the conductivity of PANI nanostructures in the range $1.87\times10^{-11} - 2.22\times10^{-10}$ ohm^{-1}cm^{-1} (Ameen, et al., 2008).

4.5 Doping with binary dopant: ZrO₂/AgI & ZrO₂/PbI₂

In this binary dopant mixture ZrO_2 is a luminescent material which usually makes no contribution in the conductivity of PANI. On the other hand, AgI is a non luminescent material which enhances the conductivity of PANI. The binary effects of these two materials have improved the optical and the electrical conductivity of PANI nanostructures in the range of $9.28\times10^{-10} - 3.763\times10^{-6}$ ohm^{-1}cm^{-1} (Ameen, et al., 2007). Shaoxu et al reported the synthesis, characterization and the thermal analysis of PANI-ZrO_2 nanocomposites (Wang, et al., 2006). The increased ZrO_2 concentration in the binary dopant mixture exhibits a high PL emission intensity which improves optoelectronic properties of PANI. Ameen et al reported the effect of ZrO_2/PbI₂ doping on the electrical properties of PANI and the conductivity improves from 2.70×10^{-13} to 8.81×10^{-4} ohm^{-1}cm^{-1} (Ameen, et al., 2009).

5. Applications of PANI in heterostructure devices

5.1 Electrophoretic deposition of PANI for heterostructure devices

5.1.1 Electrophoretically deposited PANI/ZnO heterostructure diodes

The fabrication of inorganic/organic heterostructure diodes is reported by the electrophoretic process for depositing organic materials (PANI, p-type) film on inorganic n-type ZnO NPs thin film substrate with top Pt thin layer contact by Ameen et al (Ameen, et al., 2009). The pristine ZnO film is composed of well crystalline ZnO nanoparticles of size ~50-70 nm, shown in Fig. 3 (a). The penetration of the PANI molecules into the crystalline ZnO NPs thin film upon electrophoretic deposition display the accumulation of ZnO NPs and thus, the size of ZnO nanoparticles get increased to ~90 nm (Fig. 3 (b)). This assemblage confirms the substantive interaction and incorporation of electrophoretic deposited PANI into the crystalline ZnO nanoparticles thin film substrates.

The FTIR studies (Fig. 4 (I)) shows the considerable shifting of the peaks might due to the bonding between the hydroxyl groups and the imine groups of the PANI molecules. Noticeably, the UV-visible spectra of PANI/ZnO thin film (Fig. 4 (II)) shows that the peak (a) retains its position, considerable large blue shift for peak (b) is noticed and small red shift

Fig. 3. Surface FE-SEM images of (a) synthesized ZnO nanoparticles (b) electrophoretically deposited PANI/ZnO thin film. Reprinted with permission from [Ameen S. et al, 2009], Superlatt. Microstruc. 46 (2009) 872. © 2009, Elsevier Ltd.

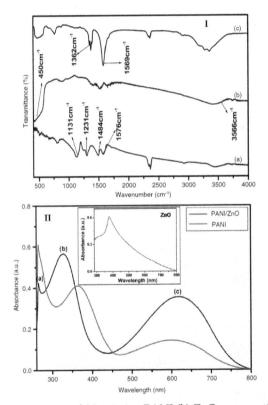

Fig. 4. (I): Typical FTIR spectrum of (a) pristine PANI (b) ZnO nanoparticles and (c) electrophoretically deposited PANI/ZnO thin film. (II): UV-Vis spectra of pristine PANI and electrophoretically deposited PANI/ZnO thin film. Inset shows the UV-Vis spectrum of ZnO nanoparticles. Reprinted with permission from [Ameen S., 2009], Superlatt. Microstruc. 46 (2009) 872. © 2009, Elsevier Ltd.

in peak (c) are noticed after the deposition of PANI and thus, ascribed to the selective interactions between ZnO and the quinoid ring of ES which facilitates the charge transfer from quinoid unit of ES to ZnO via highly reactive imine group. This interaction between ZnO and PANI contributes to the decrease in the degree of orbital overlap between the л electrons of the phenyl rings with the lone pair of the nitrogen atom in the PANI molecules and found to form a strong hydrogen bonding. Consequently, the extent of conjugation of PANI decreases, thus resulting in the increased intensity of the peak in PANI/ZnO thin film substrate and conclusively, the absorption of PANI/ZnO thin film increases over the whole range of visible light.

The I-V characteristics exhibit almost the symmetrical behavior both in the reverse and the forward bias and the current increases linearly with the increased applied voltage. The ohmic behavior is due to the formation of ohmic contacts at the interfaces of PANI and ZnO layers. Additionally, the presence of PANI minimizes the width of the depletion layer at the interface of ZnO nanoparticles and thus contributes to the typical ohmic system. The PANI/ZnO heterostructure reveals that the forward bias current increases by a factor two with an increase in the applied voltage. It is attributed that the formation of polarons and bipolarons in PANI increases rapidly and ultimately, contributes to the higher current in PANI at the high applied voltage.   A schematic diagram and I-V characteristics of PANI/ZnO p-n heterostructure diode is shown in Fig. 5 (a).

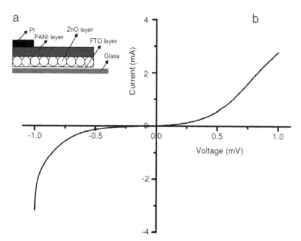

Fig. 5. (a) A schematic representation and (b) I-V characteristics of p-n heterostructure diode of Pt/PANI/ZnO. Reprinted with permission from [Ameen S., 2009], Superlatt. Microstruc. 46 (2009) 872. © 2009, Elsevier Ltd.

Current-voltage (I-V) measurements of PANI/ZnO heterostructure was carried out at 298 K with an applied voltage ranges from -1 V to +1 V. The obtained I-V characteristics exhibit weak rectifying with the non-linear nature of Pt/PANI/ZnO heterostructure diode where Pt thin layer might form a Schottky contact with PANI/ZnO layer as shown in Fig. 5 (b). The Pt layer on PANI/ZnO heterostructure might have originated a barrier between the interfaces of PANI/ZnO layer and Pt layer, and is responsible for the non-linear nature of I-V characteristics. The Pt/PANI/ZnO heterostructure diode displays the rectification ratio (I_f/I_r) of ~4.2 which is obtained from the forward (I_f) and reverse (I_r) current at 0.5 V. This low I_f/I_r value indicates that the PANI/ZnO heterostructure exhibits a weak rectifying nature and attains low turn-on voltage of ~0.15 V and delivers reasonably high current of ~0.08 mA with broader breakdown voltage (~0.52 V) with very high leakage current of ~0.6 mA.

5.1.2 Electrophoretically deposited PANI/TiO₂ heterostructure diodes

PANI has been electrophoretically deposited (EPD) on the surface of TiO_2 nanoparticulate thin film for the fabrication of heterostructure devices by Ameen et al (Ameen, et al., 2011). The enhancements in the electron transport and current density have been explained in terms of the increased charge carriers mobility with improved optical absorption by the heterostucture diodes. Fig. 6 (a) shows the morphology of well crystalline TiO_2 nanoparticles thin film of size ~20-25 nm. However, upon EPD, the size of TiO_2 nanoparticles increased to ~70-80 nm (Fig. 6 (b)) might due to the penetration of PANI molecules into the pores of mesoporous TiO_2 nanoparticulate thin film. Additionally, the uniform deposition and the penetration of PANI on TiO_2 nanoparticulate thin film could be seen in the Fig. 6 (c).

Fig. 6. Surface FESEM images of (a) TiO_2 nanoparticles (b) electrophoretically deposited PANI/TiO_2 nanoparticulate thin film and (c) cross-section image of electrophoretically deposited PANI/TiO_2 nanoparticulate thin film. Reprinted with permission from [Ameen S., 2011], J. Nanosci. Nanotech. 11 (2011) 1559. © 2011, American Scientific Publishers.

The shifting of FTIR peaks, as shown in Fig. 7 (a) indicates the existence of the hydrogen bonding between the surface hydroxyl groups of TiO_2 nanoparticles and the imine groups of PANI molecules (Somani, et al., 1999 & Nivetal, et al., 2003). The UV-Vis spectra of pristine PANI and PANI/TiO_2 nanoparticulate thin film is shown in (Fig. 7 (b)). The pristine PANI exhibits two characteristics bands at 331 and 625 nm which correspond to л–л* transitions of the phenyl rings of PANI and n-л* transitions respectively. In PANI/TiO_2 nanoparticulate thin film substrates, the peak at 269 nm retains its position, while the bands at 625 nm and

331 nm exhibit the blue shift to 599 nm and 321 nm respectively. These noticeable shifting and the intensity changes of the bands are ascribed to the bonding and the interaction between TiO_2 and the quinoid ring of PANI due to the hydrogen bonding in the form of $NH \cdots O - Ti$ in $PANI/TiO_2$ nanoparticulate thin film. The extensive XPS studies, as shown in Fig. 8 have been examined to investigate the bonding between PANI and TiO_2 nanoparticles in $PANI/TiO_2$ nanoparticulate thin film. The four major peaks at ~285.6 eV, 400.5 eV, 530.9 eV and 459.6/465.4 eV, are ascribed to carbon (C 1s), nitrogen (N 1s), oxygen (O 1s) and titanium (Ti 2p) respectively. The XPS studies favor the interaction of EPD PANI with TiO_2 nanoparticles by the existence of hydrogen bonding between the imine group of PANI and the surface hydroxyl of TiO_2 nanoparticles and thus, the electrophoretic deposition of PANI is a promising technique to construct the highly efficient $PANI/TiO_2$ nanoparticulate thin film.

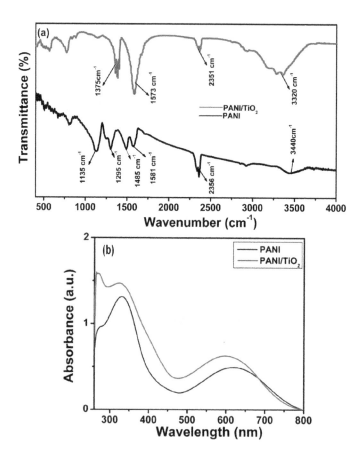

Fig. 7. (a) FT-IR spectra and (b) UV-Vis spectra of pristine PANI and electrophoretically deposited $PANI/TiO_2$ nanoparticulate thin film. Reprinted with permission from [Ameen S., 2011], J. Nanosci. Nanotech. 11 (2011) 1559. © 2011, American Scientific Publishers.

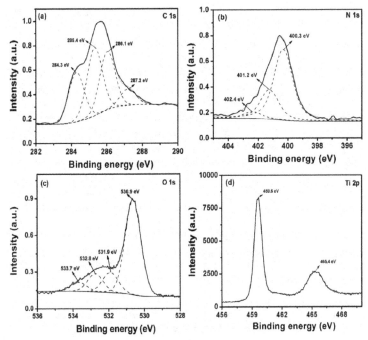

Fig. 8. (a) Resolved C1s, (b) N1s, (c) O1s and (d) Ti 2p XPS spectra of electrophoretically deposited PANI/TiO₂ nanoparticulate thin film. Reprinted with permission from [Ameen S., 2011], J. Nanosci. Nanotech. 11 (2011) 1559. © 2011, American Scientific Publishers.

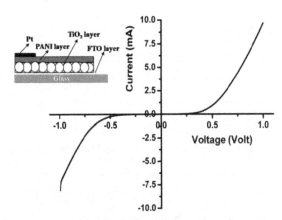

Fig. 9. Pt/PANI/TiO₂ nanoparticulate diode. Insets show the schematic representation of the PANI/TiO₂ nanoparticulate heterostructure diode. Reprinted with permission from [Ameen S., 2011], J. Nanosci. Nanotech. 11 (2011) 1559. © 2011, American Scientific Publishers.

The current–voltage (I–V) characteristics of EPD PANI/TiO₂ nanoparticulate heterostructure was carried out at different temperatures (25-150⁰C) with an applied voltage ranges from −1V to +1V, as shown in Fig. 9. A schematic diagram of Pt/PANI/TiO$_2$ nanoparticulate heterostructure diode is shown in the inset of Fig. 9. The I–V characteristic of Pt/PANI/TiO$_2$ nanoparticulate heterostructure diode exhibits a rectifying and nonlinear behavior which might originate by the formation of schottky contact/barrier between Pt layer and EPD PANI/TiO$_2$ layer in Pt/PANI/TiO$_2$ nanoparticulate heterostructure diode. Moreover, the ratio of forward (I_f) and reverse current (I_r), called as rectification ratio (I_f/I_r) of the device is estimated as 7.0 at the bias voltage of 0.5 V. The I_f/I_r value indicates that the EPD PANI/TiO$_2$ nanoparticulate heterostructure shows a weak rectifying nature. The EPD PANI/TiO$_2$ heterostructure diode attains turn-on voltage of ~ 0.40 V and derives quite high current of ~0.25 mA at 298 K. A broader breakdown voltage (~0.51 V) and good leakage current of ~0.9 mA are achieved by Pt/PANI/TiO$_2$ nanoparticulate heterostructure diode. The low turn on voltage and high leakage current might result from the generation of high density of minority charge carriers from EPD PANI to n-type TiO$_2$. In a reverse bias, Pt/PANI/TiO$_2$ nanoparticulate heterostructure diode shows slightly high break down voltage (~0.55 V) which could be explained by the hopping effect originated from the conjugated bonding of PANI and the geometry/ morphology of TiO$_2$ nanoparticles. Therefore, the uniform and controlled thickness of PANI layer onto TiO2 nanoparticulates thin film substrate by the electrophoretic deposition provides the improved performance of the p–n heterostructure devices.

5.2 Plasma enhanced polymerization of PANI for heterostructure devices
5.2.1 Fabrication of PANI/TiO$_2$ heterostructure device
The formation of an inorganic/organic heterojunction structure by depositing the plasma polymerized PANI on n-type nanocrystalline titania (TiO$_2$) thin film substrate has been reported by Ameen et al (Ameen, et al., 2009). Fig. 10 (a) demonstrates a porous TiO$_2$ formed on the FTO substrates. The film is composed of the well crystalline TiO$_2$ nanoparticles of size ~25 nm, as shown in Fig. 10 (c). It is seen in Fig. 10 (b) that the PANI molecules are well penetrated into the porous TiO$_2$ thin film upon the plasma enhanced polymerization. The accumulation of TiO$_2$ nanoparticles occurs after the deposition of PANI on the surface of TiO$_2$ thin film, and could be seen in Fig. 10 (d). This assemblage confirms the internalization of plasma deposited PANI into porous TiO$_2$ nanoparticles thin film substrates. The I-V characteristics for PANI/TiO$_2$ heterojunction structure, as shown in our reported work (Ameen, et al., 2009,), indicates that no barrier is apparently present because the I-V characteristic curve is almost linear. The forward bias current increases with an increased applied voltage and have attained almost the linear curve. The proficient current in PANI/TiO$_2$ is attributed to well penetration of PANI molecules into the pores of TiO$_2$ nanoparticles layers which improves the degree of contact with TiO$_2$ and thus decreases the series resistance of the cell while increasing the current. The efficient charge movement at the junction of PANI and TiO$_2$ interfaces has made the heterojunction structure which behaves as a typical ohmic system. Moreover, in the reverse bias, the lower current is related to the decrease of the depletion layer because of the presence of the charge carriers at TiO$_2$ nanoparticles thin film.

5.2.2 Plasma deposited PANI on single and bilayer of TiO$_2$ thin film for the fabrication of PANI/TiO$_2$ heterostructure diodes
The organic/inorganic thin film is obtained by plasma enhanced p-type PANI deposited on the individual single and bilayer n-type TiO$_2$ thin film electrodes for the fabrication of p-n

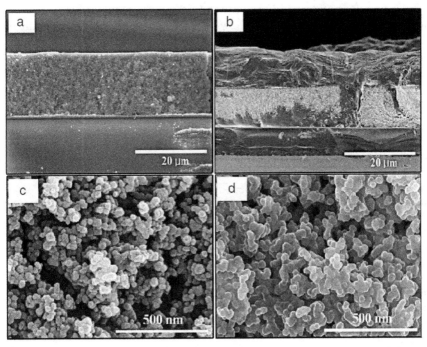

Fig. 10. Cross-sectioned FE-SEM images of (a) pristine TiO_2 (b) Plasma enhanced deposited PANI/TiO_2 thin film and surface FESEM images of (c) pristine TiO_2 (d) Plasma enhanced deposited PANI/TiO_2 thin film. Reprinted with permission from [Ameen S., 2009], Superlatt. Microstruc. 46 (2009) 745. © 2009, Elsevier Ltd.

heterostructure by Ameen et al (Ameen, et al., 2011). The single layer TiO_2 electrode exhibits non uniform and less penetration of PANI molecules by plasma enhanced polymerization into the pores of TiO_2 thin film. However, the extent of PANI deposition has drastically enhanced by the additional layer of TiO_2 on the single layered electrode due to the uniform and well penetration of PANI into the pores of TiO_2 nanoparticles.

The XPS characterizations suggest that the plasma polymerized PANI are interacted to TiO_2 by the formation of hydrogen bonding between imine group of PANI and surface hydroxyl of TiO_2. UV-Vis of PANI/TiO_2 thin film electrode exhibit the peaks at 324 nm corresponds to л–л* transitions centered on the benzenoid and quinoid units. This peak appears with high intensity in PANI/bilayer TiO_2 thin film electrode which indicates the enhanced deposition of PANI molecules on the surface of TiO_2. It could be attributed that highly porous surface of bilayer TiO_2 thin film absorbs large amount of PANI than the single layered TiO_2 thin film, as described in previous report (Ameen, et al., 2011).

The fabricated p-n heterostructure devices are comprised of plasma polymerized PANI (p-type) deposited on the single and bilayerTiO_2 thin film (n-type) coated on FTO glass substrates with a top Pt thin layer contact. The I-V curve of the fabricated PANI/single layer TiO_2 device attains low turn-on voltage of ~ 0.4 V and delivers reasonably high current of ~ 0.08 mA at 298 K, as described in previous report (Ameen, et al., 2011). However, a breakdown voltage (~0.57 V) with high leakage current of ~0.6 mA is obtained by Pt/PANI/single layer TiO_2

heterostructure device. On comparison, Pt/PANI/bilayer TiO_2 heterostructure device shows the lower turn on voltage and low breakdown voltage with high leakage current than single layer TiO_2 based heterostructure device. Here, the second layer of TiO_2 in PANI/bilayer TiO_2 provides a highly porous surface for the high loading of PANI molecules during the plasma enhanced polymerization and thus, produces high density of the minority charge carriers to n-type TiO_2 thin film layer and results to the high leakage current with small turn-on voltage and breakdown voltage. It is attributed that the additional TiO_2 thin film layer might increase the n-type character of the device. On the other hand, both the devices exhibit slight high reverse break down voltage (~0.58 V in single layer TiO_2 and ~0.5 V in bilayer TiO_2) which could be understood by considering the molecular geometry of PANI chains and geometry/morphology of TiO_2 nanoparticles deposited on the FTO substrates which likely generate the hopping effect. The current densities reported by Sadia et al in the forward bias are considerably better than the data reported elsewhere on PANI based or other conducting polymer based heterostructure device (Dhawale, et al., 2008, Nadarajah, et al., 2008 & Maeng, et al., 2008). Under light illumination, both the devices constitute the no current flowing condition i.e. called open circuit voltage condition (V_{OC}) and the maximum current condition called short circuit current (J_{SC}) at zero voltage condition. This phenomena is absent in I-V characteristics under dark condition, which means that the fabricated devices could absorb photons for the generation of current and having the photovoltaic behavior under light illumination and thus, Pt/PANI/bilayer TiO_2 heterostructure device is reported to achieve high J_{SC} (~0.32 mA) with low V_{OC} (~0.343 V) as compared with Pt/PANI/single layer TiO_2 heterostructure device (J_{SC} = ~0.20 mA and V_{OC} = ~0.462 V).

6. Applications of PANI in dye sensitized solar cells (DSSCs)

Organic solar cells have attracted considerable attention in recent years due to several advantages such as low cost, processing at low temperature, flexible and large area production etc (Thompson, et al., 2008, Gunes, et al., 2007 & Bundgoard, et al., 2007). Usually, the organic materials are not good in carrier transport and thus, the power conversion efficiency is limited by the low dissociation probability of excitons and the inefficient hopping carrier transport (Shaw, et al. 2008 & Sirringhaus, et al., 1999). Among various organic solar cells, polymer solar cells have demonstrated an impressive scale up on power conversion efficiency and improvement in lifetime. Right now a remarkable 8% power conversion efficiency has been reached at laboratory scale, 6–8% for normal and 3–4% for inverted configurations (Waldauf, et al., 2006, Steim, et al., 2010, 2008 & Ameri, et al., 2009). Still, much work must be done in order to overcome the main difficulties observed for the technology i.e. enhancement of efficiency and lifetime. Thus, the need for novel materials acting as acceptors and donors, and the understanding of the interplay between interfaces (electrodes, optical spaces) are currently under an intense investigation. Semiconductor oxides have been applied extensively like antireflection coatings and scattering layers (Lee, et al., 2008), as interfacial layers in organic solar cells (Steim, et al., 2010, 2008) and as part of the active constituents of the device acting as electron transport materials in and hybrid DSSCs (Cantu, et al., 2010, 2006, 2007, 2006,Valls, et al., 2010, 2009). The semiconductor nanostructures are therefore proposed to combine with the organic materials to provide not only a large interface area between organic and inorganic components for exciton dissociation but also a fast electron transport in semiconductors. Hybrid solar cells are a mixture of nanostructures of both organic and inorganic materials.

Therefore, they combine the unique properties of inorganic semiconductor nanoparticles with properties of organic/polymeric materials (Arici, et al., 2003). In addition to this, low cost synthesis, processability and versatile manufacturing of thin film devices make them attractive (Sariciftci, et al., 1992 & Yu, et al., 1995). Also, inorganic semiconductor nanoparticles might have high absorption coefficients and particle size induced tunability of the optical band-gap. Thus, the hybrid solar cell is becoming interesting and attractive in recent years.

6.1 PANI/TiO₂ thin film electrodes for the performance of DSSCs

The TiO$_2$/PANI and dye absorbed TiO$_2$/PANI electrodes are prepared by plasma enhanced polymerization using aniline monomer for the fabrication of DSSCs by Ameen et al (Ameen, et al., 2009). The TiO$_2$ thin film, prior to PANI deposition exhibits the particles of size ~20–25 nm, as shown in (Fig. 11 (b)). However, the proper attachment of plasma polymerized PANI on TiO$_2$ thin film substrates (Fig. 11 (d)) and increases the size of TiO$_2$ nanoparticles to ~50–60 nm indicate the effects of the penetration of PANI into TiO$_2$ nanoparticles. The current density-voltage (J–V) performance of solar cell FTO/TiO$_2$/Dye/PANI/Pt and FTO/TiO$_2$/PANI/Pt are shown in Fig.12 respectively. The electric contact of the hole conductor and the pore filling extent of the hole conductor into the dye-sensitized TiO$_2$ film are the two important factors which determine the photovoltaic behaviors of the devices. It has been found that the solar cell based on TiO$_2$/Dye/PANI electrode executes great improvement in the overall conversion efficiency with the incorporation of dye layer on TiO$_2$/PANI electrode and dramatically enhanced the photovoltaic properties such as open-circuit voltage (V$_{OC}$), short-circuit current (J$_{SC}$) and fill factor (FF) as compared to TiO$_2$/PANI electrode based DSSC. These improvements are resulted from the formation of TiO$_2$/PANI thin film, where the photon generated electrons could freely travel at the interface of the PANI and TiO$_2$ without decay, and dissociate into free charge carriers effectively.

Fig. 11. (a) Cross-section and (b) surface view FESEM images of TiO$_2$ deposited on FTO substrate. (c) and (d) show the cross-section and surface view FESEM images of Plasma enhanced deposited PANI/TiO$_2$ film. Reprinted with permission from [Ameen S. et al, 2009], J. Alloys Compd.164 (2009) 382. © 2009, Elsevier Ltd.

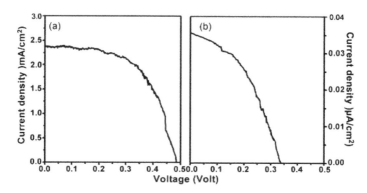

Fig. 12. J–V curve of fabricated solar cell (a) FTO/TiO$_2$/Dye/PANI/Pt (b) FTO/TiO$_2$//PANI/Pt. Reprinted with permission from [Ameen S. et al, 2009], J. Alloys Compd.164 (2009) 382. © 2009, Elsevier Ltd.

The DC conductivity vs 1000/T plots for the fabricated solar cells of FTO/TiO$_2$/Dye/PANI/Pt and FTO/TiO$_2$//PANI/Pt are shown in Fig.13. It is cleared that the conductivity of TiO$_2$/PANI is higher as compared to the pristine PANI. The room temperature conductivity of pristine PANI (28.36×10^{-3} ohm^{-1} cm^{-1}) is lower than the room temperature conductivity of PANI/TIO$_2$ thin film (75×10^{-3} ohm^{-1} cm^{-1}). The increase in conductivity might due to the increase of charged carrier concentration with improved mobility of the charged carriers. The rougher the surface, large-area defects and pinholes on PANI films exist and therefore, result to the high conduction of film. The effective electrical conductivity is thus, due to the retardation of both the transport of photoelectrons from counter electrode to hole conductor and the regeneration of the dye by the reduced-state hole conductors. Such improvements could suppress the recombination of photoelectrons, injected into the conduction band of TiO$_2$ from PANI and retards the internal resistance of the device.

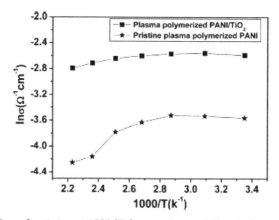

Fig. 13. Plots of DC conductivity vs 1000/T for pristine PANI and plasma enhanced deposited PANI/TiO$_2$ thin film. Reprinted with permission from [Ameen S. et al, 2009], J. Alloys Compd.164 (2009) 382. © 2009, Elsevier Ltd

AC impedance of FTO/TiO$_2$//PANI/Pt (Fig.14(a)) and TiO$_2$/Dye/PANI (Fig.14(b)) thin film electrode based DSSCs under the illumination of 100mW/cm^2 (AM1.5) by applying a 10mV AC signal at the frequency range from 10 Hz to 100 kHz are also measured by Sadia et al. A very high $R_{P/TCO}$ of 52.4Ω and R_{CT} of 3700Ω was observed for TiO$_2$/PANI electrodes based cells. Comparatively, TiO$_2$/Dye/PANI based device showed the low $R_{P/TCO}$ (35.8Ω) and R_{CT} (81.9Ω) due to the influence of dye layer which is placed between the TiO$_2$ and PANI layer of the electrode. It is found that the value of R_{CT} in TiO$_2$/Dye/PANI based device is very low as compared to the R_{CT} of TiO$_2$/PANI based device. Thus, it explains the high electron transfer at the junction of TiO$_2$ and PANI in TiO$_2$/Dye/PANI based device, resulting in the high photocurrent density and overall conversion efficiency.

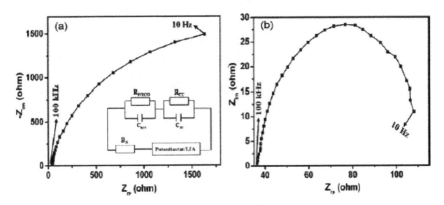

Fig. 14. AC impedance of (a) FTO/TiO$_2$//PANI/Pt and (b) TiO$_2$/Dye/PANI thin film electrode based DSSCs at the frequency range from 10 Hz to 100 kHz. Inset shows the equivalent circuit model of the device. Reprinted with permission from [Ameen S. et al, 2009], J. Alloys Compd.164 (2009) 382. © 2009, Elsevier Ltd.

6.2 PANI thin film electrode as counter electrode for DSSCs

Conducting polymers generally show p-type semiconducting behavior, good specific capacitance and catalytic activity for the reduction of I$_3$- ion. Owing to their specialty, the conducting polymers are used as counter electrode (CE) materials for DSSCs. In 2008, Saito et al reported the preparation of the counter electrode using p-toluenesulfonate (PEDOTTsO) or polystyrenesulfonate (PEDOT-PSS) doped poly (3, 4-ethylenedioxythiophene) (PEDOT) for DSSCs (Saito et al., 2004). They achieved a comparably high conversion efficiency of 4.6% with J$_{SC}$ =11.2 mA/cm^2, V$_{OC}$ = 670 mV and FF = 0.61 with the fabricated DSSCs. Consecutively, the composite films of graphene and PEDOT-PSS deposited on indium tin oxide (ITO) substrates by spin coating and applied as counter electrodes of DSSCs. The prepared CE exhibited high transmittance (>80%) at visible wavelengths and high electrocatalytic activity as compared to the above system with the similar conversion efficiency (4.5%). Q. Li et al group manufactured a microporous polyaniline (PANI) and applied as CE for the fabrication of DSSCs (Li, et al., 2008). The advanced microporous PANI CE with particle diameter of 100 nm possessed the lower charge-transfer resistance and higher electrocatalytic activity for the redox reaction. DSSC fabricated with microporous PANI CE exhibited reasonably high overall conversion efficiency of 7.15% i.e. compared to DSSC with Pt CE. This report could open the utilization of the simple preparation technique with low cost and excellent photoelectric properties of PANI based counter electrode as appropriate alternative CE materials for DSSCs. Furthermore, J. Wu et al prepared polypyrrole (PPy) nanoparticle and deposited on a fluorine-doped tin oxide (FTO) glass for the construction of PPy counter electrode and applied to DSSC (Wu, et al., 2008). The fabricated DSSC achieved a very high conversion efficiency of 7.66% owing to its smaller charge transfer resistance and higher electrocatalytic activity for the I_2/I^- redox reaction. After this significant breakthrough, K. M. Lee et al developed poly (3, 4-alkylenedioxythiophene) based CE by electrochemical polymerization on FTO glass substrate for DSSC (Lee, et al., 2009). A high conversion efficiency of 7.88% was acquired by the fabricated DSSC which attributed to the increased effective surface area and good catalytic properties for I_3^- reduction. Progressively, the nanostructured polyaniline films were grown on FTO glass using cyclic voltammetry (CV) method at room temperature and applied as counter electrode for DSSCs. They found that the controlled thickness of nanostructured polyaniline (>70 nm) by the used method increased the reactive interfaces, which conducted the charge transfer at the interface and low resistance hinders electronic transport within the film. The fabricated DSSCs achieved a high overall conversion efficiency of 4.95% with very high J_{SC} of 12.5 mA/cm². Importantly, the nanostructured PANI electrode showed the 11.6% improvement in J_{SC} as compared to DSSC with an electrodeposited platinum counter electrode (Zhang, et al., 2010). Recently, Ameen et al synthesized the undoped and sulfamic acid (SFA) doped PANI nanofibers (NFs) via template free interfacial polymerization process and deposited on FTO substrates using spin coating to prepare counter electrode for DSSCs (Ameen, et al., 2010).

6.3 Sulfamic acid doped PANI Nanofibers counter electrode for DSSCs
Ameen et al developed a simple interfacial polymerization method for the synthesis of PANI nanofibers (NFs) and its doping with sulfamic acid (SFA) to increase the conductivity (Ameen, et al., 2010). These undoped and SFA doped PANI NFs were applied as new counter electrodes materials for the fabrication of the highly efficient DSSCs. The selection of SFA was based on its exclusively important properties such as high solubility, easy handling, nonvolatile stable solid acid, and low corrosiveness. The proposed doping mechanism for PANI with SFA is shown in Fig. 15. PANI NFs exhibit well-defined fibrous morphology with the diameter of 30 nm (Fig. 16 (b)) and the diameter of PANI NFs has considerably increased to ~40 nm after doping with SFA, as shown in Fig. 16 (a). The chemical doping of SFA causes some aggregation of PANI NFs, and therefore, the formation of voids into the fibrous network of PANI NFs are noticed. The TEM images of PANI NFs (Fig. 16 (c)) and SFA-doped PANI NFs (Fig. 16 (d)) justifies the doping effect on the morphology of PANI NFs. The entrapped SFA into the fibers of PANI results to the increase of average diameter by ~40 nm as compared to undoped PANI NFs.

The UV-Vis of SFA doped PANI-NFs, as shown in Fig. 17 (a), exhibits a slight blue shift of the peak at 296 nm from 298 nm and a considerably large red shift at 380 nm from 358 nm which indicates the interactions between SFA dopants and the quinoid ring of emeraldine salt (ES) and facilitate the charge transfer between the quinoid unit of ES and the dopant via highly reactive imine groups. The CV curves (Fig. 17 (b)) of SFA-doped PANI NFs electrode attains a reasonably high anodic peak current (I_a) of 0.24 mA/cm² and cathodic peak current (I_c) of -0.17 mA/cm² with a considerably high value of switching point (0.22 mA/cm²).

However, the undoped PANI NFs electrode exhibits a low I_a of 0.21 mA/cm² and I_c of -0.2 mA/cm² with a low switching point (0.17 mA/cm²). These results suggest that the high peak current might increase the redox reaction rate at SFA-doped PANI NFs counter electrode, which may attribute to its high electrical conductivity and surface area.

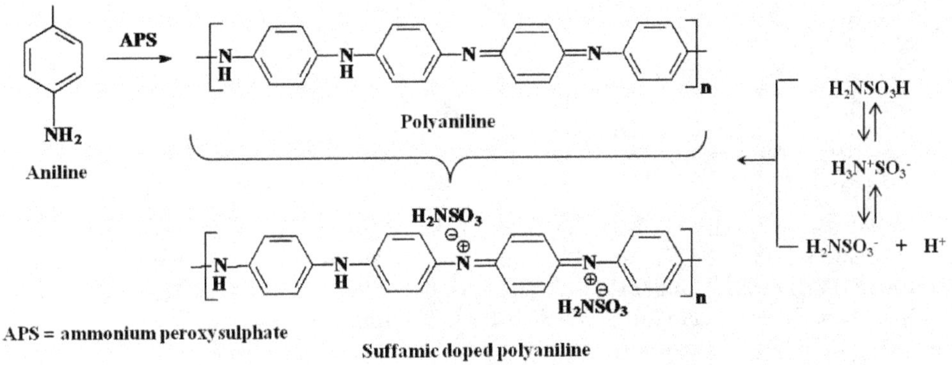

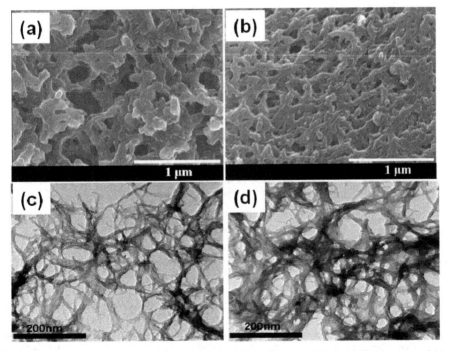

Fig. 15. Proposed mechanism of sulfamic doping into PANI NFs.

Fig. 16. FESEM images of (a) SFA doped PANI NFs and (b) PANI NFs. TEM images of (c) PANI and (d) SFA doped PANI NFs. Reprinted with permission from [Ameen S. et al, 2010], J. Phys. Chem. C 114 (2010) 4760. © 2010, ACS Publications Ltd.

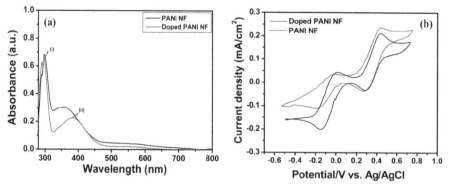

Fig. 17. (a) UV-vis spectra of PANI NFs and SFA-doped PANI NFs. (b) Cyclic voltammetry of iodide species on PANI NFs and SFA doped PANI NFs electrodes in acetonitrile solution with 10 mM LiI, 1 mM I_2, and 0.1M $LiClO_4$. Reprinted with permission from [Ameen S. et al, 2010], J. Phys. Chem. C 114 (2010) 4760. © 2010, ACS Publications Ltd.

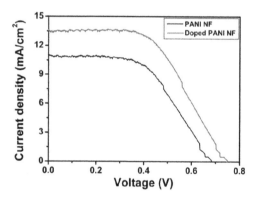

Fig. 18. J-V curve of fabricated solar cell of PANI NFs and SFA doped PANI NFs as counter electrodes under light illumination of 100 mW/cm². Reprinted with permission from [Ameen S. et al, 2010], J. Phys. Chem. C 114 (2010) 4760. © 2010, ACS Publications Ltd.

The Fig. 18 shows that the DSSCs fabricated with SFA-doped PANI NFs counter electrode achieve a high conversion efficiency (η) of 5.5% with a high short circuit current (J_{SC}) of 13.6 mA/cm² open circuit voltage (V_{OC}) of 0.74 V, and fill factor (FF) of 0.53. The conversion efficiency increases by ~27% and thus, after SFA doping of PANI NFs the conversion efficiency reaches the value of 5.5% than that of DSSC fabricated with PANI NFs counter electrode (4.0%). Further, the SFA-doped PANI NFs counter electrode has significantly increased the J_{SC} and V_{OC} of ~20% and ~10%, respectively, as compared to the DSSC fabricated with PANI NFs counter electrode. It indicates that the SFA doping has increased the fast reaction of I^-/I_3^- species at counter electrode and therefore, the superior photovoltaic properties such as η, J_{SC}, and V_{OC} of the cell are attributed to the sufficiently high conductivity and electrocatalytic activity of doped PANI NFs, which alleviates the reduction of I_3^- at the thin SFA-doped PANI NFs layers. Importantly, the IPCE curves of DSSCs fabricated with PANI NFs counter electrode exhibit the low IPCE of ~54% in the absorption range of 400-650 nm.

The IPCE value has prominently increased by ~70% with the SFA doped PANI NFs counter electrode-based DSSCs. It is noteworthy that the IPCE of the device is considerably enhanced by ~24% upon SFA doping on PANI NFs-based counter electrodes. The enhanced IPCE in DSSCs with SFA-doped PANI NFs electrode results in the high J_{SC} and photovoltaic performance, which are related to its high electrical conductivity and the higher reduction of I_3^- to I^- in the electrolyte at the interface of PANI NFs layer and electrolyte.

7. Fabrication of DSSCs with metal oxide nanomaterials photoanodes

In DSSCs, the choice of semiconductor is governed by the conduction band energy and density of states which facilitate the charge separation and minimizing the recombination. Secondly, the high surface area and morphology of semiconductor are important to maximize the light absorption by the dye molecules while maintaining the good electrical connectivity with the substrate (Baxtera, et al., 2006).The semiconducting metal oxides such as TiO_2, ZnO and SnO_2 etc have shown good optical and electronic properties and are accepted as the effective photoelectrode materials for DSSCs. These metal oxide nanostructures present discrete morphologies of nanoparticles (Ito, et al., 2008) nanowires (Law, et al., 2005 & Feng, et al., 2008) and nanotubes (Macak, et al., 2005 & Mor et al., 2005) which are the key component in DSSCs for the effective dye adsorption and the efficient electron transfer during the working operation of DSSCs. To improve the light harvesting efficiency, the metal oxide nanostructures must possess high surface to volume ratio for high absorption of dye molecules. These metal oxide nanostructures are usually prepared by the methods like hydrothermal synthesis (Zhang, et al., 2003 & Wang et al., 2009) template method (Ren, et al., 2009 & Tan, et al., 2008) electrodeposition (Tsai, et al., 2009) and potentiostatic anodization (Chen, et al., 2009 & kang, et al., 2009) and are important for improving the photovoltaic properties of DSSCs such as J_{SC}, V_{OC}, FF and conversion efficiency. Out of these, TiO_2 has been intensively investigated for their applications in photocatalysis and photovoltaic (Regan, et al., 1991 & Duffie, et al., 1991). Particularly in DSSCs, the porous nature of nanocrystalline TiO_2 films provides the large surface for dye-molecule adsorption and therefore, the suitable energy levels at the semiconductor–dye interface (the position of the conduction-band of TiO_2 being lower than the excited-state energy level of the dye) allow for the effective injection of electrons from the dye molecules to the semiconductor. Compared with other photovoltaic materials, anatase phase TiO_2 is outstanding for its stability and wide band gap and thus, widely used in the devices (Gratzel, et al., 2001). On the other hand, ZnO nanomaterials are chosen as an alternative material to TiO_2 photoanode due to its wide-band-gap with higher electronic mobility which would be favorable for the efficient electron transport, with reduced recombination loss in DSSCs. Studies have already been reported on the use of ZnO material photoanode for the application in DSSCs. Although the conversion efficiencies of ZnO (0.4–5.8%) is comparably lower than TiO_2 (11%) but still ZnO is a distinguished alternative to TiO_2 due to its ease of crystallization and anisotropic growth. In this part of the chapter, the various nanostructures of TiO_2 and ZnO have been briefly summarized for the application for DSSCs.

7.1 Various TiO₂ nanostructures photoanodes for DSSCs
7.1.1 Photoanodes with TiO₂ nanotubes
TiO_2 nanotubes (NTs) arrays are generally synthesized by the methods like electrochemical approach (Zwilling, et al., 1999 & Gong, et al., 2001) layer-by-layer assembly (Guo, et al.,

2005) template synthesis, sol–gel method (Martin, et al., 1994, Limmer, et al., 2002 & Lakshmi, et al., 1997) etc and are the effective photoanode for the fabrication of DSSCs. The reported methods for the synthesis of TiO_2 NTs provide low yield and demand advanced technologies with the high cost of templates (anodic aluminum oxide, track-etched polycarbonate or the amphiphilic surfactants). A. J. Frank obtained the bundle-free and crack-free NT films by using the supercritical CO_2 drying technique and found that the charge transport was considerably increased with the decreased of NTs bundles which created the additional pathways through the intertube contacts. However, J. H. Park et al. reported a simple and inexpensive methodology for preparing TiO_2 NTs arrays on FTO glass and applied as photoanodes for DSSCs which exhibited the significantly high overall conversion efficiency of 7.6% with high J_{SC} of 16.8 mA/cm^2, V_{OC} of 0.733 V and a fill factor (FF) of 0.63. The enhanced photovoltaic performance was attributed to the reduced charge recombination between photoinjected electrons in the substrate via tubular morphology of TiO_2 photoanode (Park, et al., 2008).

7.1.2 Photoanodes with TiO_2 nanorods

The Highly crystalline TiO_2 nanorods (NRs) with lengths of ~100-300 nm and diameters of ~20-30 nm were grown by J. Jui et al using the hydrothermal process with cetyltrimethylammonium bromide surfactant solution (Jiu, et al., 2006). In this synthesis, the length of nanorods was substantially controlled and maintained by the addition of a tri-block copolymer poly-(ethylene oxide) 100-poly (propylene oxide) 65-poly (ethylene oxide) 100 (F127) and polymer decomposed after sintering of TiO_2 nanorods at high temperatures. The fabricated DSSCs attained a high overall conversion efficiency of 7.29% with considerably high V_{OC} of 0.767 V and fill factor of 0.728. The enhancement in the photovoltaic properties was attributed to increase the ohmic loss and high electron transfer through TiO_2 NRs. As compared to P-25 based DSSCs, the less amount of dye was absorbed by the TiO_2 NRs photoanode might due to the larger size of the nanorods and therefore, result a slightly lower photocurrent density of 13.1 mA/cm^2. B. Liu group proposed a hydrothermal process to develop the oriented single-crystalline TiO_2 NRs or nanowires on a transparent conductive substrate (Liu, et al., 2009). The DSSCs fabricated with $TiCl_4$ generated 4 μm-long rutile TiO_2 NRs electrode and demonstrated relatively low light-to-electricity conversion efficiency of 3% with J_{SC} ~6.05 mA/cm^2, V_{OC} of ~0.71 V, and FF of 0.7. The device delivered the improved IPCE of ~50% at the peak of the dye absorption. The improved V_{OC} and FF revealed that the $TiCl_4$ treatment decreased the surface recombination. Conclusively, TiO_2 NRs improved the dye adsorption and the optical density through the surface of oriented NRs.

7.1.3 Photoanodes with TiO_2 nanowires

Single-crystal-like anatase TiO_2 nanowires (NWs) as compared to NRs and NTs morphology are extensively applied as photoanode for the fabrication of DSSCs. The perfectly aligned morphology of TiO_2 NWs and networks of NWs could be achieved by the solution, electrophoretic and hydrothermal process due to the "oriented attachment" mechanism. The aligned TiO_2 network with single-crystal anatase NWs conducted the high rate of electron transfer and achieved significantly high overall conversion efficiency of 9.3% with high J_{SC} of 19.2 mA/cm^2, V_{OC} of 0.72 V, and FF of 0.675. The improved photovoltaic performance was ascribed to the network structure of single-crystal-like anatase NWs which acquired a high surface to volume ratio and thus, presented the high IPCE of ~ 90%. Recently, J. K. Oh et al reported the branched TiO$_2$ nanostructure photoelectrodes for DSSCs with TiO$_2$ NWs as a seed material (Oh, et al., 2010). The prepared TiO$_2$ electrode possessed a three-dimensional structure with rutile phase and showed high conversion efficiency of 4.3% with high J$_{SC}$ of 12.18 mA/cm^2. Compared to DSSCs with TiO$_2$ NWs, the cell performance and J$_{SC}$ was enhanced by 2 times, which was due to the increased specific surface area and the roughness factor. However, the lower FF was originated from the branches of TiO$_2$ electrodes, resulting in the reduction of grain boundaries.

7.2 Various ZnO nanostructures photoanodes for DSSCs
7.2.1 Photoanodes with ZnO nanoparticles
The techniques like vapor liquid solid, chemical vapor deposition, electron beam evaporation, hydro thermal deposition, electro chemical deposition and thermal evaporation etc are generally applied for the synthesis of ZnO nanostructures. Out of these, the chemical solution method is the simplest procedure for achieving uniform ZnO nanoparticles (NPs) thin films and delivers almost the same performance as that of nanocrystalline TiO$_2$ with similar charge transfer mechanism between the dye and semiconductor. The synthesis of ZnO NPs is reported by the preparation of ZnO sols with zinc acetate as precursor and lithium hydroxide to form homogeneous ethanolic solutions (Spanhel, et al., 1991 & Keis, et al., 2001). Several researchers have fabricated DSSCs using sol–gel-derived ZnO NPs films and reported the low conversion efficiencies with values generally around 0.4–2.22% (Redmond, et al., 1994, Rani, et al., 2008 & Zeng, et al., 2006). Highly active ZnO nanoparticulate thin film through a compression method was prepared for high dye absorption by Keis et al for the fabrication of DSSCs (Keis, et al., 2002, 2002). The morphology of ZnO NPs, synthesized by a sol–gel route exhibited an average size of 150 nm. The thin film photoelectrodes were prepared by compressing the ZnO NPs powder under a very high pressure and the DSSCs fabricated with the obtained film achieved a very high overall conversion efficiency of 5% under the light intensity of 10 mWcm^2.

7.2.2 Photoanodes with ZnO nanosheets and other nanostructures
ZnO nanosheets (NSs) are quasi-two-dimensional structures that could be fabricated by a re-hydrothermal growth process of hydrothermally grown ZnO NPs (Suliman, et al., 2007). M. S. Akhtar et al prepared sheet-spheres morphology of ZnO nanomaterials through citric acid assisted hydrothermal process with 5 M NaOH solution (Akhtar, et al., 2007). The high conversion efficiency and high photocurrent of ZnO NSs based DSSCs was attributed to the effective high light harvesting by the maximum dye absorption via ZnO NSs film surface which promoted a better pathway for the charge injection into the ZnO conduction layer. Sequentially, C. F. Lin et al fabricated a prepared ZnO nanobelt arrays on the FTO substrates by an electrodeposition method and applied as photoelectrode for the fabrication of DSSCs (Lin, et al., 2008). Y. F. Hsu et al had grown a 3-D structure ZnO tetrapod nanostrcutures, comprised of four arms which were extended from a common core (Hsu, et al., 2008 & Chen, et al., 2009). The length of the arms was adjusted within the range of 1–20 mm, while the diameter was tuned from 100 nm to 2 μm by changing the substrate temperature and the oxygen partial pressure during vapor deposition.

7.2.3 Photoanodes with ZnO nanowires
Law et al designed ZnO nanowire (NWs) arrays to increase the electron diffusion length and was applied as photoelectrode for the fabrication of DSSCs (Law, et al., 2005 & Greene, et al.,

2006). The grown ZnO nanowires arrays films exhibited the relatively good resistivity values between the range of 0.3 to 2.0 Ω cm for the individual nanowires with an electron concentration of 1 - 5 x 10^{18} cm^3 and a mobility of 1–5 cm^{2}V^{-1}s^{-1}. The overall conversion efficiencies of 1.2-1.5% were obtained by DSSCs fabricated with ZnO nanowires arrays with short-circuit current densities of 5.3–5.85 mA/cm^2, open-circuit voltages of 0.610–0.710 V, and fill factors of 0.36–0.38. Another group synthesized ZnO NWs by the use of ammonium hydroxide for changing the supersaturation degree of Zn precursors in solution process (Regan, et al., 1991). The length-to-diameter aspect ratio of the individual nanowires was easily controlled by changing the concentration of ammonium hydroxide. The fabricated DSSCs exhibited remarkably high conversion efficiency of 1.7% which was much higher than DSSC with ZnO nanorod arrays (Gao, et al., 2007). C. Y. Jiang et al reported the flexible DSSCs with a highly bendable ZnO NWs film on PET/ITO substrate which was prepared by a low-temperature hydrothermal growth at 85 °C (Jiang, et al., 2008). The fabricated composite films obtained by immersing the ZnO NPs powder in a methanolic solution of 2% titanium isopropoxide and 0.02 M acetic acid was treated with heat which favored the good attachment of NPs onto NWs surfaces (Jiang, et al., 2008). Here, the conversion efficiency of the fabricated DSSCs was achieved less as compared to DSSCs based on NPs.

7.2.4 Photoanodes with ZnO nanorods

A. J. Cheng et al synthesized aligned ZnO nanorods (NRs) on indium tin oxide (ITO) coated glass substrate via a thermal chemical vapor deposition (CVD) (Cheng, et al., 2008) at very high temperature which affected the crystalline properties of ZnO NRs. The rapid large-scale synthesis of well-crystalline and good surface area of hexagonal-shaped ZnO NRs was carried out by A. Umar et al at very low temperature (70°C) for the application of DSSCs (Umar, et al., 2009). A high overall light to electricity conversion efficiency of 1.86% with high fill factor (FF) of 74.4%, high open-circuit voltage (V_{OC}) of 0.73V and short-circuit current (J_{SC}) of 3.41mA/cm^2 was achieved by fabricated DSSCs. M. S. Akhtar et al reported the morphology of ZnO flowers through hydrothermal process using Zinc acetate, NaOH and ammonia as capping agent. The photoanode was prepared by spreading the ZnO paste on FTO substrate by doctor blade technique for the fabrication of DSSCs (Akhatr, et al., 2007). Unfortunately, the DSSC presented a very low conversion efficiency of 0.3% with high FF of 0.54. The low performance might attribute to the low dye absorption on the surface of ZnO due to the less uniformity of the thin film with low surface to volume ratio. Furthermore, a flower like structures comprised with nanorods/nanowires can be assumed to deliver a larger surface area and a direct pathway for electron transport with the channels arisen from the branched to nanrods/nanowire backbone. Recently, hydrothermally grown ZnO nanoflower films accomplished improved overall conversion efficiency of 1.9% with high J_{SC} of 5.5mA cm^2, and a fill factor of 0.53 (Jiang, et al., 2007) which is higher than nanorod arrays films based DSSC of the conversion efficiency 1.0%, J_{SC} 4.5 mA/cm^2, and FF 0.36.

7.2.5 Photoanodes with ZnO nanotubes

L. Vayssiers et al grown the ZnO microtubes arrays by thermal decomposition of a Zn^{2+} amino complex at 90°C in a regular laboratory oven (Vayssieres, et al., 2001). The synthesized ZnO microtubes arrays possessed a high porosity and large surface area as compared to ZnO NWs arrays. A. B. F. Martinson et al fabricated the ZnO nanotubes (NTs) arrays by coating anodic aluminum oxide (AAO) membranes via atomic layer deposition

(ALD) and constructed the DSSCs which showed a relatively low conversion efficiency of 1.6% due to the less roughness factor of commercial membranes (Martinson, et al., 2007). In continuity, Ameen et al reported the aligned ZnO NTs, grown at low temperature and applied as photoanode for the performances of DSSCs (Ameen, et al., 2011). The ZnO seeded FTO glass substrate supported the synthesis of highly densely aligned ZnO NTs whereas, non-seeded FTO substrates generated non-aligned ZnO NTs. The non-aligned ZnO NTs photoanode based fabricated DSSCs reported the low solar-to-electricity conversion efficiency of ~0.78%. However, DSSC fabricated with aligned ZnO NTs photoanode showed three times improved solar-to-electricity conversion efficiency than DSSC fabricated with non-aligned ZnO NTs. Fig. 19 shows the surface FESEM images of ZnO NTs deposited on non-seeded and ZnO seeded FTO substrates. Fig. 19 (a & b) exhibits the highly densely aligned ZnO NTs, substantially grown on ZnO seeded FTO substrates. Importantly, the ZnO NTs possess a hexagonal hollow structure with average inner and outer diameter of ~150nm and ~300 nm, respectively, as shown in Fig. 19 (c & d). However, non-seeded FTO substrates (Fig. 19 (e)) obtain the random and non-aligned morphology of NTs with the average diameter of 800 nm. The high resolution image clearly displays the typical hexagonal hollow and round end of the NTs (Fig. 19 (f)). Fig. 20(a) of TEM image reveals hollow NT morphology with the outer and inner diameter of ~250nm and ~100 nm, respectively. SAED patterns (Fig. 20 (c)) exhibits a single crystal with a wurtzite hexagonal phase which is preferentially grown in the [0001] direction. It is further confirmed from the HRTEM image of the grown ZnO NTs, presented in Fig. 20(b). HRTEM image shows well-resolved lattice fringes of crystalline ZnO NTs with the inter-planar spacing of ~0.52nm. Additionally, this value corresponds to the d-spacing of [0001] crystal planes of wurtzite ZnO. Thus, the synthesized ZnO NTs is a single crystal and preferentially grown along the c-axis [0001].

The XRD peaks (Fig 21 (a)) of grown aligned ZnO NTs on the seeded substrates appear at the same position but with high intensity might due to high crystalline properties of aligned morphology of ZnO NTs. The UV-Vis spectra as shown in Fig 21 (b)) exhibit a single peak which indicates that the grown ZnO NTs do not contain impurities. Moreover, the aligned morphology of ZnO NTs attains high absorption, indicating the higher crystalline properties than non-aligned ZnO NTs.

The Raman spectra of non- aligned and aligned ZnO NTs is shown in Fig 22 (a). The grown ZnO NTs exhibits a strong Raman peak at ~437cm^{-1} corresponds to E_2 mode of ZnO crystal and two small peaks at ~330cm^{-1} and ~578cm^{-1} are assigned to the second order Raman spectrum arising from zone-boundary phonons $3E_{2H}$–E_{2L} for wurtzite hexagonal ZnO single crystals and E_1 (LO) mode of ZnO associated with oxygen deficiency in ZnO nanomaterials respectively (Exarhas, et al., 1995). Compared to non-aligned ZnO NTs, the stronger E_2 mode and much lower E_1 (LO) mode indicates the presence of lower oxygen vacancy. The Raman active E_2 mode with high intensity and narrower spectral width is generally ascribed to the better optical and crystalline properties of the materials (Serrano, et al., 2003) and thus, the grown aligned ZnO NTs results high crystallinity of ZnO crystals with less oxygen vacancies. Fig 22 (b) depicts the PL spectra of grown non-aligned and aligned ZnO NTs. An intensive sharp UV emission at ~378nm and a broader green emission at ~581nm are attributed to the free exciton emission from the wide band gap of ZnO NTs and the recombination of electrons in single occupied oxygen vacancies in ZnO nanomaterials (Vanheusden, et al., 1996). The high intensity and less broaden green emission indicates that the aligned ZnO NTs exhibits less oxygen vacancies and considerable stoichiometric phase structure formation. Thus, the PL spectra suggest that ZnO seeding on FTO substrates might improve surface-to-volume ratio and optical properties of ZnO NTs.

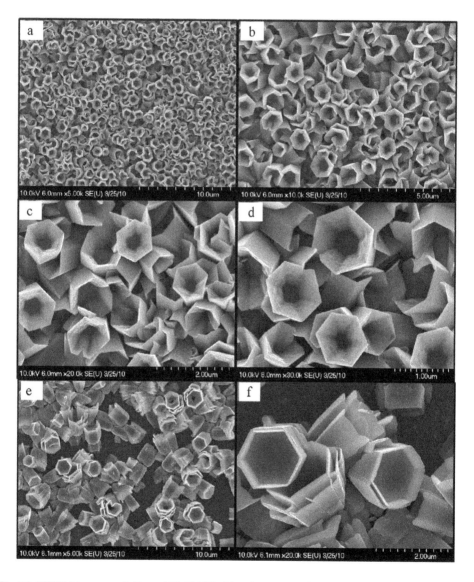

Fig. 19. FESEM images of aligned ZnO NTs (a) at low magnification and (b–d) at high magnification. (e) non-aligned ZnO NTs images at low magnification and (f) at high magnification. Reprinted with permission from [Ameen S. et al, 2011], Electrochim. Acta, 56 (2011) 1111. ©2011, Elsevier Ltd.

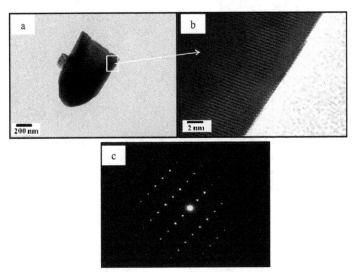

Fig. 20. (a) TEM, (b) HR-TEM and (c) corresponding SAED images of grown ZnO NTs. Reprinted with permission from [Ameen S. et al, 2011], Electrochim. Acta, 56 (2011) 1111. © 2011, Elsevier Ltd.

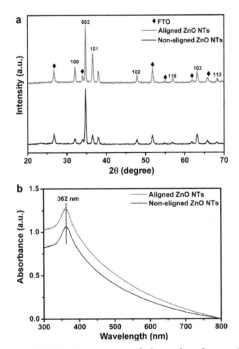

Fig. 21. (a) XRD pattern and (b) UV–Vis spectra of aligned and non-aligned ZnO NTs. Reprinted with permission from [Ameen S. et al, 2011], Electrochim. Acta, 56 (2011) 1111. © 2011, Elsevier Ltd.

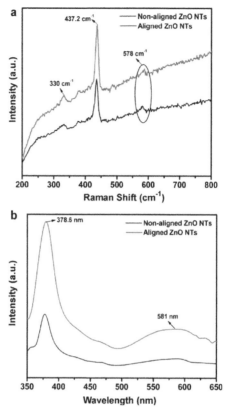

Fig. 22. (a) Raman spectra and (b) photoluminescence spectra of aligned and nonaligned ZnO NTs. Reprinted with permission from [Ameen S. et al, 2011], Electrochim. Acta, 56 (2011) 1111. © 2011, Elsevier Ltd.

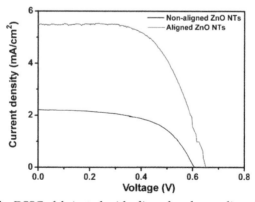

Fig. 23. J–V curve of the DSSCs fabricated with aligned and non-aligned ZnO NTs photoanode. Reprinted with permission from [Ameen S. et al, 2011], Electrochim. Acta, 56 (2011) 1111. © 2011, Elsevier Ltd.

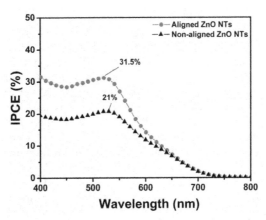

Fig. 24. IPCE curves of the DSSCs fabricated with aligned and non-aligned ZnO NTs photoanode. Reprinted with permission from [Ameen S. et al, 2011], Electrochim. Acta, 56 (2011) 1111. © 2011, Elsevier Ltd.

Fig 23 shows that DSSCs fabricated with aligned ZnO NTs photoanode achieve high solar-to-electricity conversion efficiency of 2.2% with a high short circuit current (J_{SC}) of 5.5 mA/cm^2, open circuit voltage (V_{OC}) of 0.65 V, and fill factor (FF) of 0.61. Compared with non-aligned ZnO NTs photoanode based DSSC, the aligned ZnO NTs photoanode has appreciably enhanced the conversion efficiency by three times with significantly improved J_{SC}, V_{OC} and FF. The DSSC fabricated with non-aligned ZnO NTs pho-to anode executes relatively low of 0.78 % with J_{SC} of 2.2mA/cm^2, V_{OC} (0.60 V) and FF of 0.57. The enhanced photovoltaic performances and the improved J_{SC} are mainly related to the highly dense morphology of aligned ZnO NTs and also, high dye absorption which leads to improved light harvesting efficiency. The aligned morphology might result from the sufficiently high surface area of ZnO NTs and thus, execute reasonably high charge collection and the transfer of electrons at the interface of ZnO NTs and electrolyte layer. While, low efficiency of non-aligned ZnO NTs might associate to low surface area of ZnO NTs and non-uniform surface which might result to low light harvesting efficiency and increases the recombination rate between the electrolyte and the FTO substrate. Ameen et al has reported that the performance of DSSCs with grown aligned ZnO NTs photoanode is significantly higher than the reported DSSCs with aligned ZnO nanorods, nanowires and nanotubes based photoanode (Pasquier, et al., 2006 & Singh, et al., 2010). Importantly, the aligned ZnO NTs based DSSC achieves a maximum IPCE value of ~31.5% at ~520nm as shown in Fig 24. The considerably low IPCE (~21%) is obtained with non-aligned ZnO NTs photoanode based DSSC and thus, the aligned ZnO NTs photoanode presents approximately two times improved IPCE compared to non-aligned ZnO NTs photoanode. The enhancement in IPCE imputes the influence of highly ordered aligned ZnO NTs morphology with high surface area which might improve the light scattering capacities and provides the better interaction between the photons and the dye molecules (Tachibana, et al., 2002).

8. Conclusions

The chapter summarizes the synthesis of PANI through simple and novel polymerization techniques, the concept of doping, types of dopants and the application of PANI in heterostructure devices, diodes and DSSCs. Additionally, the recent surveys on various metal oxide nanomaterials nanomaterials have been thoroughly carried out in terms of their synthesis, morphology and applications in photovoltaic devices. The effective polymerization procedures for PANI particularly, electrophoretic and plasma enhanced deposition are the most promising techniques for optimizing the uniformity, penetration, thickness, electrical conductivity and form the uniform PANI thin films for the high performance and high quality of p-n heterostructure devices and diodes. The choices of dopants are crucial to define the conductive, electrical properties and performances of heterostructure devices such as diodes and DSSCs. The review analyzes various organic/inorganic acids as efficient dopants to enhance the conducting properties of PANI, which confirm that the electronic and optical properties of PANI could be easily controlled and tailored by the oxidizing/reducing agents and acid/base doping during the polymerization procedures. In the second part, the unique and the versatile properties of metal oxides nanostructures especially TiO_2 and ZnO show significant influences on the performances of electrical, electrochemical, and photovoltaic devices by delivering high surface to volume ratio for high absorption of dye molecules, which leads high light harvesting efficiency and increases the electron transfer as well as photocurrent density during the operation of DSSCs. Moreover, various sizes and shapes like nanorods, nanowires and nanotubes of metal oxides nanomaterials particularly TiO_2 and ZnO have been reviewed evidently in terms of morphology and the photovoltaic properties of DSSCs such as J_{SC}, V_{OC}, FF and conversion efficiency. Conclusively, the doping, and method of PANI deposition techniques improve the electrical conductivity and the electrocatalytic activity of the devices and exhibit the direct effect on the performances of DSSCs.

9. References

Grätzel, M., *Solar Energy Conversion by Dye-Sensitized Photovoltaic Cells*, Inorg. Chem., 44 (2005) 6841-6851.

Grätzel, M., *Conversion of sunlight to electric power by nanocrystalline dye-sensitized solar cells*, J. Photochem. Photobiol. A, 164 (2004) 1-14.

Frank, A. J., & Kopidakis, N., J. V. D. Lagemaat, *Electrons in nanostructured TiO_2 solar cells: transport, recombination and photovoltaic properties*, Coord., Chem. Rev., 248 (2004) 1165–1179.

Gratzel, M., *Perspectives for Dye-sensitized Nanocrystalline Solar Cells*, Prog. Photovolt. Res. Appl., 8 (2000) 171-185.

Huang, J., & Kaner, R. B., *The intrinsic nanofibrillar morphology of polyaniline*, Chem. Commun., (2006) 367-376.

Zhang, D., & Wang, Y., *Synthesis and applications of one-dimensionalnano-structured polyaniline: An overview*, Mater. Sci. Eng. B, 134 (2006) 9-19.

Wan, M., *A Template-Free Method Towards Conducting Polymer Nanostructures*, Adv. Mater., 20 (2008) 2926-2932.

Wan, M., *Some Issues Related to Polyaniline Micro-/Nanostructures*, Macromol. Rapid Commun., 30 (2009) 963-975.

Li, D., Huang, J., & Kaner, R.B., *Polyaniline Nanofibers: A Unique Polymer Nanostructure for Versatile Applications*, Acc. Chem. Res., 42 (2009)135-145.

Stejskal J. & Gilbert R. G., *Polyaniline. Preparation of a Conducting Polymer*, Pure Appl. Chem., 74 (2002) 857–867.

Friend, R. H., Gymer, R.W., & Holmes, A. B., *Electroluminescence in conjugated polymers,* Nature (London), 397 (1999) 121-128.

Park J. K. & Kwon O-P., *Enhanced electrical conductivity of polyaniline film by a low magnetic field,* Synth. Met. 160 (2010) 728-731.

Gerard, M., Chaubey, A., & Malhotra, B. D., *Application of conducting polymers to biosensors,* Biosensor Bioeleton., 17 (2002) 345-359.

Xia, L., Wei, Z., & Wan, M., *Conducting polymer nanostructures and their application in biosensors,* J. Coll. Interf. Sci., 341 (2010) 1-11.

Smith, J. D. S., *Intrinsically Electrically Conducting Polymers. Synthesis, Characterization and their Applications,* Prog. Polym. Sci., 23 (1998) 57-79.

Wessling, B., *From conductive polymers to organic metals,* Chem. Innov., 31 (2001) 34-40.

Cho, Y. H., Cho, M.S., Choi, H. J., & Jhon, M. S., *Electrorheological characterization of polyaniline-coated poly(methyl methacrylate)suspensions,* Colloid Polym. Sci., 280 (2002) 1062-1066.

Choi, H. J., Cho, M. S., Kim, J.W., Kim, C.A., & Jhon, M. S., *A yield stress scaling function for electrorheological fluids,* Appl. Phys. Lett., 78 (2001) 3806-3809.

Ameen, S., Akhtar, M. S., & Husain, M., *Polyaniline and Its Nanocomposites: Synthesis, Processing, Electrical Properties and Applications,* Sci. Adv. Mater., 2 (2010) 441-462.

Huang, J. X., Virji, S., Weiller, B. H., & Kaner, R.B., *Polyaniline Nanofibers: Facile Synthesis and Chemical Sensors,* J. Am. Chem. Soc., 125 (2003) 314-315.

Virji, S., Huang, J. X., Kaner, R. B., & Weiller, B. H., *Polyaniline Nanofiber Gas Sensors: Examination of Response Mechanisms,* Nano Lett., 4 (2004) 491-496.

King, R. C. Y., & Rousse, l. F., *Morphological and electrical characteristics of polyaniline nanifibers,* Synth. Met., 153 (2005) 337-340.

Huang, J. X., & Kaner, R. B., *A General Chemical Route to Polyaniline Nanofibers,* J. Am. Chem. Soc., 126 (2004) 851-855.

Jing, X. L., Wang, Y.Y., Wu, D., She, L., & Guo, Y., *Polyaniline Nanofibers Prepared with Ultrasonic Irradiation,* J. Polym. Sci., Part A: Polym. Chem., 44 (2006) 1014-1019.

Jing, X. L., Wang, Y. Y., Wu, D., & Qiang, J.P., *Sonochemical synthesis of polyaniline nanofibers,* Ultrason. Sonochem. 14 (2007) 75-80.

Dhaoui, W., Zarrouk, H., & Pron, A., *Spectroscopic properties of thin layers of sulfamic acid-doped polyaniline and their application to reagentless determination of nitrite,* Synth. Metals, 157 (2007) 564-569.

Dhaoui, W., Bouzitoun, M., Zarrouk, H., Ouada, H. B., & Pron, A., *Electrochemical sensor for nitrite determination based on thin films of sulfamic acid doped polyaniline deposited on Si/SiO_2 structures in electrolyte/insulator/semiconductor (E.I.S.) configuration,* Synth. Metals, 158 (2008) 722-726.

Ameen, S., Ali, V., Zulfequar, M., Haq, M. M., & Husain, M., *Electrical conductivity and dielectric properties of sulfamic acid doped polyaniline,* Curr. Appl. Phys., 7 (2007) 215-219.

Ameen, S., Akhtar, M. S., Kim, Y. S., Yang, O. B., & Shin, H. S., *Sulfamic Acid-Doped Polyaniline Nanofibers Thin Film-Based Counter Electrode: Application in Dye-Sensitized Solar Cells,* J. Phys. Chem. C, 114 (2010) 4760-4764.

Ameen, S., Ali, V., Zulfequar, M., Haq, M. M., & Husain, M., *Preparation and measurements of electrical and spectroscopic properties of sodium thiosulphate doped polyaniline,* Curr. Appl. Phys., 9 (2009) 478-483.

Ameen, S., Ali, V., Zulfequar, M., Haq, M. M., & Husain, M., *Photoluminescence, FTIR, and Electrical Characterization of Samarium (III) Chloride-Doped Polyaniline*, J. Appl. Polym. Sci., 112 (2009) 2315-2319.

Ameen, S., Ali, V., Zulfequar, M., Haq, M. M., & Husain, M., *Preparation and measurements of electrical and spectroscopic properties of praseodymium (III) chloride-doped polyaniline*, Physica E, 40 (2008) 2805-2809.

Ameen, S., Ali, V., Zulfequar, M. Haq, M. M., & Husain, M., *DC Conductivity and Spectroscopic Characterization of Binary Dopant (ZrO₂/AgI)-Doped Polyaniline*, J. Polym. Sci.: Part B: Polym. Phys., 45 (2007) 2682-2687.

Wang, S., Tan, Z., Li, Y., Sun, L., & Zhang, T., *Synthesis, characterization and thermal analysis of polyaniline/ZrO₂ composites*, Thermochim. Acta, 441 (2006) 191–194.

Ameen, S., Lakshmi, G. B. V. S., & Husain, M., *Synthesis and characterization of polyaniline prepared with the dopant mixture of (ZrO₂/PbI₂)*, J. Phys. D: Appl. Phys. 42 (2009) 105104-105109.

Ameen, S., Akhtar, M. S., Ansari, S. G., Yang, O. B., & Shin, H. S., *Electrophoretically deposited polyaniline/ZnO nanoparticles for p-n heterostructure diodes*, Superlatt. Microstr., 46 (2009) 872-880.

Ameen, S., Akhtar, M. S., Kim, Y.S., Yang, O. B., & Shin, H. S., *Diode Behavior of Electrophoretically Deposited Polyaniline on TiO2 Nanoparticulate Thin Film Electrode*, J. Nanosci. Nanotech., 11(2011) 1559-1564.

Somani, P. R., Marimuthu, R., Mulik, U. P., Sainkar, S. R., & Amalnerkar, D. P., *High piezoresistivity and its origin in conducting polyanilinerTiO₂ Composites*, Synth. Met., 106 (1999) 45-52.

Niu, Z., Yang, Z., Hu, Z., Lu, Y., & Han, C. C., *Polyaniline-Silica Composite Conductive Capsules and Hollow Spheres*, Adv. Funct. Mater., 13 (2003) 949-954.

Ameen, S., Ansari, S. G., Song, M., Kim, Y. S., & Shin, H. S., *Fabrication of polyaniline/TiO₂ heterojunction structure using plasma enhanced polymerization technique*, Superlatt. Microstruc. 46 (2009) 745-751.

Ameen, S., Akhtar, M. S., Kim, Y. S., Yang, O. B., & Shin, H. S., *Electrical and Structural Characterization of Plasma Polymerized Polyaniline/TiO₂ Heterostructure Diode: A Comparative Study of Single and Bilayer TiO₂ Thin Film Electrode*, J. Nanosci. Nanotech. 11 (2011) 3306–3313.

Dhawale, D. S., Alunkhe, R. R. S., Patil, U. M., Gurav, K. V., More, A. M., & Lokhande, C. D., *Room temperature liquefied petroleum gas (LPG) sensor based on p-polyaniline/n-TiO₂ heterojunction*, Sens. Actua. B, 134 (2008) 988–992.

Nadarajah, A., Word, R. C., Meiss, J.,& Konenkamp, R., *Flexible Inorganic Nanowire Light-Emitting Diode*, Nano Lett., 8 (2008) 534-537.

Maeng, J., Jo, M., Kang, S. J., Kwon, M. K., Jo, G., Kim, T. W., Seo, J., Hwang, H., Kim D. Y.,& Park S. J., T. Lee. *Transient reverse current phenomenon in a p-n heterojunction comprised of poly (3,4-ethylene-dioxythiophene):poly(styrene-sulfonate) and ZnO nanowall*, Appl. Phys. Lett., 93 (2008) 123109-123111.

Thompson, B. C., & Frechet, J. M. J., *Polymer–Fullerene Composite Solar Cells*, Angew. Chem.-Int. Ed., 47(2008) 58-77.

Gunes, S., Neugebauer, H., & Sariciftci, N. S., *Conjugated Polymer-Based Organic Solar Cells*, Chem. Rev., 107(2007)1324-1338.

Bundgaard, E., & Krebs, F. C., *Low band gap polymers for organic photovoltaics*, Sol. Energy Mater. Sol. Cells, 91(2007) 954-985.

Shaw, P. E., Ruseckas, A., & Samuel, I. D. W., *Exciton Diffusion Measurements in Poly(3-hexylthiophene)*, Adv. Mater., 20 (2008) 3516-3520.

Sirringhaus, H., Brown, P. J., Friend, R. H., Nielsen, M. M., Bechgaard, K., Langeveld V. B. M. W., Spiering, A. J. H., Janssen, R. A. J., Meijer, E. W., Herwig, P., & Leeuw, D. M. D., *Two-dimensional charge transport in self-organized, high-mobility conjugated polymers*, Nature, 401(1999) 685-688.

Waldauf, C., Morana, M., Denk, P., Schilinsky, P., Coakley, K., Choulis, S. A., & Brabec, C. J., *Highly efficient inverted organic photovoltaics using solution based titanium oxide as electron selective contact*, Appl. Phys. Lett., 89 (2006) 233517-233519.

Steim, R., Kogler, F. R., & Brabec, C. J., *Interface materials for organic solar cells*, J. Mater. Chem., 20 (2010) 2499-2512.

Steim, R., Choulis, S. A., Schilinsky, P., & Brabec, C. J., *Interface modification for highly efficient organic photovoltaics*, Appl. Phys. Lett., 92 (2008) 093303-093305.

Ameri, T., Dennler, G., Lungenschmied, C., & Brabec, C. J., *Organic tandem solar cells: A review*, Ener. Environ. Sci., 2 (2009) 347-363.

Lee Y. J., Ruby, D. S., Peters, D.W., McKenzie, B. B., & Hsu, J. W. P., *ZnO Nanostructures as Efficient Antireflection Layers in Solar Cells*, Nano Lett., 8 (2008) 1501-1505.

Cantu, M. L., IddikiM, K. S, Rojas, D. M., Amade, R., & Pech, N. I. G., *Nb-TiO₂/polymer hybrid solar cells with photovoltaic response under inert atmosphere conditions*, Sol. Energy Mater. Sol. Cells, 94 (2010)1227-1234.

Cantu, M. L., Norrman, K., Andreasen, J.W., & Krebs, F.C., *Oxygen Release and Exchange in Niobium Oxide MEHPPV Hybrid Solar Cells*, Chem. Mater., 18 (2006) 5684-5690.

Cantu, M. L., Norrman, K., Andreasen, J.W., Pastor, N. C., & Krebs, F.C., *Detrimental Effect of Inert Atmospheres on Hybrid Solar Cells Based on Semiconductor Oxides*, J. Electrochem. Soc., 154 (2007) B508.

Cantu, M. L., & Krebs, F.C., *Hybrid solar cells based on MEH-PPV and thin film semiconductor oxides (TiO₂, Nb₂O₅, ZnO, CeO₂ and CeO₂-TiO₂): Performance improvement during long-time irradiation*, Sol. Energy Mater. Sol. Cells, 90 (2006) 2076-2086.

Valls, I. G., & Cantu, M. L., *Dye sensitized solar cells based on vertically-aligned ZnO nanorods: effect of UV light on power conversion efficiency and lifetime*, Ener. Environ. Sci., 3 (2010) 789-795.

Valls, I. G., & Cantu, M. L., *Vertically-aligned nanostructures of ZnO for excitonic solar cells: a review*, Ener. Environ. Sci., 2 (2009) 19-25.

Arici, E., Meissner, D., Schaffler, F., & Sariciftci, N.S., *Core/shell nanomaterials in photovoltaics*, Int. J. Photoenergy, 5 (2003) 199-208.

Sariciftci, N.S., Smilowitz, L., Heeger, A. J., & Wudl, F., *Photoinduced Electron Transfer from a Conducting Polymer to Buckminsterfullerene*, Science, 258 (1992) 1474-1476.

Yu, G., Gao, J., Hummelen, J.C., Wudl, F., & Heeger, A.J., *Polymer Photovoltaic Cells: Enhanced Efficiencies via a Network of Internal Donor-Acceptor Heterojunctions*, Science, 270 (1995) 1789-1791.

Ameen, S., Akhtar, M. S., Kim, G.S., Kim, Y. S., Yang, O. B., & Shin, H. S., *Plasma-enhanced polymerized aniline/TiO₂ dye-sensitized solar cells*, J. Alloys Compd., 487 (2009) 382-386.

Saito, Y., Kitamura, T., Wada, Y., & Yanagida, S., *Application of Poly (3,4-ethylenedioxythiophene) to Counter Electrode in Dye-Sensitized Solar Cells*, Chem. Lett. 31 (2002) 1060-1063., Y. Saito, W. Kubo, T. Kitamura, Y. Wada, S. Yanagida, *I⁻/I₃⁻ redox reaction behavior on poly(3,4-ethylenedioxythiophene) counter electrode in dye-sensitized solar cells*, J. Photochem. Photobiol. A. Chem. 164 (2004) 153-157.

Li, Q., Wu, J., Tang, Q., Lan, Z., Li, P., Lin, J., & Fan, L., *Application of microporous polyaniline counter electrode for dye-sensitized solar cells*, Electrochem. Commun. 10 (2008) 1299-1302.

Wu, J., Li, Q., Fan, L., Lan, Z., Li, P., Lin, J., & Hao, S., *High-performance polypyrrole nanoparticles counter electrode for dye-sensitized solar cells*, J. Power Sour. 181 (2008) 172-176.

Lee, K. M., Chen, P. Y., Hsu, C. Y., Huang, J. H., Ho, W. H., Chen, H. C., & Ho, K. C., *A high-performance counter electrode based on poly(3,4-alkylenedioxythiophene) for dye-sensitized solar cells*, J. Power Sour. 188 (2009) 313-318.

Zhang, J., Hreid, T., Li, X., Guo, W., Wang, L., Shi, X., Su, H., & Yuan, Z., *Nanostructured polyaniline counter electrode for dye-sensitised solar cells: Fabrication and investigation of its electrochemical formation mechanism*, Electrochim. Acta, 55 (2010) 3664-3668.

Baxtera, J. B., & Aydil, E. S., *Dye-sensitized solar cells based on semiconductor morphologies with ZnO nanowires*, Sol. Energy Mater. Sol. Cells, 90 (2006) 607-622.

Ito, S., Murakami, T. N., Comte, P., Liska, P., Grätzel, C., & Nazeeruddin, M. K., *Fabrication of thin film dye sensitized solar cells with solar to electric power conversion efficiency over 10%*, Thin Solid Films, 14 (2008) 4613-4619.

Law, M., Greene, L. E., Johnson, J.C., Saykally, R., & Yang, P.D., *Nanowire dye-sensitized solar cells*, Nat. Mater., 6 (2005) 455-459.

Feng, X. J, Shankar, K., Varghese, O. K., Paulose, M., Latempa, T. J., & Grimes, C. A., *Vertically Aligned Single Crystal TiO2 Nanowire Arrays Grown Directly on Transparent Conducting Oxide Coated Glass: Synthesis Details and Applications*, Nano Lett., 11(2008)3781-3786.

Macak, J. M., Tsuchiya, H., & Schmuki, P., *High-Aspect-Ratio TiO2 Nanotubes by Anodization of Titanium*, Angew. Chem. Int Ed, 14 (2005) 2100-2102.

Mor, G. K., Shankar, K., Paulose, M., Varghese, O. K., & Grimes, C. A., *Enhanced Photocleavage of Water Using Titania Nanotube Arrays*, Nano Lett., 1 (2005) 191-195.

Zhang, D. S., Yoshida, T., & Minoura, H., *Low-Temperature Fabrication of Efficient Porous Titania Photoelectrodes by Hydrothermal Crystallization at the Solid/Gas Interface*, Adv Mater., 10 (2003) 814-817.

Wang, H., Liu, Y., Li, M., Huang, H., Zhong, M.Y., & Shen, H., *Hydrothermal growth of large-scale macroporous TiO2 nanowires and its application in 3D dye-sensitized solar cells*, Appl. Phys. A, 97 (2009) 25-29.

Ren, X., Gershon, T., Iza, D. C, Rojas, D. M., Musselman, K., & Driscoll, J. L. M., *The selective fabrication of large-area highly ordered TiO2 nanorod and nanotube arrays on conductive transparent substrates via sol–gel electrophoresis*, Nanotechnology, 36 (2009) 365604-9.

Tan, L. K., Chong, M. A. S., & Gao, H., *Free-Standing Porous Anodic Alumina Templates for Atomic Layer Deposition of Highly Ordered TiO2 Nanotube Arrays on Various Substrates*, J. Phys. Chem. C, 112 (2008) 69-73.

Tsai, T.Y., & Lu, S.Y., *A novel way of improving light harvesting in dye-sensitized solar cells – Electrodeposition of titania*, Electrochem. Commun., 11 (2009) 2180-2183.

Chen, Q.W., & Xu, D.S., *Large-Scale, Noncurling, and Free-Standing Crystallized TiO2 Nanotube Arrays for Dye-Sensitized Solar Cells*, J. Phys. Chem. C, 113 (2009) 6310-6314.

Kang, T. S., Smith, A. P., Taylor, B. E., & Durstock, M. F., *Fabrication of highly-ordered TiO2 nanotube arrays and their use in dye-sensitized solar cells*, Nano. Lett., 9 (2009) 601-606.

Regan, B. O., & Graetzel, M., *A low-cost, high-efficiency solar cell based on dye-sensitized colloidal TiO2 films*, Nature, 353 (1991) 737-740.

Duffie, J., & Beckman, W., *Solar Engineering of Thermal Processes*. Wiley: U.S.A., 1991

Graetzel, M., *Photoelectrochemical cells*, Nature, 414 (2001) 338-344.

Zwilling, V., Aucouturier, M., & Ceretti, E. D., *Anodic oxidation of titanium and TA6V alloy in chromic media. An electrochemical approach*, Electrochem. Acta, 45 (1999) 921-929.

Gong, D., Grimes, C. A., Varghese, O. K., Hu, W., Singh, R. S., Chen, Z., & Dickey E.C., *Titanium oxide nanotube arrays prepared by anodic oxidation*, J. Mater. Res., 16 (2001) 3331-3334.

Guo, Y. G., Hu, J. S., Liang, H. P., Wan, L. J., & Bai, C.L., *TiO₂-Based Composite Nanotube Arrays Prepared via Layer-by-Layer Assembly*, Adv. Funct. Mater., 15 (2005) 196.

Martin, C. R., *Nanomaterials: A Membrane-Based Synthetic Approach*, Science, 266 (1994) 1961-1966.

Limmer, S. J., Seraji, S., Wu, Y., Chou, T. P., Nguyen, C., & Cao, G. Z., *Template-Based Growth of Various Oxide Nanorods by Sol–Gel Electrophoresis*, Adv. Funct. Mater., 12 (2002) 59-64.

Lakshmi, B. B., Dorhout, P. K., & Martin, C. R., *Sol–Gel Template Synthesis of Semiconductor Nanostructures*, Chem. Mater., 9 (1997) 857-862.

Park, J. H., Lee, T.W., & Kang, M.G., *Growth, detachment and transfer of highly-ordered TiO₂ nanotube arrays: use in dye-sensitized solar cells*, Chem. Commun., (2008) 2867-2869.

Jiu, J., Isoda, S., Wang, F., & Adachi, M., *Dye-Sensitized Solar Cells Based on a Single-Crystalline TiO2 Nanorod Film*, J. Phys. Chem. B, 110 (2006) 2087-2092.

Liu, B., & Aydil, E. S., *Growth of Oriented Single-Crystalline Rutile TiO₂ Nanorods on Transparent Conducting Substrates for Dye-Sensitized Solar Cells*, J. Am. Chem. Soc., 131(2009) 3985-3990.

Oh, J. K., Lee, J. K., Kim, H.S., Han, S. B., & Park, K.W., *TiO₂ Branched Nanostructure Electrodes Synthesized by Seeding Method for Dye-Sensitized Solar Cells*, Chem. Mater., 22 (2010) 1114-1118.

Spanhel, L., & Anderson, M. A., *Semiconductor clusters in the sol-gel process: quantized aggregation, gelation, and crystal growth in concentrated zinc oxide colloids*, J. Am. Chem. Soc., 113(1991) 2826-2833.

Keis, K., Vayssieres, L., Rensmo, H., Lindquist, S. E., & Hagfeldt, A., *Photoelectrochemical properties of nano-to microstructured ZnO electrodes*, J. Electrochem. Soc., 148 (2001) A149-153.

Redmond, G., Fitzmaurice, D., & Graetzel, M., *Visible Light Sensitization by cis-Bis(thiocyanato)bis(2,2'-bipyridyl-4,4'-dicarboxylato)ruthenium (II) of a Transparent Nanocrystalline ZnO Film Prepared by Sol-Gel Techniques*, Chem. Mater., 6(1994) 686-691.

Rani, S., Suri, P., Shishodia, P. K., & Mehra, R. M., *Synthesis of nanocrystalline ZnO powder via sol–gel route for dye-sensitized solar cells*, Sol. Energ. Mat. Sol. C, 92 (2008) 1639-1645.

Zeng, L. Y., Dai, S. Y., Xu, W. W., & Wang, K. J., *Dye-Sensitized Solar Cells Based on ZnO Films*, Plasma Sci. Tech., 8 (2006) 172-176.

Keis, K., Bauer, C., Boschloo, G., Hagfeldt, A., Westermark, K., Rensmo, H., & Siegbahn, H., *Nanostructured ZnO electrodes for dye-sensitized solar cell applications*, J. Photochem. Photobiol. A: Chem., 148(2002) 57-64.

Keis, K., Magnusson, E., Lindstrom, H., Lindquist, S. E., & Hagfeldt, A., *A 5% efficient photoelectrochemical solar cell based on nanostructured ZnO electrodes*, Sol. Energ. Mat. Sol. Cells, 73(2002) 51-58.

Suliman, A. E., Tang, Y. W., & Xu, L., *Preparation of ZnO nanoparticles and nanosheets and their application to dye-sensitized solar cells*, Sol. Energ. Mat. Sol. Cells, 91(2007) 1658-1662.

Akhtar, M. S., Khan, M. A., Jeon, M. S., & Yang, O. B., *Controlled synthesis of various ZnO nanostructured materials by capping agents-assisted hydrothermal method for dye-sensitized solar cells*, Electrochim. Acta, 53(2008) 7869-7874.

Lin, C. F., Lin, H., Li, J. B., & Li, X., *Electrodeposition preparation of ZnO nanobelt array films and application to dye-sensitized solar cells*, J. Alloys Compd., 462 (2008)175-180

Hsu, Y. F., Xi, Y. Y., Yip, C. T., Djurisic, A. B., & Chan, W. K., *Dye-sensitized solar cells using ZnO tetrapods*, J. Appl. Phys., 103(2008) 083114-083117.

Chen, W., Zhang, H. F., Hsing, I. M., & Yang, S. H., *A new photoanode architecture of dye sensitized solar cell based on ZnO nanotetrapods with no need for calcination*, Electrochem. Commun., 11(2009) 1057-1060.

Law, M., Greene, L. E., Johnson, J. C., Saykally, R., & Yang, P. D., *Nanowire dye-sensitized solar cells*, Nat. Mater., 4 (2005) 455-459.

Greene, L. E., Yuhas, B. D., Law, M., Zitoun, D., & Yang, P. D., *Solution-Grown Zinc Oxide Nanowires*, Inorg. Chem., 45(2006) 7535-7543.

Gao, Y. F., Nagai, M., Chang, T. C., & Shyue, J. J., *Solution-Derived ZnO Nanowire Array Film as Photoelectrode in Dye-Sensitized Solar Cells*, Cryst. Grow. Des. 7 (2007) 2467-2471.

Jiang, C. Y., Sun, X. W., Tan, K. W., Lo, G. Q., Kyaw, A. K. K., & Kwong, D. L., *High-bendability flexible dye-sensitized solar cell with a nanoparticle-modified ZnO-nanowire electrode*, Appl. Phy. Lett., 92(2008) 143101-143103.

Cheng, A. J., Tzeng, Y., Zhou, Y., Park, M., Wu, T.H., Shannon, C., Wang, D., & Lee, W., *Thermal chemical vapor deposition growth of zinc oxide nanostructures for dye-sensitized solar cell fabrication*, Appl. Phys. Lett., 92 (2008) 092113-092115

Umar, A., Al-Hajry, A., Hahn, Y.B., & Kim, D.H., *Rapid synthesis and dye-sensitized solar cell applications of hexagonal-shaped ZnO nanorods*, Electrochim. Acta, 54 (2009) 5358-5362.

Jiang, C. Y., Sun, X. W., Lo, G. Q., Kwong, D. L., & Wang, J. X., *Improved dye-sensitized solar cells with a ZnO-nanoflower photoanode*, Appl. Phy. Lett. 2007, 90, 263501-263503.

Vayssieres, L., Keis, K., Hagfeldt, A., & Lindquist, S. E., *Three-Dimensional Array of Highly Oriented Crystalline ZnO Microtubes*, Chem. Mater., 13(2001) 4395-4398.

Martinson, A. B. F., Elam, J. W., Hupp, J. T., & Pellin, M. J., *ZnO Nanotube Based Dye-Sensitized Solar Cells*, Nano Lett., 7(2007) 2183-2187.

Ameen, S., Akhtar, M. S., Kim, Y.S., Yang, O. B., & Shin, H. S., *Influence of seed layer treatment on low temperature grown ZnO nanotubes: Performances in dye sensitized solar cells*, Electrochim. Acta, 56 (2011) 1111-1116.

Exarhos, G. J., & Harma, S. K. S, *Influence of processing variables on the structure and properties of ZnO films*, Thin Solid Films, 270 (1995) 27-32.

Serrano, J., Manjon, F. J., Romero, A. H., Widulle, F., Lauck, R., & Cardona, M., *Dispersive Phonon Linewidths: The E_2 Phonons of ZnO*, Phys. Rev. Lett., 90 (2003) 055510-0555113.

Vanheusden, K., Warren, W. L., Seager, C. H., Tallant, D. R., Voigt, J. A., & Gnade, B.E., *Mechanisms behind green photoluminescence in ZnO phosphor powders*, J. Appl. Phys., 79 (1996) 7983-7990.

Pasquier, A. D., Chen, H., & Lu, Y., *Dye sensitized solar cells using well-aligned zinc oxide nanotip arrays*, Appl. Phys. Lett., 89 (2006) 253513-253515.

Singh, D. P., *Synthesis and Growth of ZnO Nanowires*, Sci. Adv. Mater., 2 (2010) 245-271.

Tachibana, Y., Hara, K., Sayama, K., & Arakawa, H., *Quantitative Analysis of Light-Harvesting Efficiency and Electron-Transfer Yield in Ruthenium-Dye-Sensitized Nanocrystalline TiO_2 Solar Cells*, Chem. Mater., 14 (2002) 2527-2535.

Physical and Optical Properties of Microscale Meshes of Ti_3O_5 Nano- and Microfibers Prepared via Annealing of C-Doped TiO_2 Thin Films Aiming at Solar Cell and Photocatalysis Applications

N. Stem[1], E. F. Chinaglia[2] and S. G. dos Santos Filho[1]

[1]*Universidade de São Paulo/Escola Politécnica de Engenharia Elétrica (EPUSP)*
[2]*Centro Universitário da FEI/Departamento de Física*
Brazil

1. Introduction

Dye-sensitized nanocrystalline solar cells (DSSC) or photoeletrochemical solar cells were firstly described by Gratzel and O'Reagan in the early 1990s (Sauvage et. al., 2010) and they have reached the global photovoltaic market since 2007. Later on, the investments in nanotechnology enabled the rapid development of DSSC cells with nanostructured thin films. According to a review performed by Hong Lin et. al. (Lin et. al., 2009) the numbers of papers focusing on the development of the DSSC cells increased in last decade, being mainly originated in countries such as Japan, China, South Korea, Swiss and USA, where there is an enlarged integration of nanotechnology, electrochemical and polymers research and finantial supported projects like National Photovoltaic Program by Department of Energy (DOE) and NEDO's New Sunshine from USA and Japan, respectively. Some research groups of the institutions (Kim et. Al., 2010),which have recently obtained efficiencies around 10%, are EPFL (11.2% in 2005) and AIST (10% in 2006). They have used the N719 colorant in devices with area $0.16cm^2$ and $0.25cm^2$. On the other hand, Sharp, Tokyo University and Sumitomo Osaka Cell have used the black dye colorant in devices with areas of approximately $0.22cm^2$, providing the efficiencies of about 11.1%, 10.2% and 10% in the years 2006, 2006 and 2007, respectively. In 2006, Tokyo University has also reached the efficiency of 10.5% in devices with $0.25cm^2$ area, but using β-diketonide colorant.

Initially, the DSSC (Sauvage et. al., 2010) were based on a nanocrystalline semiconductor (pristine titanium dioxide) coated with a monolayer of charge-transfer dye, with a broad absorption band (generally, polypyridyl complexes of ruthenium and osmium), to sensitize the film. The principle of operation of these devices can be divided into: a) the photo-current generation that occurs when the incident photons absorbs in the dye, generates electron-hole pairs and injects electrons into the conduction band of the semiconductor ($Ru^{2+} -> Ru^{3+} + e^-$), and b) the carrier transport that occurs because of the migration of these electrons through the nanostructured semiconductor to the anode (Kim et. al., 2010). Thus, since this device requires an electrode with a conduction band with a lower level than the dye one, the main desired properties for the electrode are optimized band structure and good electron injection efficiency and diffusion properties (Wenger, 2010).

Since Ru has become scarce and its purification and synthesis is too complex for production in large scale, new outlets for doping the titanium dioxide became necessary. Among the materials usually adopted for the electrode, TiO_2, ZnO, SnO_2, Nb_2O_5 and others have been employed (Kong et al., 2007), besides nanostructured materials. For instance, in a previous work, H. Hafez et. al. (Hafez et. al., 2010) made a comparison between the J-V curves of three different structures for the TiO_2 electrodes combined with N719 dye for dye-sensitized cells: a) pure nanorod with adsorbed dye of 2.1x 10^{-5}mol.cm^{-2}; b) pure nanoparticle with adsorbed dye of 3.6x10^{-5}mol.cm^{-2} and c) a mix between nanorods and nanoparticles with adsorbed dye of 6.2x10^{-5}mol.cm^{-2}. These cells presented the incident photon-to-current conversion efficiency, IPCE (at λ=575nm) of approximately 63.5%, 70.0% and 88.9%, and the efficiencies, 4.4%; 5.8% and 7.1%, respectively. A higher efficiency of 7.1% was found for a mixed structure of nanorods and nanoparticles and the efficiencies found for either pure nanoparticules or nanorods were around 5.8% and 4.4%, respectively.

Despite showing lower efficiency compared with the crystalline silicon solar cells, this thin film technology has been pointed as a potential solution to reduce costs of production. Also, they can be engineered into flexible sheets and are mechanically robust, requiring no special protection from environmental events like hail strikes. Other major points of DSSC technology is the fact that it is less sensitive to impurities compared with the conventional crystalline ones because the constituents used are low cost and abundant. Furthermore, differently from the Si-based modules, the performance of dye PV modules increases with temperature. For instance, comparing the Si-based modules with the dye PV modules, Pagliaro et. Al. (2009) showed for temperature variying from 25°C to 60°C that the percentage of power efficiency decreased approximately 40% for the silicon-based one and increased approximately 30% for the STI titania cells (Pagliaro et. al., 2009). Another important characteristic is associated with the color that can vary by changing the dye, being possible to be transparent, which is useful for application on windows surface. However, degradation under heat and UV light are the main disavantages and, in addition, the sealing can also be a problem because of the usage of solvents in the assembling, which makes necessary the development of some gelators combined with organic solvents. The stability of the devices is another important parameter to be optimized (Fieggemeier et. al., 2004), and the competitive light-to-energy conversion efficiencies must be tested. Recently, Wang et. al. (Wang et. al., 2003) have proved that it is possible to keep the device stable under outdoor conditions during 10 years in despite of the complexity of the system.

2. An overview of the techniques for producing titanium oxide nanofibers

The study of titania nanotubes (Ou & Lien, 2007) started in the nineties, with the development of the formation parameters of several processes (temperature, time interval of treatment, pressure, Ti precursors and alkali soluters, and acid washing). With the evolution of the characterization techniques, the thermal and post-thermal annealings were studied, and optimized for the several types of applications (photocatalysis, littium battery, and dye sensitized solar cells). The hydrothermal treatments have also been modified either physically or chemically depending on the desired application and on the desired stability after post-hydrothermal treatment and post-acid treatments.

Focusing on nanostructured materials developed for solar cells and photocatalysis, titanium dioxide (TiO_2) is one of the most promising due to its high efficiency, low cost and photostability (Kim et. al. , 2007) (Varghese et. al., 2003). Some resources have been used for enlarging efficiency and for reducing costs. The enhanced porosity of the nanofibers, nanobelts or nanorods of these new structures, which can be used as photoanodes, were proved to have a better response than titanium-dioxide nanoparticles, because of their structure that facilitates the chemical adsorption for polymer electrolytes (Varghese et. al., 2003). There is a wide variety of methods for producing nanofibers and nanotubes techniques, such as sol-gel techniques combined with low cost processes such as arc-plasma evaporation, electrospinning techniques, and hydrothermal methods (Chen and Mao, 2007), (Nuansing et. al., 2006) and (Park et. al., 2010) .

Another resource usually used for enhancing efficiency is the doping (Chennand and Mao, 2007) (Valentini et. al., 2005) , either with non-metallic elements (N, C, S or P) or halogens, in order to reduce bandgap and to shift the adsorption band edge to the visible-light range. And, for producing nanostructured materials, several precursor seeds have been successfully used including alkalines (Kukovecz et. al., 2005), carbon (Puma et. al., 2008) and (Varghese et. al., 2003) and water vapor (Yamamoto et. al., 2008), which also have the role as dopants. For instance, Khan et. al. (Khan et. al., 2009) showed that hydrothermally synthesized titanium dioxide doped with Ru, provided a significantly decrease in the energy bandgap and showed an increase (>80% higher after 140min) in their photocatalytic activity to degrade methylene blue (MB) under visible light compared with undoped tubes. Concomitantly, Zhang et. al. (Zhang et. al., 2010) report the doping of TiO_2 with transition metal ions, specially Fe(III) and Cr(III) as a good tool for improving photocatalytic properties.

According to previous works (Reyes-Garcia et. al., 2009) (Konstantinova et al., 2007), concerning with photocatalytic properties, carbon has been shown as one of the most proeminent dopant for titanium dioxide because it can provide a significant reduction of the optical band gap and the appearance of some C states in the mid-gap. For example, the energy of oxygen vacancies can be reduced from 4.2eV to 3.4eV (interstitional position in the titanium dioxide lattice) and to 1.9eV (substitutional one) for anatase phase and, from 4.4eV to 2.4eV for rutile phase for both positions, interstitial and substitutional. As a result, it has been showed that the photosensitization property is enhanced (Valentini et. al., 2005).

The hydrothermal route and calcination have been the most used techniques by varying time, atmosphere and temperature of annealing. In a previous work (Suzuki & Yoshikawa, 2004) , nanofibers of TiO_2 were synthesized by hydrothermal method (150 ºC for 72 h) using natural rutile sand as the starting material and calcination at 700ºC for 4 h. On the other hand, pure rutile phase TiO_2 nanorods (Chen et al., 2011) were also successfully synthesized under hydrothermal conditions, showing an increase of the photocatalytic activity for the times ranging from 1 to 15h because of the increase of the crystal domain. The best performance of DSSC measured under "1 sun condition" gave a current density <7.55 mA/cm², an open circuit voltage <0.70V, a fill factor <60%, and an energy conversion efficiency <3.16%. Meanwhile, Hafez et. al. (Hafez et. al., 2010) processed anatase TiO_2 nanorods by hydrothermal method and proved that the efficiency could increase from 5.8% to 7.1% if the DSSC electrodes were changed from nanoparticles to nanorods (Wang et. al., 2003). Wu et. al. (2009) proved that the use of ethanol as precursor for producing H-titanate nanotubes in inert N_2 atmosphere. Depending on the calcination temperature, the nanostructure could be altered, presenting either nanotubes, or nanowires or nanorods for calcination temperatures of 400ºC, 500ºC and 600ºC, respectively. It is believed that during the calcination in N_2, the decomposed products of ethanol were not burnt out because there was not observed oxygen in the environment. Thus, the residual carbon either remainded in the TNTs or it doped the titanium dioxide by forming different nanostructures and, therefore, acting as seeds. Tryba (Tryba, 2008) has also demonstrated that the carbon-based coating of TiO_2, prepared by the calcination of TiO_2 with carbon precursor (polyvinylalcohol, poly (terephthalate ethylene), or hydroxyl propyl cellulose (HPC)) at high temperatures 700°C – 900°C retarded the phase transformation from anatase to rutile and increased the photoactivity, but the carbon coating reduced the UV radiation once it reached the surface of the TiO_2 particles and altered the absorbed light.

This work is focused on the development of new technique for producing carbon-doped TiO_2 thin films on silicon substrates together with Ti_3O_5 fiber meshes and on the investigations about the properties of this novel material. The innovation of the proposed technique relies on the fact that thermal evaporation is the most common method to fabricate single crystalline nanowires on silicon substrate by means of the Vapor-Liquid-Solid (VLS) mechanism (Dai et. al., 2002), (Yin et. al., 2002) and (Pan et. al., 2001). On the other hand, it is not an useful process for growing TiO_2 nanowires because Ti precursor can react with silicon to form Ti-Si alloys before nucleation and growth of TiO_2 nanowires (Wu et. al., 2005). Also, it is too difficult the production of titania nanowires by thermal treatment of Ti on Si substrate because $TiSi_2$ phases is favored before nucleation of titanium oxide nanowires in inert gas or high vaccum (Xiang et. al., 2005). On the other hand, a recent study has shown that single crystalline rutile TiO_2 nanowires could be obtained by annealing TiO_2 nanoparticles on silicon substrates at high temperature in air without catalysts (Wang et. al., 2009). Although it is possible to obtain titania nanowires on silicon by thermal annealing, there is a complete lack of information in literature about the effect of carbon as dopant on the physical and electrical properties of TiO_2 nanowires produced by thermal annealing of TiO_2 on silicon substrates. C-doped TiO_2 can evolve to lower oxides of titanium like Ti_4O_7, Ti_3O_5, and Ti_2O_3 after thermal annealing at 1000-1100°C in vacuum or argon. This process is known as carbothermal reduction of titanium dioxide in presence of carbon and can produce TiC powders of submicron size at a very high temperature of 1500°C (Sen et. al, 2011) and (Swift & Koc, 1999).

Thus , in the following, the formation mechanism of nano- and microfibers of Ti_3O_5 produced by annealing of carbon-doped TiO_2 thin films on silicon substrates at 900-1000°C for 120min in wet $N_2(0.8\%H_2O)$ is presented. The effects of concentration of carbon, concentration of water vapor and temperature on the formation of the nano and microfibers are addressed.

3. Nanofibers formation mechanism

Generally speaking, the formation of titania nanotubes has been explained by the sheet roll-up mechanism. In this process the nanosheet-like features produced after thermal treatment composed of highly distorted TiO_6 octahedra are believed to be formed by scrolling up, such that the driving force gets high enough because of the saturation of the undercoordinated sites or dangling bonds. In this structure, each Ti_{4+} ion is surrounded by an octahedron of six O_{2-} ions, and the distortion is generated with the aid of thermal treatment and precursor seeds (Chen & Mao, 2007) and (Kukovecz et. al., 2005). According to the previous work of Bavykin et. al. (Bavykin et. al., 2006) and (Bavykin et. al., 2009), the nanotubes are believed to be thermodinamically less stable than the nanofibers due to their increased surface area and the higher stress in the crystal lattice.

Figure 1 presents a simplified scheme of the possible formation mechanism of the nanofibers: a) starting from carbon-doped titanium dioxide crystals; b) after thermal annealings at temperatures lower than 900ºC, it might occur delamination and the nanosheets are dettached; c) as the driving force is increased, the hollow nanofibers are formed, being composed by the distorted TiO₆ octahedra; d) after the hydrothermal annealing performed at 1000ºC , the nanofibers probably are filled in because of the –OH bonds.

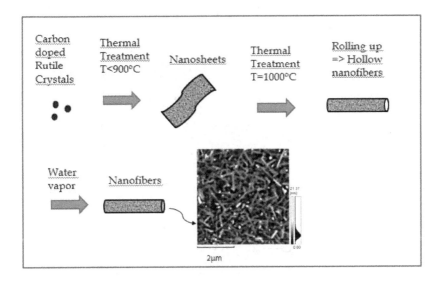

Fig. 1. The carbon doped crystals after thermal treatment are dettached in nanosheets. Increasing the temperature up to 1000ºC, the sheet roll-up forming hollow nanofibers. Then, the nanofibers are filled in, probalby due to the presence of water vapor during annealing.

4. Details of sample preparation and cleaning monitoring

The initial wafer cleaning is a quite important to drop out: a) contaminant films, b) discrete particles, and c) adsorbed gases. While the RCA 1 is responsible for the organic compound dropping (such as condensed organic vapors from lubrificants, greases, photoresist, solvent residues or components from plastic storage containers), RCA 2 is responsible for the metallic (heavy metals, alkalis, and metal hydroxides) compound dropping.

Thus, a common cleaning for P-type Si (100) consists of the following sequence: a) RCA 1: 4 parts deionized (DI) water H₂O, 1 part 35% ammonium hydroxide (NH₄OH) , 1 part 30% hydrogen peroxideH₂O₂ (heated at 75ºC during 15 min); b) RCA2: 4 parts DI water (H2O), 1 part 35% hydrogen chloride (HCl), 1 part 30% hydrogen peroxide (H₂O₂) (heated at 80ºC during 15min) (Santos Filho et. al., 1995), (Kern, 1990) and (Reinhardt & Kern, 2008). According to S. G. Santos et. al. (Santos Filho et. al., 1995), the typical impurities found on the wafer surface analyzed by TRXFA after the conventional standard cleaning are up to 10¹⁰ atoms/cm², and the drying with the aid of isopropyl alchoholis was shown to be efficient in removing a high percentage of particles of almost all measurable sizes (submicron and larger), as presented at table 1. Thus, after the deposition in order to perform the thermal annealings the samples were previously boiled in ultrapure isopropanol alcohol during 15 min, followed by rinsing in DI water during 5 min.

Elemental analysis were performed by using EDS technique, indicating the presence of the elements Ti, O, C or another contaminant before and after hydrothermal treatment. The EDS spectra presented show the obtained peaks for: a) as-deposited film, and b) for sample 1E (annealed at 1000°C) where the K_α line peaks of carbon, oxygen, silicon and titanium are indicated. The L line peak of the titanium (not shown) is superimposed to the K line of the oxygen.

Element	TXRFA Convencional 10^{10} atoms/cm^2
S	<LD
K	<LD
Ca	70±30
Ti	40±20
Cr	20±10
Mn	<LD
Fe	45±8
Co	<LD
Ni	<LD
Cu	10±8
Zn	54±4

Table 1. TRXFA performed after the initial cleaning and drying at isopropyl alchoholis (Santos Filho et. al., 1995).

After the cleaning process, TiO_2 (rutile phase) and C were co-deposited on bare silicon by e-beam evaporation using the EB3 Multihearth Electron Beam Source from Edwards and targets with 99.99% of purity from Sigma Aldrich. The carbon contents were fixed at two different concentrations: 1.5%wt or about 3.0%wt (Stem et al., 2010); (Stem et al., 2011). Then samples were boiled in a neutral ambient (isopropanol alcohol) aiming at the remotion of possible contaminants.

The deposition pressure was controlled in the range of (2.3×10^{-6} – 4.6×10^{-6}) Torr; the e-beam co-deposition current used was 150mA for a fixed time of 1min in order to produce a thickness close to 200nm.

After the co-deposition, hydrothermal annealing was performed in resistance-heated furnace with an open horizontal quartz tube; samples were introduced by a quartz boat. The temperature was adjusted in the range of 700°C to 1000°C for the following gases (2L/min): ultrapure N_2 or wet N_2 (0.8%H_2O), for 120min. As reported by Shannon et. al. (Shannon et. al., 1964), the presence of water can greatly promote the formation of oxygen vacancies, which increases the diffusivity of oxygen ions through TiO_2 layer and reduces diffusivity of titanium interstitials. In addition, wet inert gas plays a crucial role in triggering the much higher growth rate of titanium oxide nanowires (Liu et. al., 2010). A brief summary of the sample preparation is presented at figure 2. In this figure, the AFM analysis of the samples just after the initial cleaning, the as-deposited film and after thermal annealing are shown. The EDS spectra of the as-deposited film and after annealing are also presented.

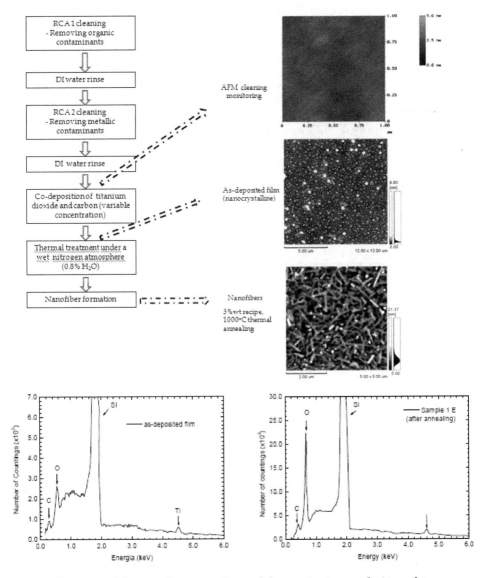

Fig. 2. Brief scheme of the sample preparation and the monitoring analysis: surface morphology by AFM technique and elemental analysis by EDS technique. The EDS spectra are not normalized; and therefore, only qualitative.

5. Producing meshes of Ti_3O_5 nano and microfibers

It is well known that is not easy to obtain titanium oxide nanowires by thermal treatment of Ti on Si, because $TiSi_2$ phases are favored over the nucleation of titanium dioxide nanowires in an inert gas or under high vacuum (Wu et. al., 2005), (Xiang et. al., 2005). In case of TiO_2 on Si, only when the high vacuum or inert gas was replaced by an oxygen-rich gas, TiO_2 nanowires could be formed on Si (Bennett et. al., 2002).

Figure 3a shows the obtained XRD spectra of titanium oxide thin films doped with 1.5%wt and 3.0%wt of carbon, respectively, and annealed at 700ºC (1G), 900ºC (1Fx and 1F) and 1000ºC (1Ex and 1E). The annealed films are primarily amorphous with a low content of crystalline Ti_3O_5 and rutile, except for the sample 1E where the higher crystallinity is demonstrated by high intensity peaks (about 772 times higher than the lowest intensity found for sample 1G) and for sample 1G where Ti_3O_5 was not be identified. However, when temperature reaches an intermediate value for the 3%wt carbon recipe, about 900ºC (as for sample 1F), the intensity of Ti_3O_5 and rutile increased in the amorphous film. On the other hand, for films doped with 1.5%wt of carbon recipe, only crystalline phase of Ti_3O_5 was observed at 700-900ºC, while Ti_3O_5 and rutile are observed at 1000ºC. .

Figure 3b is an ampliation of the XRD pattern shown in figure 3a of sample 1E, with the scale of the intensity reduced and, and with 2θ varying from 55 to 58 degrees when annealing to view the high intensity peaks and the peak deconvolution. It could be demonstrated that region is composed by three superposed peaks: Ti_3O_5 (<-5 1 2> and <-6 0 1>) and rutile (<220>), respectively.

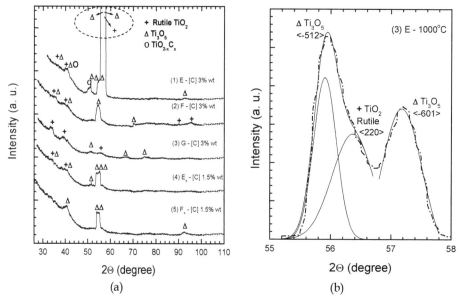

(a)   (b)

Fig. 3. (a) Typical XRD spectra for the 3%wt recipe: samples 1G (700ºC), 1F (900ºC) and 1E (1000ºC), and for the 1.5%wt recipe: samples $1F_x$ (900ºC) and $1E_x$ (1000ºC); (b) ampliation of the most intense peaks of sample 1E (1000ºC) (dashed region of figure 3 a) and peak deconvolution, detailing the superposed peaks.

All of the crystalline orientations for Ti$_3$O$_5$ fitted well with the XRD patterns of λ-Ti$_3$O$_5$ (Monoclinic, C2/m E, a = 9.757Å, b = 3.802Å, c = 9.452Å) (Grey & Madsen, 1994). In addition, TiO$_{2-x}$C$_x$ was also identified with the aid of XRD powder patterns , which is an evidence that carbon occupies positions in the crystalline phase of the titanium dioxide (interstitial and substitutional) and introduces defects, electron and hole trapping centers because of the presence of carbon and carbonate-type species (Reyes-Garcia et. al., 2008). Therefore, after annealing at 1000°C (sample 1E), the structure becomes predominantly crystalline, being formed by λ-Ti$_3$O$_5$ and rutile with carbon incorporation.

In order to shed further light on the influence of the carbon content, film morphology was evaluated by dynamic mode technique (AFM of Shimadzu). Figure 4 shows the obtained AFM images of nano- and micro-fibers prepared by annealing at different temperatures in wet N$_2$ (0.8%H$_2$O) for 3 wt%-doped TiO$_2$ thin films on a silicon substrate: a) top view of sample 1G; b) the correspondent statistics performed for figure 5 a); c) top view of sample 1F; d) top view of sample 1E; e) 3D view of sample 1E and (f) the correspondent statistics for figure 4d.

As a result of the performed analysis, the average RMS roughness of the as-deposited film was (2.3$\pm$0.5)nm and increased to (10$\pm$2)nm after annealing at 700°C in nitrogen+water vapor, being about four times higher. The observed "islands", as shown in Figure 4(a), presenting a diameter range of 19.05nm and 158.6nm.

On the other hand, as the temperature increases to 900°C, a threshold temperature, the morphology starts evoluting from small "islands" to micro scale meshes of fibers, with length varying from 0.79µm to 2.06µm and widths lower than 0.400µm (range: 0. 100 to 0.400µm). In this case, the RMS roughness decreased to (5.8$\pm$0.7)nm (Figure 4(c)) and, in place of "islands", needle-like nanofibers and embedded fibers were formed on the surface and below it.

Finally, after annealing at 1000°C, the film morphology was completely changed, as shown in Figure 4d (top view) and in figure 4e (3D view). In this case, micro scale meshes of fibers randomly distributed were observed with length ranging from 0.1 to 1.1µm (shown in figure 4 f) and average width of (0.170$\pm$20) µm. Also, the average RMS roughness decreased from (5.8$\pm$0.7)nm to (3.3$\pm$0.2)nm.

In contrast, when the carbon concentration was decreased below 2%wt, nano- and microfibers were not observed (AFM images not shown) on the samples prepared by annealing at different temperatures (700-1000°C) in pure N$_2$ or wet N$_2$ (0.8%H$_2$O).

Figure 5a shows the FTIR analysis of C-doped TiO$_2$ samples1.5%wt (1F$_x$ and 1E$_x$) and 3.0%wt (1G, 1F and 1E) that have been annealed at 700°C, 900°C and 1000°C. A broad absorption peak at 1096cm^{-1} and this peak represents Si-O-Si stretching bond, while the Si-O-Si bending peak is also shown at 820cm^{-1} (Yakovlev et. al., 2000) and (Erkov et. al., 2000), both can be associated to silicon oxidation during the thermal annealing in water vapor atmosphere. Also, Ti-O-Ti stretching vibration of the rutile phase was observed at 614.4cm^{-1} for all samples (Yakovlev et. al., 2000) and (Erkov et. al., 2000), corroborating the XRD analysis, where a change in the cristallinity was demonstrated, evoluting from an armophous structure to a crystalline one (rutile). The higher intensity of this band is likely to be due to the increase in the amount of rutile when the carbon content is higher (3%wt). For this carbon content, Ti-O stretching at 736.5cm^{-1} (Yakovlev et. al., 2000) progressively increases as the annealing temperature increases from 700°C to 1000°C, which indicates progressive transition from an amorphous TiO$_2$ to a crystalline structure of λ-Ti$_3$O$_5$ and rutile. In addition, a band is observed at 781 cm^{-1} only for sample 1E, which was annealed at 1000°C, as shown in detail in figure 5b. Richiardi et al.(Richiardi et. al., 2001) shows this band to be due to symmetric stretching of Ti-O-Si and Si-O-Si bonds, which corroborates a quantitative mixture of SiO_2 and TiO_2 at the interface; where TiO_2 is more likely rutile since it is at the interface as established by Raman analysis (not shown).

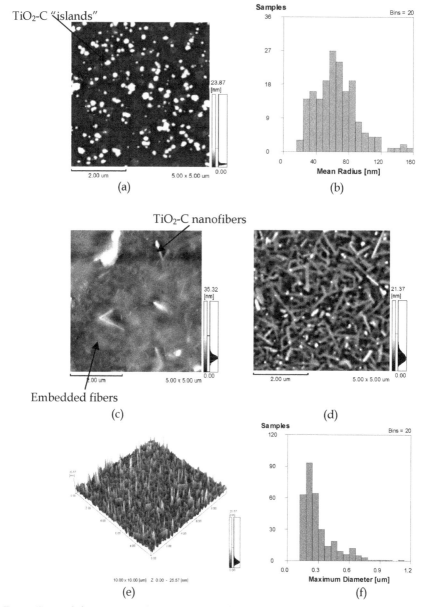

Fig. 4. Typical dynamic-mode AFM images for: (a) sample 1G annealed at 700°C; (b) statistics of (a); (c) sample 1F annealed at 900°C; (d) sample 1E (top view); (e) sample 1E (3D view) and (f) statistics of (d).

Physical and Optical Properties of Microscale Meshes of Ti$_3$O$_5$ Nano- and Microfibers Prepared via
Annealing of C-Doped TiO$_2$ Thin Films Aiming at Solar Cell and Photocatalysis Applications
159

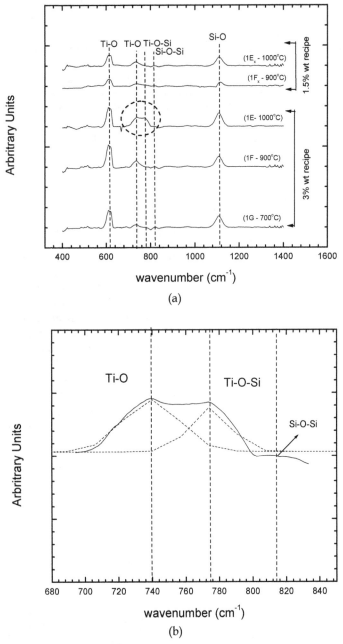

Fig. 5. a) Typical FTIR spectra as function of the wave number for the 3%wt recipe: samples
1G (700ºC), 1F (900ºC) and 1E (1000ºC), and for the 1.5%w recipe: samples 1F$_x$ (900ºC) and
1E$_x$ (1000ºC) and b) larger view of FTIR curve.

Aiming to evaluate stoichiometry and the carbon content after thermal treatments, the aerial concentrations of oxygen and titanium were obtained from Rutherford Backscattering Spectrometry (RBS) by fitting rump-code simulation (Climent-font et. al., 2002) to the experimental spectra. Using the extracted aerial concentrations (cm^{-2}), stoichiometry of the titanium oxide was determined admitting a weighted composition of $aTiO_x + bSiO_2$, where a, b and x are calculated parameters. The carbon content was obtained by EDS analysis because the detection limit was lower than the value reported to RBS analysis (Wuderlich et. al., 1993). Also, EDS has sufficient sensitivity to distinguish carbon content of 1.5%wt from 3.0wt% (detection limit of about 0.1wt%) analysis (Wuderlich et. al., 1993). Figure 6 illustrates the experimental RBS spectrum and the fitted simulation for the sample 1E.

Table 2 presents the average concentration of carbon [C], the stoichiometry and the aerial silicon-oxide concentration [SiO_2] extracted from the EDS and RBS analyses according to the procedure described in the experimental section.

For the 3.0%wt carbon concentration in table 2, the SiO_2-layer thickness ranged from 16.2 nm ($\approx 7.5 \times 10^{16}$ atoms/cm^2) to 19.4 nm ($\approx 9.0 \times 10^{16}$ atoms/cm^2) for temperatures varying from 700°C to 1000°C. In this case, as predicted by the band at 1096 cm^{-1}, the higher the temperature, the higher the aerial silicon oxide concentration, which is consistent with the increase of the band at 1096 cm^{-1} in Figure 5. However, the oxygen stoichiometric coefficient of TiO_x decreased from 2.0 to 1.7 (see table 1) when the temperature was increased from 700 to 900°C. Assuming the presence of crystalline Ti_3O_5 and rutile, as illustrated by the XRD results, $TiO_{1.70}$ fits well with 25% TiO_2 and 75% Ti_3O_5 at 1000°C. Moreover, TiO_2 is consistent with predominantly amorphous TiO_2 at 700°C (sample 1G), as illustrated by the XRD results. Finally, $TiO_{1.85}$ (sample 1F) fits well with 75% TiO_2 and 25% Ti_3O_5 at 900°C (sample 1F) and is also consistent with a predominantly amorphous TiO_2, as illustrated by the XRD results.

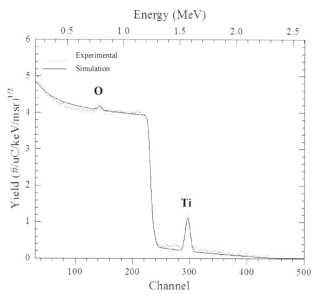

Fig. 6. Typical RBS spectrum of the sample 1E (3%w recipe).

For the 1.5%wt carbon concentration in table 2 the oxygen stoichiometric coefficient is close to 1.80 for the thermal treatments of 900°C and 1000°C. In this case, $TiO_{1.80}$ fits well with 66% TiO_2 and 33% Ti_3O_5, which is consistent with a predominantly amorphous TiO_2 with a low concentration of Ti_3O_5, as illustrated by the XRD results. In the latter case (sample 1Ex), the diffusion of the oxygen species might have been prevented, if compared to sample 1E, possibly due to a denser bulk of TiO_2 at 1000°C, which might have also slightly decreased the growth rate of the SiO_2 layer (Koch, 2002) .

Recipe	Sample	Temperature (°C)	[C] (%wt)	Stoichiometry	$[TiO_x]$ (10^{16}/cm²)	$[SiO_2]$ (10^{16}/cm²)
3.0%wt	1G	700	3.4±1.2	$TiO_{2.00}$	4.3	7.5
	1F	900	3.2±0.9	$TiO_{1.85} =$ $0.75TiO_2 + 0.25$ Ti_3O_5	7.0	8.0
	1E	1000	3.4±0.6	$TiO_{1.70}$ $=0.25TiO_2 +$ $0.75\ Ti_3O_5$	5.7	9.0
1.5%wt	1Fx	900	1.5±0.4	$TiO_{1.80} =$ $0.66TiO_2 + 0.33$ Ti_3O_5	4.3	8.0
	1Ex	1000	1.7±0.2	$TiO_{1.80} =$ $0.66TiO_2 + 0.33$ Ti_3O_5	3.6	8.5

Table 2. Average concentration of carbon [C] as obtained from EDS and, stoichiometry and aerial silicon oxide concentration $[SiO_2]$ after fitting rump-code simulation to the experimental spectra using weighted compositions of $aTiO_x + bSiO_2$. TiOx layer is divided into two different layers rutile TiO_2 and Ti_3O_5 according to XRD spectra from figure 3, except for sample G where rutile TiO_2 is dominant.

Figures 7a and 7b show the diffuse reflectance spectra and the solar spectrum for AM1.5G (ASTMG173) (Stem, 2007) and (ASTM, 2005), respectively. Figure 7a allow to infer that it is evident that the film annealed at 700°C has a less significant amount of absorption in the visible region with the absorption band limited at a wavelength below 460 nm. In this case, titanium oxide is predominantly amorphous, and the literature corroborates this limited band below 460 nm (Wang et. al., 2007). However, when the annealing temperature was increased to 900°C or 1000°C, samples 1F and 1E adsorbed a much larger light fraction in the visible region, which can be attributed to a structural change of the samples associated with a phase transition to rutile, $TiO_{2-x}C_x$ and Ti_3O_5. In this case, both positions, substitutional and interstitial, carbon significantly impacts the optical properties in the range of 500 to 800 nm because of the formation of complex midgap states (Reyes-Garcia et. al., 2008) and (Wang et. al., 2007).

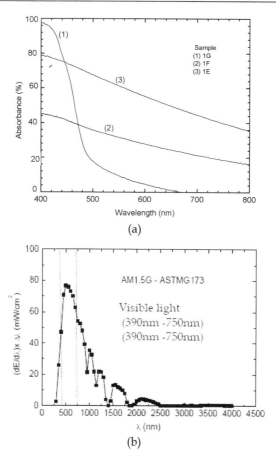

Fig. 7. (a) Absorbance curves as function of wavelength for samples processed with the 3%wt carbon recipe (1G, 1F and 1E). Their correspondent optical band-gap extracted from the curve is also presented. (b) Solar spectral irradiance as function of the wavelength, λ (nm) for AM1,5G spectrum (ASTM G173-03) (Stem, 2007), (ASTM, 2005).

Aiming at evaluating the photo catalytic properties of the developed material, the photoluminescence spectrum were obtained as function of the wavelength. Figure 8(a) shows the room temperature photoluminescence (PL) emission of the samples 1G(700ºC), 1F (900ºC) and 1E(1000ºC) in which the vertical scale of the intensity was normalized using the silicon peak at 515nm for the three spectra. Based on this normalization, the PL emission of the samples 1G and 1F are significantly lower in area compared to sample 1E. In addition, figures 8b, 8c and 8d show the obtained spectrum for each studied case and peaks deconvolutions based on Gaussian distributions, respectively.

Basically, three characteristic band peaks are obtained: a) sample 1G: at approximately 2.2eV and 2eV; b) sample 1F: at approximately 2.2eV and 1.9eV and c) sample 1E: at approximately 2.2eV, 2.0eV and 1.9eV; which are close to one another and they are distant from the optical band gap reported on rutile (3.05eV) (Wang et. al., 2009) and on Ti_3O_5 (4.04eV) (Wouter et. al., 2007). On the other hand, Enache et al. (Enache et. al., 2004) report

that PL can reveal the nature of the defects involved in C-doped titanium oxides, showing
that the broad peak at ~ 2.0eV is correlated to the amount of disorder due to the increase in
the number of defects, oxygen vacancies or titanium interstitials (Enache et. al., 2004).
Meanwhile, the broadband at ~1.90eV is believed to be associated to the presence of ionic
point defects, or to excitons bound to these defects (Enache et. al, 2004) and the broadband
at ~2.2 eV is attributed to self-trapped excitons (Enache et. al., 2004).

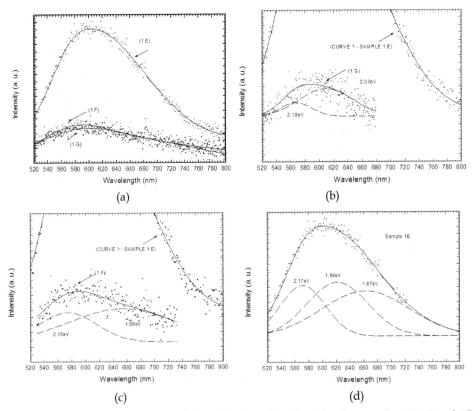

Fig. 8. PL measurements as wafunction of the wavelength: a) for the samples 1E, 1F and 1G;
and peak deconvoution for samples b) 1E; c) 1F and d) 1G.

Thus, analyzing the deconvolutions (figures 8b, 8c and 8d) it can be observed that in figure
8 (b), sample G has as dominant the band centered at 2.0eV (about 63.8%) and a minor band
centered at about 2.2eV, representing about 36.2% of total area. According to XRD spectra
presented at figure 3a, the sample G is practically amorphous presenting small peaks
associated to rutile TiO$_2$, thus it can be inferred that band peak at ~ 2.0eV to the number of
defects, oxygen vacancies or titanium interstitials in rutile TiO$_2$ (as discussed item) mainly
due to carbon doping and the band center at 2.2eV, attributed to some to self-trapped
excitons (Enache et. al., 2004).
However, as the hydrothermal temperature annealing increases to 900$_o$C (sample 1F), the
nanofibers started to be formed, and XRD peaks corresponding to Ti$_3$O$_5$ become dominant
and the band corresponding to ~ 2.0eV (tentatively associated to rutile TiO$_2$) practically

vanishes. In this sample, the band centered at 2.2 eV (some to self-trapped excitons) is about 35.6% of the total area, practically equal the one presented for sample 1G. Meanwhile, the start of nanofibers formation promoted the generation of a new band, compared to sample G spectrum, centered at about 1.9eV (about 64.4% of the total area) being believed to be associated to of ionic point defects, or to excitons bound to these defects (Enache et. al., 2004). These defects might be provenient from the vacancies produced by carbon doping; however, this fact needs further investigation afterwards.

As the temperature goes to 1000°C the nanofibers are formed, and two high intensity peaks were identified in XRD spectrum, rutile TiO_2 and Ti_3O_5. Analyzing the deconvolution of PL spectrum of sample 1E, three bands could be identified, being centered at 2.2eV, 2.0eV and 1.9eV, representing about 21.4%, 34.5% and 44.1% of total area, respectively. The band centered at 2.2eV, initially associated to some to self-trapped excitons in samples 1G and 1F, had its area increased significantly, about three times than for the other cases. On the other hand, the band centered at 2.0eV, that was vanished in the beginning of the nanofibers formation (sample F), became intense with the increase in the amount of disorder due to the random distribution of nano- and microfibers, which can promote increasing of the density of defects, oxygen vacancies and titanium interstitials on carbon doped rutile TiO_2 and λ-Ti_3O_5 (Monoclinic, C2/m E, a = 9.757Å, b = 3.802Å, c = 9.452Å). However, it should be pointed out that this disorder is not correlated to the cristallinity of the film as demonstrated by XRD spectra. The mentioned disorder also promoted an increase in the broadband centered at ~1.90eV, as mentioned previously, believed to be associated to the presence of ionic point defects, or to excitons bound to these defects.

In order to compare the peak areas of the studied PL spectrum, obtained based on the peak deconvolution presented at figure 8, the normalized areas for each samples are presented as functions of the characteristical band, 1.90eV, 2.00 eV and 2.20eV in figure 9. Analyzing this figure, it can be easily identified the growth of the three bands for sample 1E for the three characteristic bands.

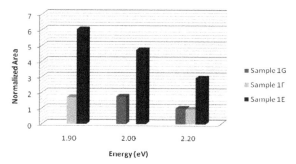

Fig. 9. Normalized areas for each studied sample as function of the normalized areas resulting from the peak deconvolution presented at figure 8.

6. Inferring about the reaction mechanisms to form the nanofibers

In order to infer a possible reaction mechanism model for producing nanofibers for the technique, the system can be divided into three groups: a) rutile carbon doped reactions; b) carbothermal reaction; c) TiO_2 behavior under nitrogen atmosphere and d) TiO_2 behavior under water vapor (an oxygen atmosphere (Richards, 2002) and hydrogen atmosphere), as presented at Table 3. The required energy to form reactions or the Gibbs potentials is presented. Thus, the reactions that present a negative free energy are expected to occur spontaneously and the positive ones require adsorption of energy. Therefore, only the most probable or spontaneously reactions will be considered (the most negative Gibbs potential). According to Valentini et. al.(Valentini et. al., 2005), the reactions that might occur in rutile titania and the correspondent required energy are represented for the equations (1)-(3) in table 3. Equation (1) stands for pure rutile material and (2)-(3) for carbon-doped titanium, occupying interstitial and substitutional positions, respectively. The energy required to interstitial reaction to occur is associated to the sum of the required energies to break the C-O and Ti-O bonds, while the required energy to substitutional reactions to occur is most probably associated to the tendency of carbon atoms trap electrons from the oxygen vacancy.

However, when high annealing temperatures are considered, carbothermal reactions (Sen et. al., 2011) and the interaction between TiO$_2$/Si (Richards, 2002) also become important. In particular, in carbothermal reaction, titanium dioxide is believed to react with carbon in order to obtain Ti$_3$O$_5$ and CO (equation (4)) in table 3. On the other hand, as the adopted atmosphere for the annealings in the proposed technique of this chapter consists of wet Nitrogen (0.8% water vapor), the dominant reactions between the interface TiO$_2$/Si are the ones obtained for nitrogen atmosphere, equation (5), so that Ti$_3$O$_5$ and SiO$_2$ are products of the expected reactions, as for the carbothermal reaction.

Focusing on the small percentage of water vapor present at the annealing atmosphere, it can be inferred that the water vapor dissociates at oxygen and hydrogen. Thus, all most probable reactions on TiO$_2$/Si interface point out to form Ti$_3$O$_5$, corroborating the XRD spectrum, AFM and FTIR spectra presented in the figures 3, 4 and 5.

Another point to be considered is that the hydrogen present in the atmosphere are expected to promote a kind of a redox reaction (Iowaki, 1983), when the hydrogen penetrates the film, forming oxygen vacancies and electrons are trapped as shown at equation (8). On the other hand, hydrogen is also adsorbed on neighboring oxygen, forming a hydroxyl group and Ti$_3$$^+$ that is not removed from surface, as shown in equation (9).

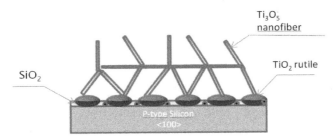

Fig. 10. Inferred scheme about nanofibers formation.

In order to understand how nano- and microfibers are formed on the silicon substrate, a schematic mechanism is proposed and illustrated in Figure 10. Initially, the amorphous TiO$_2$ would change from the amorphous to rutile phase, the carbon presence is believed to favor rutile phase (Binh, 2011). Rutile subsequently reacts with Si to form Ti$_3$O$_5$ (equations (4) and (5)). When the heating budget and carbon concentration are larger enough, Ti$_3$O$_5$ nano- and microfibers are formed to reach minimum free energy. The reactions presented in table 3 compete against each other to reach the minimum value for Gibbs potential, G$_o$. The equilibrium structure based on the competition of strain energy and surface energy would be either nanowires, or nanofibers.

		E(eV)
Carbon doping rutile titania (Valentini et. Al., 2005)	$TiO_2(s) \rightarrow TiO_{2-x} + xV_o + x/2\ O_{2(g)}$ (pure material) $\quad$ (1)	4.4
	$TiO_2C_x(s) \rightarrow TiO_{2-x}C_x(s) + xV_o + x/2\ O_{2(g)}$ (interstitial) $\quad$ (2)	2.4
	$TiO_{2-x}C_x(s) \rightarrow TiO_{2-2x}C_x(s) + xV_o + x/2\ O_{2(g)}$ (substitutional) (3)	2.4
Carbothermal Reactions (Sen et. al., 2011)		G_o (kJ)
	$3TiO_2 + C \rightarrow Ti_3O_5 + CO$ $\quad$ (4)	-156.9
TiO$_2$/ Si under nitrogen atmosphere (Richards, 2002)	Possible TiO$_2$/Si Reactions	G_o (kJ)
	$6TiO_2 + Si \rightarrow 2Ti_3O_5$ $\quad$ (5)	-156.9
TiO$_2$/ Si under oxygen atmosphere (Richards, 2002)	Possible TiO$_2$/Si Reactions	G_o (kJ)
	$3TiO_2 + O_2 \rightarrow Ti_3O_5 + 3/2O_2$ $\quad$ (6)	398.1
	$TiO_2 + O_2 + Si \rightarrow TiO_2 + SiO_2$ $\quad$ (7)	-961.5
TiO$_2$ under hydrogen Atmosphere (Iowaki, 1983),	$Ti^{4+}O^{2-}Ti^{4+}O^{2-} \xrightarrow{+H_2(g)} \overset{H^-}{\underset{|}{Ti^{3+}O}}\ \overset{H^-}{\underset{|}{Ti^{3+}O}} \xrightarrow{-H_2O(g)} Ti^{3+}O^{2-}Ti^{4+}\square^-$ $\quad$ (8) $Ti^{3+}O^{2-}Ti^{4+}\square^- \underset{}{\overset{\frac{1}{2}H_2(g)}{\rightleftharpoons}} \overset{H^-}{\underset{|}{Ti^{3+}O}}\ Ti^{3+}\square^-.$ $\quad$ (9) (I)	

Table 3. Possible involved reactions for the obtaining of the nanofiber.

7. Conclusions

In this chapter a review about the methods for producing nanofibers were presented and a new process for achieving the λ-Ti$_3$O$_5$ nano- and microfibers from C-doped TiO$_2$ thin films was also presented. Initially, the condition to form the nanofibers needs carbon (3%wt) as precursor seed followed by thermal treatment in nitrogen+water vapor (0.8%wt) environment at 1000°C during 120min. In this case, microscale meshes of fibers randomly distributed were observed with length ranging from 0.1 to 1.1μm and average width of (170±20)nm. The nano- and microfibers formation was characterized at different temperatures, including the initial stages at 900°C. From Raman and FTIR Spectroscopy techniques, it was shown that rutile is an inner layer located at the interface mesh/Si that is away from the surface so that the meshes of nano- and microfibers are predominantly composed of λ-Ti$_3$O$_5$ grown from the reaction with Si to form Ti$_3$O$_5$ and SiO$_2$. On the other hand, it was noteworthy that the microscale mesh of nano- and microfibers showed increased photoluminescence compared to amorphous TiO$_2$ with a broad peak in the visible if compared with samples built up of carbon-doped rutile titanium dioxide and samples with the nanofibers at the initial stage.

8. Acknowledgements

The authors would like to thank LSI-EPUSP staff and Nelson Ordonez for the support with the E-beam equipment; Msc. André Borges Braz from LCT-EPUSP, Msc Juliana Livi Antonossi from LCT-EPUSP, Dr. Marcel D.L. Barbosa from LamFi-IFUSP and Msc. Davinson Mariano da Silva from FATEC-SP for EDS, XRD, RBS and Absorbance measurements, respectively. The FTIR measurements were performed at CCS-UNICAMP by Msc. Jair Fernandes de Souza. The authors would also like to thank Dr. Evaldo José Corat from INPE for Raman spectra discussions and, Msc Michele Lemos de Souza and Dr. Dalva Lúcia Araújo de Faria from LEM-IQUSP by the PL measurements. Nair Stem was supported by a CNPq post-doctoral scholarship under process number 151745/2008 0.

9. References

Sauvage, F; Di Fonzo, F.; Bassi, A. L.; Casari, C. S.; Russo, V.; Divitini, G.; Ducati, C.; Bottani, C. E.; Comte, P.; and Graetzel, M. (2010), Hierarchical TiO₂ Photoanode for Dye-Sensitized Solar Cells, *Nanoletters*, Vol. 10, pp 2562-2567.

Lin, H.; Wang, Wen-li; Liu, Yi-zhu and Li, Jian-bao (2009). Review article: New Trends for Solar Cell Development and Recent Progress of Dye Sensitized Solar Cells, Frontiers of Materials Science in China, Vol. 3, N° 4, pp. 345-352.

Kim, H.; Bae, S. and Bae, D., (2010) Synthesis and Characterization of Ru Doped TiO₂ Nanoparticles by A Sol-Gel and A Hydrothermal Process, *Advances in Technology of Materials and Materials Processing*, Vol. 12, N° 9, pp 1-6.

Wenger, S. (2010) Strategies to optimizing dye-sensitized solar cells: organic sensitizers, tandem device structures, and numerical device modeling, Ph. D. Thesis, École Polytechnique Fédérale de Lausanne.

Kong, F.; Dai, S., and Wang, K. (2007); Review of recent progress in Dye-sensitized solar cells, *Advances in OptoElectronics*, Vol. 2007, pp. 1-13.

Hafez, H.; Lan, Z.; Li, Q.; Wu, J. (2010); Review: High efficiency dye-sentized solar cell based on novel TiO₂ nanorod/nanoparticle bilayer electrode. *Nanotechnology, Science and Applications*, Vol. 3, N° 1, pp. 45-51.

Pagliaro, M.; Palmisano, G. and Ciriminna, R. (2009); Working principles of dye-sensitized solar cells and future applications *Photovoltaics International journal*, Vol. 2, pp. 47 – 50.

Fieggemeier, E. and Hagfeldt, A. (2004), Are dye sensitized nano-structured solar cells stable? An overview of device testing and component analyses, *International Journal of Photoenergy*, V. 06, pp. 127-140.

Wang, P.; Zakeeruddin, S. M.; Moser, J. E.; Nazeeruddin, M. K.; Sekiguchi, T. and Gratzel, M., (2003). A stable quasi-solid-state dye-sensitized solar cell with an amphilic ruthenium sensitizer and polymer gel electrolyte. *Nature Materials*, Vol. 2, pp. 402-407.

Ou, Hsin-Hung and Lo, Shang-Lien (2007) Review of titania nanotubes synthesized via the hydrothermal treatment: Fabrication, modification, and application. *Sep Purif Technol*, Vol. 58, pp 179–191.

Suzuki Y and Yoshikawa S. (2004). Synthesis and thermal analyses of TiO₂-derived nanotubes prepared by hydrothermal method. *Journal of Materials Research*, Vol.. 19, pp. 982-985.

Kim, I.; Gwak, J.; Han, S.; Singh, K. (2007), Synthesis of Pd or Pt/titanate nanotube and its application to catalytic type hydrogen gas sensor, *Sensors and Actuators B*, Vol. 128, pp. 320-325.

Varghese, O. K.; Gong, D.; Paulose, M.; Ong, K. G. Mand Grimes, C. A., (2003), Hydrogen sensing using titania nanotubes, *Sensors and Actuators B*, Vol. 93, pp. 338-344.

Chen, X. and Mao, S. S. (2007), Titanium Dioxide Nanomaterials: Synthesis, Properties, Modifications, and Applications,*Chem. Rev.*, Vol. 107, pp. 2891-2959.

Nuansing, W.; Ninmuang, S.;Jaremboon, W.; Maensiri, S.; Seraphin, S. (2006), Structural characterization and morphology of electrospun TiO_2 nanofibers, *Materials Science and Engineering B*, Vol. 131, N⁰ 1-3, pp. 147-155.

Park, S. J.; Chase, G. G.; Jerong, K. and Kim, H. Y. (2010), Mechanical Properties of titania nanofiber mats fabricated by electropsinning of sol-gel precursor, *J. Sol-gel Technol.*, Vol. 54, pp. 188-194.

D. Valentini, D.; Pacchioni, G.; Selloni, A. (2005), Theory of carbon doping of titanium dioxide, *Chem. Matter.*, Vol 17, pp. 6656-6665.

Kukovecz, A.; Hodos, M.; Horváth, E. ; Radnóczi, G.; Kónya, Z.; and Kiricsi, I. (2005). Oriented Crystal Growth Model Explains the Formation of Titania Nanotubes, *The Journal of Physical Chemistry B Letters*, Vol. 109, N⁰ 38, pp. 17781–17783.

Puma, G. L.; Bono, A.; Krishnaiah, D.; Collin, J. G. (2008). Preparation of titanium dioxide photocatalyst loaded onto activated carbon support using chemical vapor deposition: A review paper, *Journal of Hazardous Materials*, Vol. 157, N⁰ 2-3, pp. 209-219.

Yamamoto, S.; Bluhm, H.; Andersson, K.; Ketteler, G.; Ogasawara, H.; Salmeron, M. and Nilsson, A. (2008); In situ x-ray photoelectron spectroscopy studies of water on metals and oxides at ambient conditions; *Journal of Physics: Condensed Matter*, Vol. 20, No 18, pp. 184025.

Khan, MA; Han, DH; Yang, OB (2009). Enhanced photoresponse towards visible light in Ru doped titania nanotube. *Applied Surface Science*. Vol. 255, pp. 3687-3690.

Zhang, H.; Li, X. and Chen, G. (2010); chapter Fabrication of Photoelectrode Materials, Electrochemistry for the Enviroment, Springer New York, ISBN 978-0-387-68318-8, pp. 475-513.

Reyes-Garcia, A.; Sun, Y.; Reyes-GLL , K. R. and Raftery, D. (2009), Solid-state NMR and EPR analysis of carbon doped titanium dioxide photocatalystis $(TiO_{2-x}C_x)$, *Solid State Nuclear Magnetic Resonance in Catalysis*, Vol. 35, N⁰ 2, pp. 74-81.

Konstantinova, E. A.; Kokorin, A. I.; Saktivel, S.; Kisch, H. and Lips, K. (2007); Carbon-doped titanium dioxide: visible light photocatalysis and EPR investigation; *Transformation and storage of solar Energy*, Vol. 61, N⁰ 12, pp. 810-814.

Suzuki Y and Yoshikawa S. (2004). Synthesis and thermal analyses of TiO_2-derived nanotubes prepared by hydrothermal method. *Journal of Materials Research*, Vol.. 19, pp. 982-985.

Chen, Huei-Siou; Su, C.; Chen, Ji-Lian; Yang, Tsai-Yin; Hsu, Nai-Mu; and Li, Wen-Ren (2011), Preparation and Characterization of Pure Rutile TiO_2 Nanoparticles for Photocatalytic Study and Thin Films for Dye-sensitized Solar cells, *Journal of nanomaterials*, Vol. 2011, N⁰ 1, pp. 1 – 8.

Wu, Z.; Dong, F.; Zhao, W.; Wang, H.; Liu, Y. and Guan, B. (2009); The fabrication and characterization of novel carbon doped TiO_2 nanotubes, nanowires and nanorods with high visible light photocatalytic activity; *Nanotechnology*, Vol. 20, No 23, p. 235701.

Tryba, B. (2008), Increase of the photocalaytic activity of TiO_2 by carbon and iron modifications – review article, International Journal of Photoenergy, Vol. 2008, Article ID 721824, pp.1-15.

Dai, Z. R.; Gole, J. L.; Stout, J. D. and Wang, Z. L. (2002), Tin Oxide Nanowires, Nanoribbons, and Nanotubes; *J. Phys. Chem. B*, Vol. 106, Nº 6, pp 1274–1279.

Yin, Y.; Zhang, G. and Xia, Y. (2002), Synthesis and Characterization of MgO Nanowires through a Vapor-Phase Precursor Method, *Adv. Func. Mater.*, Vol. 12, pp.293-298.

Pan, Z. W., Dai, Z.R., and Wang, Z.L. (2002), Lead oxide nanobelts and phase transformation induced by electron beam irradiation, *Appl. Phys. Lett.*, Vol. 80, pp. 309-311.

Wu, J. M.; Shin, H. C. and Wu, W. T. (2005), Growth of TiO₂ nanorods by two-step thermal evaporation, *J. Vac. Sci. Technol. B*, Vol. 23, pp. 2122 – 2126.

Xiang, B., Wang, Q. X.; Zhang, X. Z.; Liu, L. Q.; Xu, J. and Yu., D. P. (2005), Synthesis and field emission properties of TiSi₂ nanowires *Appl. Phys. Lett. 86*, pp. 243103/1-243103/3 issue 24

Bennett, P. A.; Ashcroft, B.; He,Z. and Tromp, R. M. (2002), Growth dynamics of titanium silicide nanowires observed with low-energy electron microscopy, *J. Vac. Sci. Technol. B*, Vol. 20, pp. 2500-2504.

Wang, C. C.; Yu, C.-Y.; Kei, C. C.; Lee, C. T. and Perng, T. P. (2009); Formation of TiO₂ nanowires on silicon directly from nanoparticles, *Nanotechnology*, Vol. 20, pp. 1-6.

Sen, W.; Xu, B. q.; Yang, B.; Sun, H.-y.; Song, J.-x.; Wan, H.-l. and Dai, Y.-n. (2011), Preparation of TiC powders by carbothermal reduction method in vacuum, *Transactions of Nonferrous Metals Society of China*, Vol. 21, Nº 1, pp 185-190.

Swift, G. A. and Koc, R. (1999), Formation studies of TiC from carbon coated TiO₂, *Journal of Materials Science*, Vol. 34, pp. 3083 – 3093.

Bavykin, D. V.; Friedrich, J. M. and Walsh, F. C. (2006); Protonated titanates and TiO₂ nanostructured nanomaterials: synthesis, properties and applications, Vol. 18, Nº 21, pp. 2807-2824.

Bavyjkin, D. V. and Walsh, F. C. (2009); Elongated titanate nanostructures and their applications; *European Journal of Inorganic Chemistry*, Vol. 2009, Nº 8, pp. 977-997.

Santos Filho, S. G.; Hasenak, C. M.; Sanay, L. C. and Mertens, P. (1995); A less critical cleaning procedure for silicon wafer using diluted HF dip and boiling in isopropyl alcohol as final step; *J. Electrochemical Soc.*, Vol. 142, Nº 3, pp. 902 – 907.

Stem, N.; Santos Filho, S. G. (2010); Carbon-Modified Titanium Dioxide Deposited by E-Beam Aiming Hydrogen Sensing. In: *25th Symposium on Microeletronics Technology and Devices, 2010, São Paulo. 25th Symposium on Microeletronics Technology and Devices (ECS Transactions).* New Jersey, USA : Pennington, Vol. 31, pp. 433-439.

Stem, N.; Chinaglia E. F.; Santos Filho, S. G. (2011); Ti₃O₅ nano- and microfibers prepared via annealing of C-doped TiO₂ thin films aiming at solar cell and photocatalysis applications. In: *26th Symposium on Microelectronics Technology and Devices (ECS Transactions). New Jersey, USA: Pennington, 2011.*

Stem, N.; Chinaglia E. F.; Santos Filho, S. G. (2011). Microscale meshes of Ti₃O₅ nano- and microfibers prepared via annealing of C-doped TiO₂ thin films. *Materials Science & Engineering. B, Solid-State Materials for Advanced Technology*, DOI: 10.1016/j.mseb.2011.06.013, available on line.

Kern, W. (1990); The evolution of silicon wafer technology, *The Journal of Eletrochemical Society*, Vol. 137, Nº 6, pp.1887-1892.

Reinhardt, K. A. and Wern. K (2008); Handbook of Silicon Wafer Cleaning Technology, Materials Science and Process Technology Series. 2nd Edition, Willian Andrew.

Shannon, R. D. (1964), Phase transformation studies in TiO_2 supporting different defect mechanisms in vacuum-reduced and hydrogen-reduced rutile. *J Appl Phys*, Vol. 35, pp. 3414-3416.

Liu, H.; Zhang, Y.; Li, R.; Cai, M.; Sun, X. (2010), A facile route to synthesize titanium oxide nanowires via water-assisted chemical vapor deposition *J. Nanopart. Res.* (2010) DOI 10.1007/s11051-010-0041-0.

Grey, I. E.; Li, C.; Madsen, I. C. (1994) Phase Equilibria and Structural Studies on the Solid Solution $MgTi_2O_5$-Ti_3O_5 *Journal of Solid State Chemistry*, Vol. 113, N° 1, pp. 62-73.

Yakovlev, V. V.; Scarel, G.; Aita, C. R. and Mochizuki, S. (2000), Short-range order in ultrathinfilm titanium dioxide studied by Raman spectroscopy, *Appl. Phys. Lett.*, Vol. 76, N° 9, pp. 1107-1109.

Erkov, V. G.; Devyatova, S. F.; Molodstova, E. L.; Malsteva, T. V.; and Yanovskii, U. A. (2000). Si-TiO_2 interface evolution at prolonged annealing in low vacuum or N_2O ambient Applied Surface Science, Vol. 166, N° 1, pp. 51-56.

Climent-font, A; Watjen, U. and Bax, H. (2002); Quantitative RBS analysis using RUMP. On the accuracy of the He stopping in Si Nuclear Instruments and Methods in Physics Research, Section B, Vol. 71, pp. 81-86.

G. Richiardi, Al Damin, S. Bordiga, C. Lamberti, G. Spanó, F, Rivetti and A. Zecchina (2001), *J. Am. Chem. Soc.*, Vol. 123, 11409-11419.

Wuderlich, W.; Foitzik, A. H.; and Heuer, A. H. (1993); On the Quantitative EDS Analysis of Low Carbon Concentrations in Analytical TEM; *Ultramicroscopy*, Vol. 49, pp. 220 - 224.

Koch, Carl C. (2002). Nanostructured Materials - Processing, Properties and Potential Applications.. William Andrew Publishing/Noyes.New York, USA, ISBN 0-815514514.

ASTM G-173-03, Standard Table for reference solar spectrum irradiance direct normal and hemisphericalon 37° tilted surface, avaible at www.astm.org, accessed January 2005.

Stem, N. (2007), Células solares de silício de alto rendimento: otimizações teóricas e implementações experimentais utilizando processos de baixo custo, PhD Thesis, Escola Politécnica de Engenharia Elétrica da Universidade de São Paulo, http://www.teses.usp.br/teses/disponiveis/3/3140/tde-02042008-113959/fr.php.

Wang, X.; Meng, S.; Zhang, X.; Wang, H.; Zhong, W. and Du, Q. (2007). Multi-type carbon doping of TiO_2 photocatalyst, Chem. Phys. Lett., Vol. 444, pp. 292-296.

Sakthivel, S. and Kisch, H. (2003), Daylight Photocatalysis by Carbon-Modified Titanium Dioxide. *Angewandte Chemie International Edition*, Vol. 42, N° 40, pp. 4908-4911.

Wouters, Y.; Galerie, A. and Petit, J-P. (2007), Photoelectrochemical study of oxides thermally grown on titanium in oxygen or water vapor atmospheres, *Journal of the Electrochemical Society*, Vol. 154, N° 10, pp. C587-C592.

Enache, C. S.; Schoonman, J. S. and De Krol, R. V. (2004), The Photoresponse of Iron- and Carbon-Doped TiO_2 (Anatase) Photoelectrodes, *J. Electroceramics*, Vol. 13, Numbers 1-3, pp. 177-182.

Richards, B. S. (2002), Novel Uses of Titanium Dioxide for Silicon Solar Cells, *Ph.D. Thesis*, University of New South Wales.

Iowaki, T. (1983). Studies of the surface of titanium dioxide. Part 5.—Thermal desorption of hydrogen, *J. Chem. Soc., Faraday Trans. 1*, Vol. 79, pp. 137-146.

Dang, Binh H.Q.; Rahman, Mahfujur; MacElroy, J. M. Don (2011). Conversion of amorphous TiO_2 coatings into their crystallineform using a novel microwave plasma treatment. *Surface and Coatings Technology*, Article in Press

8

Dye Sensitized Solar Cells - Working Principles, Challenges and Opportunities

Khalil Ebrahim Jasim
Department of Physics, University of Bahrain
Kingdom of Bahrain

1. Introduction

Even before the industrial revolutions human life quality is greatly affected by the availability of energy. The escalated and savage consumption of conventional sources of energy are leading to forecasted energy and environmental crises. Renewable energy sources such as solar energy are considered as a feasible alternative because *"More energy from sunlight strikes Earth in 1 hour than all of the energy consumed by humans in an entire year."*(Lewis, 2007). Facilitating means to harvest a fraction of the solar energy reaching the Earth may solve many problems associated with both the energy and global environment (Nansen, 1995). Therefore, intensive research activities have resulted in attention-grabbing to the different classes of organic and inorganic based solar cells. A major study by IntertechPira stated that "The global Photovoltaic (PV) market, after experiencing a slow period, is expected to double within the next five years, reaching US$ 48 billion. Wafer-based silicon will continue as the dominant technology, but amorphous thin-film and Cadmium Telluride (CdTe) technologies will gain ground, and are expected to account for a combined 22% of the market by 2014" (www.intertechpira.com).

A solar cell is a photonic device that converts photons with specific wavelengths to electricity. After Alexandre Edmond Becquerel (French physicist) discovered the photoelectrochemical (photovoltaic) effect in 1839 (Becquerel, 1839) while he was investigating the effect of light on metal electrodes immersed in electrolyte, research in this area continued and technology developed to produce many types and structures of the materials presently used in photovoltaic (PV) technology. First and second generations photovoltaic cells are mainly constructed from semiconductors including crystalline silicon, III-V compounds, cadmium telluride, and copper indium selenide/sulfide (Hara & Arakawa, 2003; Hoffert, 1998; Zhao et al., 1999). Low cost solar cells have been the subject of intensive research work for the last three decades. Amorphous semiconductors were announced as one of the most promising materials for low cost energy production. However, dye sensitized solar cells DSSCs emerged as a new class of low cost energy conversion devices with simple manufacturing procedures. General comparison between semiconductor based solar cells and dye sensitized solar cells is presented in Table 1.

Incorporation of dye molecules in some wide bandgap semiconductor electrodes was a key factor in developing photoelecrochemical solar cells. Michael Gratzel and coworkers at the Ecole Polytechnique Federale de Lausanne (Gratzel, 2003; Nazerruddin et al., 1993; O' Regan & Gratzel, 1991) succeeded for the first time to produce what is known as "Gratzel Cell" or the dye sensitized solar cell (DSSC) to imitate photosynthesis -the natural processes plants convert sunlight into energy- by sensitizing a nanocrystalline TiO_2 film using novel Ru bipyridl complex. In dye sensitized solar cell DSSC charge separation is accomplished by kinetic competition like in photosynthesis leading to photovoltaic action. It has been shown that DSSC are promising class of low cost and moderate efficiency solar cell (see Table 2 and Figure 1) based on organic materials (Gratzel, 2003; Hara & Arakawa, 2003).

	Semiconductor solar cells	DSSC
Transparency	Opaque	Transparent
Pro-Environment (Material & Process)	Normal	Great
Power Generation Cost	High	Low
Power Generation Efficiency	High	Normal
Color	Limited	Various

Table 1. Comparison between semiconductor based solar cell and the dye sensitized solar cell DSSC.

In fact, in semiconductor p-n junction solar cell charge separation is taken care by the junction built in electric field, while in dye sensitizes solar cell charge separation is by kinetic competition as in photosynthesis (Späth et al., 2003). The organic dye monolayer in the photoelectrochemical or dye sensitized solar cell replaces light absorbing pigments (chlorophylls), the wide bandgap nanostructured semiconductor layer replaces oxidized dihydro-nicotinamide-adenine-dinucleotide phosphate (NADPH), and carbon dioxide acts as the electron acceptor. Moreover, the electrolyte replaces the water while oxygen as the electron donor and oxidation product, respectively (Lagref. et al., 2008; Smestad & Gratzel, 1998). The overall cell efficiency of dye sensitized solar cell is found to be proportional to the electron injection efficiency in the wide bandgap nanostructured semiconductors. This finding has encouraged researchers over the past decade. ZnO_2 nanowires, for example, have been developed to replace both porous and TiO_2 nanoparticle based solar cells (Law et al., 2005). Also, metal complex and novel man made sensitizers have been proposed (Hasselmann & Meyer, 1999; Isalm et al., 2000; Yang et al., 2000). However, processing and synthesization of these sensitizers are complicated and costly processes (Amao & Komori 2004; Garcia et al., 2003; Hao et al., 2006; Kumara et al., 2006; Polo & Iha, 2006; Smestad, 1998; Yanagida et al., 2004). Development or extraction of photosensitizers with absorption range extended to the near IR is greatly desired. In our approach, the use of natural dye extracts, we found that our environment provides natural, non toxic and low cost dye sources with high absorbance level of UV, visible and near IR. Examples of such dye sources are Bahraini Henna (*Lawsonia inermis* L.) and Bahraini raspberries (Rubus spp.). In this work we provide further details about the first reported operation of Henna (*Lawsonia inermis* L.) as a natural dye sensitizer of TiO_2 nanostructured solar cell (Jasim & Hassan, 2009; Jasim et al. in press 2011). We have experienced the usefulness of commercialized dye sensitized solar cell kits such as the one provided by Dyesol™ to "illustrates how interdisciplinary science can be taught at lower division university and upper division high school levels for an understanding of renewable energy as well as basic science concepts." (Smestad, 1998; Smestad & Gratzel 1998) Furthermore, it aids proper training and awareness about the role of nanotechnology in modern civilization.

Classification	η [%]	Area[a] [cm^2]	V_{oc} [V]	J_{sc} [mA · cm^{-2}]	FF [–]	Test center[b] (and date)	Producer
Silicon							
Si (crystalline)	24.7±0.5	4.00 (da)	0.706	42.2	0.83	Sandia (3/99)	UNSW PERL
Si (multicrystalline)	19.8±0.5	1.09 (ap)	0.654	38.1	0.80	Sandia (2/93)	UNSW/Eurosolare
Si (thin-film transfer)	16.6±0.4	4.02 (ap)	0.645	32.8	0.78	FhG-ISE (7/01)	U. Stuttgart
III–V cells							
GaAs (crystalline)	25.1±0.8	3.91 (t)	1.022	28.2	0.87	NREL (3/90)	Kopin
GaAs (thin film)	23.3±0.7	4.00 (ap)	1.011	27.6	0.84	NREL (4/90)	Kopin
GaAs (multicrystalline)	18.2±0.5	4.01 (t)	0.994	23.0	0.80	NREL (11/95)	RTI
InP (crystalline)	21.9±0.7	4.02 (t)	0.878	29.3	0.85	NREL (4/90)	Spire
Polycrystalline thin film							
CuInGaSe$_2$ (CIGS)	18.4±0.5	1.04 (t)	0.669	35.7	0.77	NREL (2/01)	NREL
CdTe	16.5±0.5	1.13 (ap)	0.845	26.7	0.76	NREL (9/01)	NREL
Amorphous/microcrystalline Si							
Si (nanocrystalline)	10.1±0.2	1.20 (ap)	0.539	24.4	0.77	JQA (12/97)	Kaneka
Photoelectrochemical cells							
Nanocrystalline dye	11.0±0.5	0.25 (ap)	0.795	19.4	0.71	FhG-ISE (12/96)	EPFL. LPI
Nanocrystalline dye (submodule)	4.7±0.2	141.4 (ap)	0.795	11.3	0.59	FhG-ISE (2/98)	INAP
Multijunction cells							
GaInP/GaAs	30.3	4.0 (t)	2.488	14.22	0.86	JQA (4/96)	Japan Energy
GaInP/GaAs/Ge	28.7±1.4	29.93 (t)	2.571	12.95	0.86	NREL (9/99)	Spectrolab
GaAs/CIS (thin film)	25.8±1.3	4.00 (t)	–	–	–	NREL (11/89)	Kopin/Boing
a-Si/CIGS (thin film)	14.6±0.7	2.40 (ap)	–	–	–	NREL (6/88)	ARCO

Table 2. Confirmed terrestrial cell efficiencies measured under the global AM 1.5 spectrum (1000 W · m^{-2}) at 25 °C. [a] (ap)=aperture area; (t)=total area; (da)=designated irradiance area. [b] FhG-ISE=Fraunhofer-Institute for Solar Energy system; JQA = Japan Quality Assurance (From Green & Emery, 2002).

In this chapter, we overview some aspects of the historical background, present, and anticipated future of dye sensitized solar cells. Operation principle of the dye sensitized solar cell is explained. Some schemes used in preparation and assembly of dye sensitized solar cell are presented with few recommendations that might lead to better performance and stability of the fabricated cell. The structural, optical, electrical, and photovoltaic performance stability of DSSC are discussed. The performance of nanocrystalline solar cell samples can be appreciably improved by optimizing the preparation technique, the class of the nanostructured materials, types of electrolyte, and high transparent conductive electrodes. Challenges associated with materials choice, nanostructured electrodes and device layers structure design are detailed. Recent trends in the development of

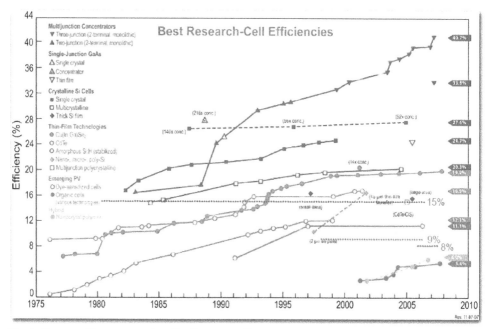

Fig. 1. Reported best research cell efficiencies (Source: National Renewable Laboratory, 2007). The Overall peak power production of dye sensitized solar cell represents a conversion efficiency of about 11%.

nano-crystalline materials for DSSCs technology are introduced. Manufacturability and different approaches suggested for commercialization of DSSC for various applications are outlined. We believe that the availability of efficient natural dye sensitizers, flexible and ink-printable conductive electrodes, and solid state electrolyte may enhance the development of a long term stable DSSCs and hence the feasibility of outdoor applications of both the dye sensitized solar cells and modules.

2. Structure of dye sensitized solar cell

The main parts of single junction dye sensitized solar cell are illustrated schematically in Figure 2. The cell is composed of four elements, namely, the transparent conducting and counter conducting electrodes, the nanostructured wide bandgap semiconducting layer, the dye molecules (sensitizer), and the electrolyte. The transparent conducting electrode and counter-electrode are coated with a thin conductive and transparent film such as fluorine-doped tin dioxide (SnO_2).

2.1 Transparent substrate for both the conducting electrode and counter electrode

Clear glass substrates are commonly used as substrate because of their relative low cost, availability and high optical transparency in the visible and near infrared regions of the electromagnetic spectrum. Conductive coating (film) in the form of thin transparent conductive oxide (TCO) is deposited on one side of the substrate. The conductive film ensures a very low electric resistance per square. Typical value of such resistance is 10-20 Ω

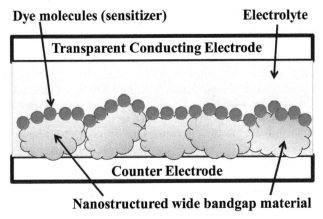

Fig. 2. Schematic of the structure of the dye sensitized solar cell.

per square at room temperature. The nanostructured wide bandgap oxide semiconductor (electron acceptor) is applied, printed or grown on the conductive side. Before assembling the cell the counter electrode must be coated with a catalyzing layer such as graphite layer to facilitates electron donation mechanism to the electrolyte (electron donor) as well be discussed later.

One must bear in mind that the transparency levels of the transparent conducting electrode after being coated with the conductive film is not 100% over the entire visible and near infrared (NIR) part of the solar spectrum. In fact, the deposition of nanostructured material reduces transparency of the electrode. Figure 3 shows a typical transmittance measurement (using dual beam spectrophotometer) of conductive glass electrode before and after being coated with nanostructured TiO$_2$ layer.

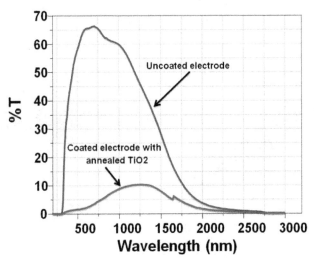

Fig. 3. Transmittance of conductive glass electrode before and after being coated with nanostructured TiO$_2$ layer.

2.2 Nanostructured photoelectrode

In the old generations of photoelectrochemeical solar cells (PSC) photoelectrodes were made from bulky semiconductor materials such as Si, GaAs or CdS. However, these kinds of photoelectrodes when exposed to light they undergo photocorrosion that results in poor stability of the photoelctrochemical cell. The use of sensitized wide bandgap semiconductors such as TiO_2, or ZnO_2 resulted in high chemical stability of the cell due to their resistance to photocorrosion. The problem with bulky single or poly-crystalline wide bandgap is the low light to current conversion efficiency mainly due to inadequate adsorption of sensitizer because of limited surface area of the electrode. One approach to enhance light-harvesting efficiency (LHE) and hence the light to current conversion efficiency is to increase surface area (the roughness factor) of the sensitized photoelectrode.

Due to the remarkable changes in mechanical, electrical, magnetic, optical and chemical properties of nanostructured materials compared to its phase in bulk structures, it received considerable attention (Gleiter, 1989). Moreover, because the area occupied by one dye molecule is much larger than its optical cross section for light capture, the absorption of light by a monolayer of dye is insubstantial. It has been confirmed that high photovoltaic efficiency cannot be achieved with the use of a flat layer of semiconductor or wide bandgap semiconductor oxide surface but rather by use of nanostructured layer of very high roughness factor (surface area). Therefore, Gratzel and his coworkers replaced the bulky layer of titanium dioxide (TiO_2) with nonoporous TiO_2 layer as a photoelectrode. Also, they have developed efficient photosensitizers (new Ru complex, see for example Figure 16) that are capable of absorbing wide range of visible and near infrared portion of the solar spectrum and achieved remarkable photovoltaic cell performance (Nazerruddin et al., 1993; O' Regan & Gratzel, 1991; Smestad & Gratzel, 1998). Nanoporusity of the TiO_2 paste (or colloidal solution) is achievable by sintering (annealing) of the deposited TiO_2 layer at approximately 450 °C in a well ventilated zone for about 15 minutes (see Figure 4). The high porosity (>50%) of the nanostructured TiO_2 layer allows facile diffusion of redox mediators within the layer to react with surface-bound sensitizers. Lindström et al. reported "A method for manufacturing a nanostructured porous layer of a semiconductor material at room temperature. The porous layer is pressed on a conducting glass or plastic substrate for use in a dye-sensitized nanocrystalline solar cell." (Lindström et al., 2001)

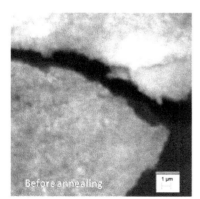 

Fig. 4. Scanning electron microscope (SEM) images for TiO_2 photoelectrode before and after annealing it at about 450°C for 15 minutes.

Because it is not expensive, none toxic and having good chemical stability in solution while irradiated, Titanium dioxide has attracted great attention in many fields other than nanostructured photovoltaics such as photocatalysts, environmental purification, electronic devices, gas sensors, and photoelectrodes (Karami, 2010). The preparation procedures of TiO_2 film is quite simple since it is requires no vacuum facilities. Nanostructured TiO_2 layers are prepared following the procedure detailed in (Hara & Arakawa, 2003; Nazerruddin et al., 1993; O' Regan & Gratzel, 1991; Smestad, 1998) "A suspension of TiO_2 is prepared by adding 9 ml of nitric acid solution of PH 3-4 (1 ml increment) to 6 g of colloidal P25 TiO_2 powder in mortar and pestle. While grinding, 8 ml of distilled water (in 1 ml increment) is added to get a white- free flow- paste. Finally, a drop of transparent surfactant is added in 1 ml of distilled water to ensure coating uniformity and adhesion to the transparent conducting glass electrode. The ratio of the nitric acid solution to the colloidal P25 TiO_2 powder is a critical factor for the cell performance. If the ratio exceeds a certain threshold value the resulting film becomes too thick and has a tendency to peel off. On the other hand, a low ratio reduces appreciably the efficiency of light absorption" (Jasim & Hassan, 2009). Our group adopted the Doctor blade method to deposit TiO_2 suspension uniformly on a cleaned (rinsed with ethanol) electrode plate. The TiO_2 layer must be allowed to dry for few minutes and then annealed at approximately 450°C (in a well ventilated zone) for about 15 minutes to form a nanoporous, large surface area TiO_2 layer. The nanostructured film must be allowed to cool down slowly to room temperature. This is a necessary condition to remove thermal stresses and avoid cracking of the glass or peeling off the TiO_2 film.

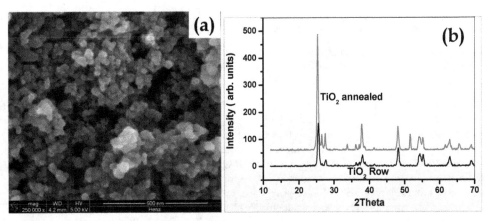

Fig. 5. (a) Scanning electron microscope (SEM) images and (b) XRD for TiO_2 photoelectrod before and after being annealed.

Scanning electron microscopy SEM (see Figure 5-a) or X-ray diffraction measurements (XRD) (see Figure 5-b) is usually used to confirm the formation of nanostructured TiO_2 layer. Analysis of the XRD data (shown in Figure 5-b) confirmers the formation of nanocrystalline TiO_2 particles of sizes less than 50 nm (Jasim & Hassan, 2009). The nanoporous structure of the TiO_2 layer suggests that the roughness factor of 1000 is achievable. In other words, a 1-cm² coated area of the conductive transparent electrode with nanostructured TiO_2 layer actually possessing a surface area of 1000 cm² (Hara & Arakawa, 2003). The formation of nanostructured TiO_2 layer is greatly affected by TiO_2 suspension preparation procedures as well as by the annealing temperature. We found that a sintered TiO$_2$ film at temperatures lower than the recommended 450°C resulted in cells that generate unnoticeable electric current even in the μA level. Moreover, nanostructured TiO$_2$ layer degradation in this case is fast and cracks form after a short period of time when the cell is exposed to direct sunlight. Recently Zhu et al. investigated the effects of annealing temperature on the charge-collection and light-harvesting properties of TiO$_2$ nanotube-based dye-sensitized solar cells (see Figure 6) and the reported "DSSCs containing titanium oxide nanotube (NT) arrays films annealed at 400 °C exhibited the fastest transport and slowest recombination kinetics. The various structural changes were also found to affect the light-harvesting, charge-injection, and charge-collection properties of DSSCs, which, in turn, altered the photocurrent density, photovoltage, and solar energy conversion efficiency" (Zhu et al. 2010).

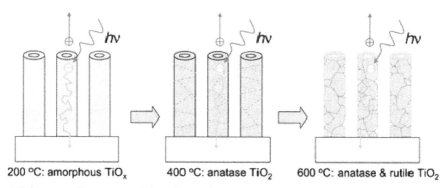

200 °C: amorphous TiO$_x$ 400 °C: anatase TiO$_2$ 600 °C: anatase & rutile TiO$_2$

Fig. 6. Schematic illustration of the effects of annealing temperature on the charge-collection and light-harvesting properties of TiO$_2$ nanotube-based dye-sensitized solar cells (From Zhu et al., 2010).

One of the important factors that affect the cell's efficiency is the thickness of the nanostructured TiO$_2$ layer which must be less than 20 μm to ensure that the diffusion length of the photoelectrons is greater than that of the nanocrystalline TiO$_2$ layer. TiO$_2$ is the most commonly used nanocrystalline semiconductor oxide electrode in the DSSC as an electron acceptor to support a molecular or quantum dot QD sensitizer is TiO$_2$ (Gratzel, 2003). Other wide bandgap semiconductor oxides is becoming common is the zinc oxide ZnO$_2$. ZnO$_2$ possesses a bandgap of 3.37 eV and a large excitation binding energy of 60 meV. Kim *et al.* reported that the nanorods array electrode showed stable photovoltaic properties and exhibited much higher energy conversion efficiency (Kim et al., 2006). Another example, Law and coworkers have grown by chemical bath deposition ZnO$_2$ nanowires 8-μm long with 100 nm diameters as photoelectrod (see Figure 7) the efficiency of a ZnO$_2$ nanowire photoelectrode DSSC is about 2.4%. This low efficiency level compared to that of nanostructured TiO$_2$ photoelectrode DSSC is probably due to inadequate surface area for sensitizer adsorption (Baxter et al., 2006; Boercker et al., 2009; Law et al., 2005). Other research groups suggested that the growth of longer, thinner, denser ZnO$_2$ nanowires is a practical approach to enhance cell efficiency (Guo et al., 2005). Investigations show that ZnO$_2$ nanorod size could be freely modified by controlling the solution conditions such as temperature, precursor concentration, reaction time, and adopting multi-step growth. Nanorod structured photoelectrode offers a great potential for improved electron transport.

It has been found that the short circuit current density and cell performance significantly increase as nanorods length increases because a higher amount of the adsorbed dye on longer nanorods, resulting in improving conversion efficiency (Kim et al. 2006). Because titanium dioxide is abundant, low cost, biocompatible and non-toxic (Gratzel & Hagfeldt, 2000), it is advantageous to be used in dye sensitized solar cells. Therefore, nanotube and nanowire-structured TiO_2 photoelectrode for dye-sensitized solar cells have been investigated (Mor et al., 2006; Pavasupree et al., 2005; Pavasupree et al., 2006; Shen et al., 2006; Suzuki et al., 2006). Moreover; SnO_2, or Nb_2O_5 employed not only to ensure large roughness factor (after nanostructuring the photoelectrode) but also to increase photgenerated electron diffusion length (Bergeron et al., 2005; Sun et al. 2006). Many studies suggest replacing nanoparticles film with an array of single crystalline nanowires (rods), nanoplants, or nanosheets in which the electron transport increases by several orders of magnitude (Kopidakis et al., 2003; Law et al., 2005; Noack et al., 2002; Tiwari & Snure, 2008; Xian et al., 2006). Incorporation of vertically aligned carbon nanotube counter electrode improved efficiency of TiO_2/anthocyanin dye-Sensitized solar cells as reported by Sayer et al. They attributed the improvement to "the large surface area created by the 3D structure of the arrays in comparison to the planar geometry of the graphite and Pt electrodes, as well as the excellent electrical properties of the CNTs." (Sayer et al., 2010).

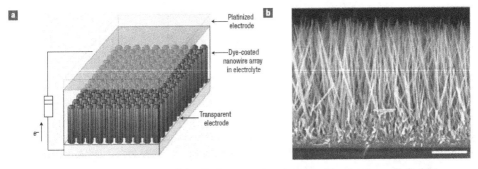

Fig. 7. (a) Schematic illustration of the ZnO nanowire dye sensitized solar cell, light is incident through the bottom electrode, and (b) scanning electron microscopy cross-section of a cleaved nanowire array. The wires are in direct contact with the transparent substrate, with no intervening particle layer. Scale bar, 5-μm (From Law et al., 2005).

2.3 Photosensitizer

Dye molecules of proper molecular structure are used to sensitized wide bandgap nanostructured photoelectrode. Upon absorption of photon, a dye molecule adsorbed to the surface of say nanostructured TiO_2 gets oxidized and the excited electron is injected into the nanostructured TiO_2. Among the first kind of promising sensitizers were Polypyridyl compounds of Ru(II) that have been investigated extensively. Many researches have focused on molecular engineering of ruthenium compounds. Nazeeruddin et al. have reported the "black dye" as promising charge transfer sensitizer in DSSC. Kelly, et.al studied other ruthenium complexes Ru(dcb)(bpy)$_2$ (Kelly, et al 1999), Farzad et al. explored the Ru(dcbH$_2$)(bpy)$_2$(PF$_6$)$_2$ and Os(dcbH$_2$)(bpy)$_2$-(PF6)$_2$ (Farzad et al., 1999), Qu et al. studied cis-Ru(bpy)$_2$(ina)$_2$(PF$_6$)$_2$ (Qu et al., 2000) , Shoute et al.

investigated the cis-Ru(dcbH$_2$)$_2$(NCS) (Shoute et al., 2003), and Kleverlaan et al. worked with OsIII-bpa-Ru (Kleverlaan et al 2000). Sensitizations of natural dye extracts such as shiso leaf pigments (Kumara et al., 2006), Black rice (Hao et al., 2006), Fruit of calafate (Polo and Iha, 2006), Rosella (Wongcharee et al., 2007), Natural anthocyanins (Fernando et al., 2008), Henna (*Lawsonia inermis* L.) (Jasim & Hassan, 2009; Jasim et al., in press 2011), and wormwood, bamboo leaves (En Mei Jin *et al.*, 2010) have been investigated and photovoltaic action of the tested cells reveals some opportunities. Calogero et al. suggested that "Finding appropriate additives for improving open circuit voltage V$_{OC}$ without causing dye degradation might result in a further enhancement of cell performance, making the practical application of such systems more suitable to economically viable solar energy devices for our society." (Calogero et al., 2009)

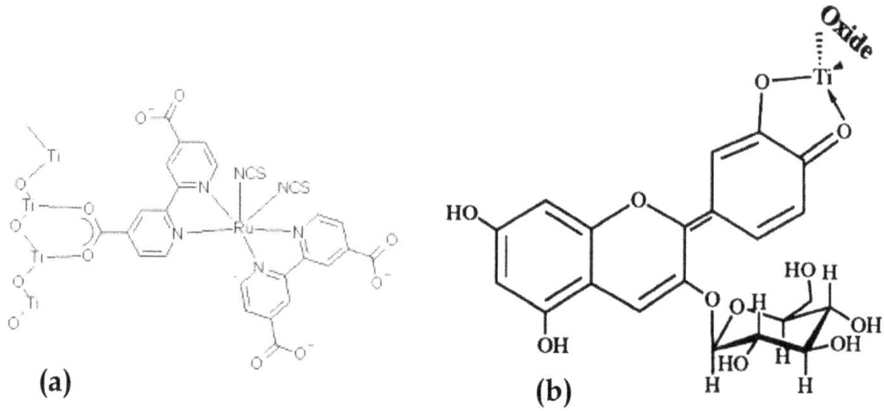

(a)　　　　　　　　(b)

Fig. 8. (a) Ruthenium based red or "N3" dye adsorbed onto a titanium dioxide surface (from Martinson et al., 2008), and (b) Proposed structure of the cyanin dye adsorbed to one of the titanium metal centers on the titanium dioxide surface (From Smestad, 1988).

Gratzel group developed many Ru complex photosensitizers (examples are shown in Figure 16). One famous example is the cis-Di(thiocyanato)bis(2,2'-bipyridyl)-4,4'-dicarboxylate) ruthenium(II), coded as N3 or N-719 dye it has been an outstanding solar light absorber and charge-transfer sensitizer. The red dye or N3 dye (structure is shown in Figure 8-a and Figure 16) is capable of absorbing photons of wavelength ranging from 400 nm to 900 nm (see Figure 16) because of metal to ligand charge transfer transition. Theoretical Study of new ruthenium-based dyes for dye sensitized solar cells by Monari et al., states "The UV/vis absorption spectra have been computed within the time-dependent density functional theory formalism. The obtained excitation energies are compared with the experimental results." (Monari et al., 2011) In fact, for dye molecule to be excellent sensitizer, it must possess several carbonyl (C=O) or hydroxyl (-OH) groups capable of chelating to the Ti$^{(IV)}$ sites on the TiO$_2$ surface as shown in Figure 8 (Tennakone et al., 1997). Extracted dye from California blackberries (Rubus ursinus) has been found to be an excellent fast-staining dye for sensitization, on the other hand, dyes extracted from strawberries lack such complexing capability and hence not suggested as natural dye sensitizer (Cherpy et al., 1997; Semistad & Gratzel, 1998; Semistad, 1988).

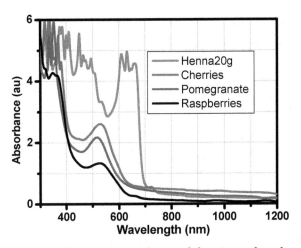

Fig. 9. Measured absorbance of some extracted natural dyes in methanol as solvent.

Commercialized dye sensitized solar cells and modules use ruthenium bipyridyl–based dyes (N3 dyes or N917) achieved conversion efficiencies above 10% (Nazerruddin, et al., 1993). However, these dyes and those chemically engineered are hard to put up and are expensive (Cherepy et al., 1997). Therefore, in attempt to develop green solar cells; our group at the University of Bahrain used Soxhlet Extractor in the extraction of natural dye solutions from abundant natural dye sources such as Bahraini Henna (*Lawsonia inermis* L.), Yemeni Henna, pomegranate, raspberries, and cherries after being dried(Jasim, submitted for publication 2011). We used methanol as solvent in each extraction process. The absorbance of the extracted dye solution has been measured using dual beam spectrophotometer (see Figure 9). Different concentrations of Henna (*Lawsonia inermis* L.) extracts have been prepared from the original extract. The light harvesting efficiency (LHE) for each concentration has been calculated from the absorbance (see Figure 10). The light harvesting efficiency is given as:

$$LHE(\lambda) = \left(1 - 10^{-A(\lambda)}\right) \times 100 \tag{1}$$

where A (λ) is the absorbance of the sample at specific wavelength.

The absorbance and hence the LHE increases with concentration of dye extract. Also, as shown in Figure 10, as Henna extract concentration increases the absorbance increases and covers broader range of wavelengths.

Since not all photons scattered by or transmitted through the nanocrystalline TiO₂ layer get absorbed by a monolayer of the adsorbed dyes molecules, the incorporation of energy relay dyes might help enhancing the light harvesting efficiency. A remarkable enhancement in absorption spectral bandwidth and 26% increase in power conversion efficiency have been accomplished with some sensitizers after energy relay dyes have been added (Harding et al., 2009). Metal free organic sensitizers such as metal free iodine reported by Horiuchi et al. demonstrated remarkable high efficiency "The solar energy to current conversion efficiencies with the new indoline dye was 6.51%. Under the same conditions, the N3 dye was 7.89%" (Horiuchi et al., 2004). Semiconductor quantum dots QDs are nanostructured crystalline semiconductors where quantum confinement effect due to their size results in

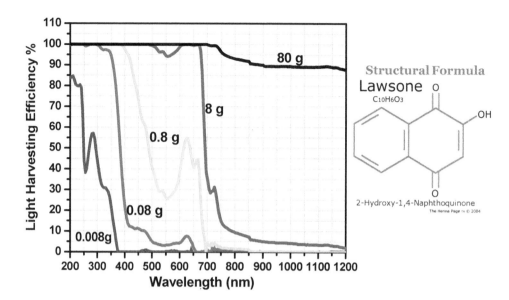

Fig. 10. Light harvesting efficiency of Henna extract at different concentrations. Data are given in grams of Henna powder per 100 ml of methanol as solvent. Also, shown the structural formula of Lawsone molecule that is responsible for the characteristic color of Henna (From www.hennapage.com).

remarkable optical linear and nonlinear behaviors. Excitonic absorption edge of quantum dots is size dependent as shown in Figure 11 for lead sulfide PbS quantum dots suspended in toluene. It is anticipated that quantum dots are alternatives of dyes as light-harvesting structures in DSSC. Light absorption produces excitons or electron-hole pairs. Excitons have an average physical separation between the electron and hole, referred to as the Exciton Bohr Radius. Usually, Bohr radius is greater the QD diameter (e.g., for PbS Boher radius is 20 nm) leading to quantum confinement effect (discrete energy levels = artificial molecule). Excitons dissociate at the QD TiO$_2$ interface. The electron is subsequently injected in the semiconductor oxide conduction band, while the hole is transferred to a hole conductor or an electrolyte. Efficient and rapid hole injection from PbS QDs into triarylamine hole conductors has been demonstrated, and IPCE (Incident Photon to Current Conversion Efficiency) values exceeding 50% have been obtained. QDs have much higher optical cross sections than molecular sensitizers, depending on their size. However, they also occupy a larger area on the surface of the nanostructured photoelectrode, decreasing the QD concentration in the film. Thus, the value of the absorption length is similar to that observed for the dye-loaded nanostructured photoelectrode. Investigations show that multiple excitons can be produced from the absorption of a single photon by a QD via impact ionization if the photon energy is 3 times higher than its band gap (Ellinson et al., 2005; Nozik, 2004; Nozik, 2005). The issue to be confronted is to find ways to collect the excitons before they recombine get lost in the cell.

Unlike dyes that absorb over relatively narrow region, semiconductor quantum dots such as PbS (see Figure 11-b) absorb strongly all photons with energy greater than the bandgap, thus a far higher proportion of light can be converted into useful energy using nanocrystals compared to dyes. Perhaps most important, dyes are disgracefully unstable and tend to photobleach over a relatively short amount of time. Quantum dots prepared with a properly designed outer shell are very stable and hence long lasting solar cells without degradation in performance are feasible. Quantum dots-sensitized solar cell produces quantum yields greater than one due to impact ionization process (Nozik, 2001). Dye molecules cannot undergo this process. Solar cells made from semiconductor QDs such as CdSe, CdS, PbS and InP showed a promising photovoltaic effect (Hoyer & Konenkamp, 1995; Liu & Kamat 1993; Plass et al., 2002; Vogel & Weller 1994; Zaban et al., 1998; Zweible & Green, 2000). Significant successes have been achieved in improving the photo-conversion efficiency of solar cells based on CdSe quantum dote light harvesters supported with carbon nanotube this is accomplished by incorporating carbon nanotubes network in the nanostructured TiO_2 layer, and accordingly assisting charge transport process network (Hasobe et al., 2006; Robel et al., 2005). Consequently, appreciable improvement in the photo-conversion efficiency of the DSSC is attainable. Recently Fuke et al., reported CdSe quantum-dot-sensitized solar cell with ~100% internal quantum efficiency. A significant enhancement in both the electron injection efficiency at the QD/TiO_2 interface and charge collection efficiency at the QD/electrolyte interface" were achieved (Fuke et al., 2010).

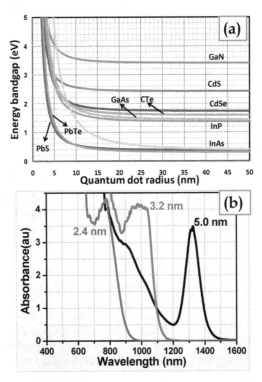

Fig. 11. (a) Calculated energy gap of some semiconductor quantum dots using the effective mass- approximation -model and (b) measured absorbance of PbS quantum dots suspended in toluene of three different sizes (radius).

2.4 Redox electrolyte

Electrolyte containing I^-/I_3^- redox ions is used in DSSC to regenerate the oxidized dye molecules and hence completing the electric circuit by mediating electrons between the nanostructured electrode and counter electrode. NaI, LiI and R_4NI (tetraalkylammonium iodide) are well known examples of mixture of iodide usually dissolved in nonprotonic solvents such as acetonitrile, propylene carbonate and propionitrile to make electrolyte. Cell performance is greatly affected by ion conductivity in the electrolyte which is directly affected by the viscosity of the solvent. Thus, solvent with lower viscosity is highly recommended. Moreover, counter cations of iodides such as Na^+, Li^+, and R_4N^+ do affect the cell performance mainly due to their adsorption on nanostructured electrode (TiO_2) or ion conductivity. It has been found that addition of *tert*-butylpyridine to the redoxing electrolyte improves cell performance (Nazeeruddin et al., 1993) (see Figure 19). Br^-/Br_3^- redox couple was used in DSSCs and promising results were obtained. The V_{oc} and I_{sc} increased for the Eosin Y-based DSSC when the redox couple was changed from I^-/I_3^- to Br^-/Br_3^- (Suri & Mehra, 2006).

The redoxing electrolyte needs to be chosen such that the reduction of I_3^- ions by injection of electrons is fast and efficient (see Figure 13). This arise from the fact that the dependence of both hole transport and collection efficiency on the dye-cation reduction and I^-/I_3^- redox efficiency at counter electrodes are to be taken into account (Yanagida, 2006). Besides limiting cell stability due to evaporation, liquid electrolyte inhibits fabrication of multi-cell modules, since module manufacturing requires cells be connected electrically yet separated chemically (Matsumoto et al., 2001; Tennakone et al., 1999). Hence, a significant shortcoming of the dye sensitized solar cells filled with liquid state redoxing electrolyte is the leakage of the electrolyte, leading to reduction of cell's lifespan, as well as the associated technological problems related to device sealing up and hence, long-term stability (Kang, *et al.*, 2003). Many research groups investigate the use of ionic liquids, polymer, and hole conductor electrolytes (see Figure 12) to replace the need of organic solvents in liquid electrolytes. Despite the reported relative low cell's efficiency of 4–7.5% (device area < 1 cm²) , these kind of electrolyte are promising and may facilitate commercialization of dye sensitize solar modules (Kawano, et al., 2004; Kuang et al., 2006; Schmidt-Mende & Gratzel, 2006; Wang et al., 2004).

Addition of polymer gel to quasi-solidify electrolytes has been investigated by many research groups (Ren et al., 2001; Kubo et al., 2001; Nogueira et al., 2001). It has been found that the addition of Poly(vinylidene fluoride-co-hexafluoropropylene) to the KI/I_2 electrolyte has improved both the fill factors and energy conversion efficiency of the cells by about 17% (Kang, et al., 2003). Gel electrolytes also are very attractive from many perspectives such as: Efficiency is a compromise between electrolyte viscosity and ionic mobility; gelled ionic liquids have an anomalously high ionic mobility despite their high viscosity, and particularly for realization of monolithic arrays inter-cell sealing (Wang, et al., 2005). Innovative classes of electrolytes such as p-type, polymeric conductor, PEDOT or PEDOT:TMA, which carries electrons from the counter electrode to the oxidized dye encouraging further investigations to optimize and/or design new ones. Recently one of the first systematic study of charge transport and recombination in solid state dye sensitized solar cell SDSCs using conjugated polymer hole transporter has been reported by Zhang et al., in this investigation organic indoline dye D131 as the sensitizer and poly(3-hexylthiophene) (P3HT) as the hole transporter a power conversion efficiency of 3.85% have been recorded. Therefore, this class of solar cells is expected to represent one of the most efficient SDSCs using polymeric hole transporter (Zhang et al, 2011).

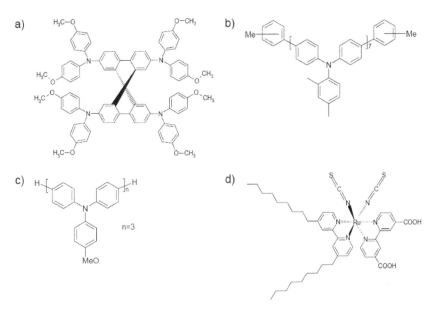

Fig. 12. (a) Chemical structure of the hole-conductor spiro-OMeTAD resulted in cells energy conversion efficiency η = 4%, (b) Chemical structure of AV-DM resulted in cells with η = 0.9%, (c) Structure of AV-OM. resulted in cells with η = 2%, (d) Structure of the Z907 dye used for all solar cells as sensitizer of the nanostructured TiO$_2$ film (From Schmidt-Mende & Gratzel, 2006).

3. How dye sensitized solar cell works

In this section we overview the following: Process during which light energy get converted to electric one, photovoltaic performance, charge injection, charge transport in the nanostructured electrode, charge recombination, and cell dark current.

3.1 Operating principle of dye sensitized solar cell

Nanocrystalline TiO$_2$ is deposited on the conducting electrode (photoelectrode) to provide the necessary large surface area to adsorb sensitizers (dye molecules). Upon absorption of photons, dye molecules are excited from the highest occupied molecular orbitals (HOMO) to the lowest unoccupied molecular orbital (LUMO) states as shown schematically in Figure 13. This process is represented by Eq. 2. Once an electron is injected into the conduction band of the wide bandgap semiconductor nanostructured TiO$_2$ film, the dye molecule (photosensitizer) becomes oxidized, (Equation 3). The injected electron is transported between the TiO$_2$ nanoparticles and then extracted to a load where the work done is delivered as an electrical energy, (Equation 4). Electrolytes containing I^-/I_3^- redox ions is used as an electron mediator between the TiO$_2$ photoelectrode and the carbon coated counter electrode. Therefore, the oxidized dye molecules (photosensitizer) are regenerated by receiving electrons from the I^- ion redox mediator that get oxidized to I_3^- (Tri-iodide ions). This process is represented by Eq. 5. The I_3^- substitutes the internally donated electron with that from the external load and reduced back to I^- ion, (Equation 6). The movement of electrons in the conduction band of the wide bandgap nanostructured semiconductor is accompanied by the diffusion of charge-compensating cations in the electrolyte layer close to the nanoparticle surface. Therefore, generation of electric power in DSSC causes no permanent chemical change or transformation (Gratzel, 2005).

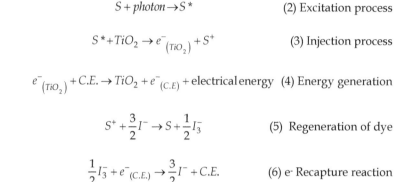

$$S + photon \rightarrow S* \qquad \text{(2) Excitation process}$$

$$S* + TiO_2 \rightarrow e^-_{(TiO_2)} + S^+ \qquad \text{(3) Injection process}$$

$$e^-_{(TiO_2)} + C.E. \rightarrow TiO_2 + e^-_{(C.E)} + \text{electrical energy} \qquad \text{(4) Energy generation}$$

$$S^+ + \frac{3}{2}I^- \rightarrow S + \frac{1}{2}I_3^- \qquad \text{(5) Regeneration of dye}$$

$$\frac{1}{2}I_3^- + e^-_{(C.E.)} \rightarrow \frac{3}{2}I^- + C.E. \qquad \text{(6) } e^- \text{ Recapture reaction}$$

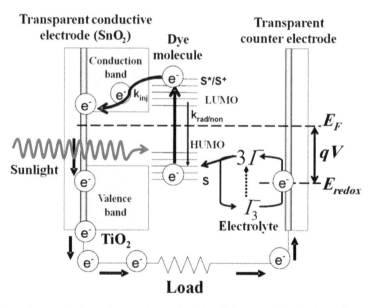

Fig. 13. Schematic illustration of operation principle of dye sensitized solar cell.

As illustrated in Fig. 13, the maximum potential produced by the cell is determined by the energy separation between the electrolyte chemical potential (E_{redox}) and the Fermi level (E_F) of the TiO_2 layer. The small energy separation between the HOMO and LUMO ensures absorption of low energy photons in the solar spectrum. Therefore, the photocurrent level is dependent on the HOMO-LUMO levels separation. This is analogous to inorganic semiconductors energy bandgap (E_g). In fact, effective electron injection into the conduction band of TiO_2 is improved with the increase of energy separation of LUMO and the bottom of the TiO_2 conduction band. Furthermore, for the HOMO level to effectively accept the donated electrons from the redox mediator, the energy difference between the HOMO and redox chemical potential must be more positive (Hara & Arakawa, 2003).

3.2 Photovoltaic performance

Figure 14 presents examples of the I-V characteristics of natural dye sensitized solar cell NDSSC with Bahraini Henna (*Lawsonia inermis* L.), pomegranate, Bahraini raspberries, and cherries. We found that nature of the dye and its concentration has a remarkable effect on the magnitude of the collected photocurrent. Under full solar spectrum irradiation with photon flux I_0 = 100 mW/cm² (Air Mass 1.5), the photon energy –to- electricity conversion efficiency is defined as (Gratzel, 2003):

$$\eta = \frac{J_{sc} \times V_{oc} \times FF}{I_0} \qquad (7)$$

where J_{sc} is the short circuit current, V_{oc} the open circuit voltage, and FF is the fill factor of the solar cell which is calculated by multiplying both the photocurrent and voltage resulting in maximum electric power delivered by the cell.

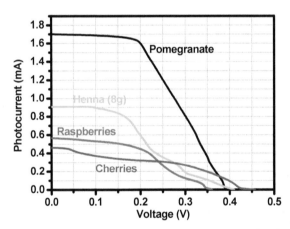

Fig. 14. Photocurrent vs. voltage curves obtained for nanostructured TiO_2 photoelectrodes sensitized with some extracted natural dyes (Jasim, submitted for publications).

Table 3 shows the electrical properties of some assembled NDSSCs. Photocurrent and voltage drop on a variable load have been recorded instantaneously while the cell is exposed to direct sun illumination. Due to light reflection and absorption by the conductive photoelectrode and the scattering nature of the nanostructured TiO_2, the measured transmittance of the photoelectrode (see Figure 3) shows an average of 10% of the solar spectrum (Air Mass 1.5) may reach the sensitizers. Since TiO_2 past is applied on the conductive electrode using doctor blade method the effective area of the irradiated part of the cell is 1.5 cm × 2 cm. Despite the variation of Bahraini Henna extract concentration the cells produced almost the same open circuit voltage V_{oc}. On the other hand, the short circuit current I_{sc} varies with Henna extract concentration. Highly concentrated Bahraini Henna extracts results in non-ideal I-V characteristics even though it possesses 100% light harvesting efficiency in the UV and in the visible parts of the electromagnetic spectrum. The dye concentration was found to influence remarkably the magnitude of the collected photocurrent. High concentration of Henna extract introduces a series resistance that ultimately reduces the generated photocurrent. On the other hand, diluted extracts reduces the magnitude of the photocurrent and cell efficiency. (Jasim et al, 2011).

Dye	V_{oc} (V)	I_{sc} (mA)	FF	% η
Bahraini Henna 80g	0.426	0.368	0.246	0.128
Bahraini Henna 8g	0.410	0.906	0.363	0.450
Bahraini Henna 0.8g	0.419	0.620	0.330	0.286
Yameni Henna 84g	0.306	0.407	0.281	0.117
Yameni Henna 21g	0.326	0.430	0.371	0.174
Yameni Henna 4.2g	0.500	0.414	0.276	0.191
Cherries in Methanol	0.305	0.466	0.383	0.181
Cherries in Methanol+ 1% HCL	0.301	0.463	0.288	0.134
Pomegranate	0.395	1.700	0.481	1.076
Raspberries	0.360	0.566	0.455	0.309

Table 3. Electrical properties of some assembled natural dye sensitized solar cells NDSSCs (From Jasim et al, 2011; Jasim, submitted for publications).

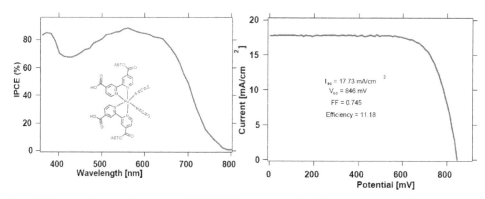

Fig. 15. Photovoltaic performance of DSSC laboratory cell (a) Photo current action spectrum showing the monochromatic incident photon to current conversion efficiency (IPCE) as function of light wavelength obtained with the N-719 sensitizer. (b) *Photocurrent density – voltage* curve of the same cell under AM 1.5 standard test conditions. (From Nazeeruddin et al., 2005).

Gratzel and coworkers reported cell efficiency of 10.4% using black dye $(RuL'(NCS)_3$ complexes) and as shown in Figure 15, cells with solar to electric power conversion efficiency of the DSSC in full AM 1.5 sun light validated by accredited PV calibration laboratories has reached over 11% (Chiba et al., 2006). Jiu et al., (Jiu, et al., 2006) have synthesized highly crystalline TiO₂ nanorods with lengths of 100-300 nm and diameters of 20-30 nm. The rod shape kept under high calcination temperatures contributed to the achievement of the high conversion efficiency of light-to-electricity of 7.29%. Reported efficiencies of nanostructured ZnO₂ photoelectrodes based cells are encouraging and many research groups are dedicating their efforts to provide cells with efficiency close to that reported for sensitized nanostructured TiO₂ photoelectrodes.

The short circuit current magnitude affects directly the incident photon-to-current conversion efficiency IPCE which is defined using the photoresponse and the light intensity as:

$$IPCE(\lambda) = \frac{1240(eV.nm) \times J_{sc}(\mu A / cm^2)}{\lambda(nm) \times I(\mu W / cm^{-2})} \qquad (8)$$

where λ is the wavelength of the absorbed photon and I is the light intensity at wavelength λ. Figures 15 and 16 present IPCE examples of some commonly used sensitizers by Gratzel and coworkers.

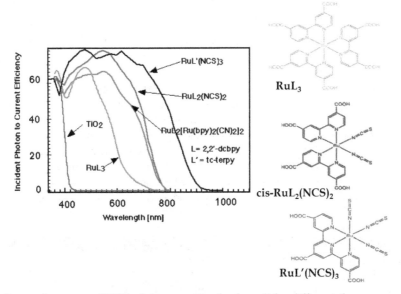

Fig. 16. Spectral response (IPCE) of dye-sensitized solar cell for different dyes compared with the spectral response of bare TiO₂ electrode and the ideal IPCE curve for a single bandgap device (From http://dcwww.epfl.ch/icp/ICP-2/solarcellE.html. and Gratzel et al., 2005).

In terms of light harvesting efficiency LHE, quantum yield of electron injection quantum yield Φ_{inj}, and collection efficiency η_c of the injected electrons at the back contact IPCE is given by:

$$IPCE = LHE \times \Phi_{inj} \times \eta_c \qquad (9)$$

Therefore, IPCE equals the LHE if both Φ_{inj} and η_c are close 100%. However, Charge injection from the electronically excited sensitizer into the conduction band of the nanostructured wide bandgap semiconductor is in furious competition with other radiative and non-radiative processes. Due to electron transfer dynamics (see Figure 17), if electron injection in the semiconductor is comparable to, or slower than, the relaxation time of the dye, Φ_{inj} will be way below 100%. This can be deduced from the definition of the quantum yield Φ_{inj} (Cherepy et al., 1997):

$$\Phi_{inj} = \frac{k_{inj}}{k_{inj} + k_{rad} + k_{nrad}} \tag{10}$$

The quantum yield approaches 100% only when the radiative and nonradiative rates (k_{rad}, k_{nrad}) (paths shown in Figure 13) are much smaller than the injection rate k_{inj}. The rate constant for charge injection k_{inj} is given by Fermi golden rule (Gratzel, 2001; Hara & Arakawa, 2003):

$$k_{inj} = \left(\frac{4\pi^2}{h} \right) |V|^2 \rho(E) \tag{11}$$

where h is Planck's constant, $|V|$ is the electron-coupling matrix element and $\rho(E)$ is the density of electronic acceptor states in the conduction band of the semiconductor. Equation (11) assumes that electron transfer from the excited dye molecules into the semiconductors is activationless and hence exhibits a temperature-independent rate. Some representative examples of electron injection rate constants k_{inj} and electronic coupling matrix elements $|V|$ measured by laser flash photolysis for some sensitizers adsorbed onto nanocrystalline TiO_2, t_f and Φ_{inj} (the excited-state lifetime and the injection quantum yield, respectively) are presented in Table 4 (Gratzel, 2001; Hara & Arakawa, 2003). The shown values of $|V|$ on Table 4 credited to the degree of overlapping of photosensitizer excited states wavefunction and the conduction band of the nanostructured photoelectrode. The distance between the adsorbed sensitizer and the nanostructured photoelectrode affect the value of the electronic coupling matrix elements.

Sensitizers	k_{inj} [s⁻¹]	$|V|$ [cm⁻¹]	t_f [ns]	Quantum yield
Ru^{II}(bpy)$_3$	2×10^5	0.04	600	0.1
Ru^{III}L$_3$ (H$_2$O)	3×10^7	0.3	600	0.6
Ru^{III}L$_3$ (EtOH)	4×10^{12}	90	600	1.0
Ru^{III}L$_2$(NCS)$_2$	10^{13}	130	50	1.0
Coumarin-343	5×10^{12}	100	10	1.0
Eosin-Y	9×10^8	2	1	0.4

Table 4. Electron injection rate constants k_{inj} and electronic coupling matrix elements $|V|$ measured by laser flash photolysis for various sensitizers adsorbed onto nanocrystalline TiO_2. In the sensitizers column, L stands for the 4,4'-dicarboxy-2,2'-bipyridyl ligand and bipy for 2,2'-bipyridyl (From Gratzel, 2001).

Advantages of tandem structure have been investigated both theoretically and experimentally as approaches to improve the photocurrent of DSSC (Durr et al., 2004). "The tandem structured cell exhibited higher photocurrent and conversion efficiency than each single DSSC mainly caused by its extended spectral response." (Kubo et al., 2004)

3.3 Charge injection, transport, recombination, and cell dark current

Kinetics of electron injection into the semiconductor photoelectrod after being excited from the photosensitizer has been investigated by many researchers using time-resolved laser spectroscopy (Hara & Arakawa, 2003). It has been found that both the configuration of the photosensitizer material and the energy separation between the conduction band level of the wideband gap semiconductor and the LUMO level of the photosensitizer are greatly affecting the electron transfer rate to the wideband gap semiconductor. Figure 17 shows a schematic illustration of kinetics in the DSSC. The shown arrows indicate excitation of the dye from the HOMO to the LUMO level, relaxation of the exited state (60 ns), electron injection from the dye LUMO level to the TiO_2 conduction band (50 fs -1.7 ps), recombination of the injected electron with the hole in the dye HOMO level (ns -ms), recombination of the electron in the TiO_2 conduction band with a hole (I_3^-) in the electrolyte (10 ms), and the regeneration of the oxidized dye by I^- (10 ns). (Hagfeldt & Gratzel, 2000).

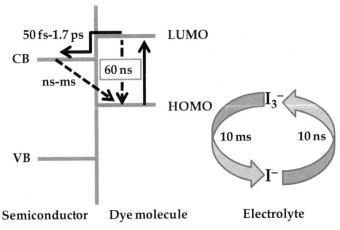

Fig. 17. Schematic illustration of kinetics in the DSSC, depicted from Hagfeldt & Gratzel, 2000.

It has been confirmed that electron injection from the excited dye such as the N_3 dye or $RuL_2(NCS)_2$ complex into the TiO_2 conduction band (CB) is a very fast process in femtosecond scale. The reduction of the oxidized dye by the redox electrolyte's I^- ions occur in about 10^{-8} seconds. Recombination of photoinjected CB electrons with oxidized dye molecules or with the oxidized form of the electrolyte redox couple (I_3^- ions) occurs in microseconds (Hara & Arakawa, 2003). To achieve good quantum yield, the rate constant for charge injection should be in the picosecond range. In conclusion, Fast recovery of the sensitizer is important for attaining long term stability. Also, long-lasting charge separation is a very important key factor to the performance of solar cells. Thus, new designs for larger conjugated dye-sensitizer molecules have been reported by investigators ,for example, Haque et al., (Haque et al., 2004) studied hybrid supermolecules that are efficiently retard the recombination of the charge-separated state and therefore assure enhanced energy conversion efficiency by extending the lifetime of light-induced charge-separated states as illustrated in Figure 18, "Hybrid supermolecule: This is the structure of the redox triad that gave the most efficient charge separation in the report by Haque and colleagues . The triad is made of a ruthenium complex anchored to nanocrystalline TiO_2 (the electron acceptor) and covalently linked to polymeric chains of triphenyl-amine groups (the electron donor). Arrows represent the direction of the electron transfer process. The first step of the electron transfer is the light-induced excitation of the chromophore (process 1). Following this an electron is readily injected from the sensitizer excited state into the conduction band of the TiO_2 semiconductor (process 2). The direct recombination of primarily separated charges (process 3) would degrade the absorbed energy into heat. In this supermolecule this is avoided through the fast reduction of the ruthenium by the linked triphenyl-amine electron donor groups (process 4). The secondary recombination process (process 5) between the injected electron and the oxidized amine radical is made increasingly slow because the positive charge can hop from one triphenylamine function to the adjacent one along the chain (process 6) and the hole moves away from the TiO_2 surface. The overall photo-initiated process thus results in unidirectional electron flow from the end of the polymeric chains to the oxide (from right to left) and a very long-lived charge-separated state" (Moser, 2005).

In TiO_2 nanoparticle DSSCs, the electrons diffuse to the anode by hopping 103-106 times between particles (Baxter et al., 2006). With each hop there is a considerable probability of recombination of the photoexcited electron with the electrolyte since both the diffusion and recombination rates are on the order of milliseconds. Hence, this allows recombination to limit the cell efficiency. On the other hand, nanowire or tube structured photoelectrode (e.g., ZnO_2) provide a direct path (express highway) to the anode, leading to increased diffusion rate without increasing the recombination rate and thus increases cell efficiency.

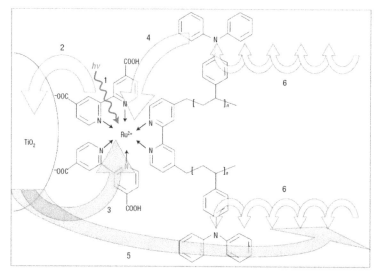

Fig. 18. Schematics of the hybrid supermolecule. The supersensitizer molecule adsorbed to a nanostructured TiO_2 surface promise to improve the photovoltaic conversion efficiency of dye sensitized solar cell (From Moser, 2005).

Dark current in DSSC is mainly due to the loss of the injected electron from nanostructured wide bandgap semiconductor (say TiO_2) to I_3^- (the hole carrier in solution electrolyte). Thus, it is a back reaction that must be eliminated or minimized. Reduction of dark current enhances the open circuit voltage of the cell, this can be deduced from the following general equation of solar cell relating the open circuit voltage V_{OC} to both the injection current I_{inj} and dark current I_{dark}:

$$V_{OC} = \frac{k_B T}{q} \ln\left(\frac{I_{inj}}{I_{dark}} + 1\right) \tag{12}$$

where k_B is the Boltzmann constant, T is the absolute temperature of the cell, and q is the magnitude of the electron charge. In fact, dark current mainly occurs at the TiO_2/electrolyte interface where no photosensitizer got adsorbed. One successful way to suppress dark current is to use one of pyridine derivatives (e.g., tert-butylpyridine TBP) as coadsorbates on the nanostructured TiO_2 surface. Figure 19 shows the current–voltage characteristics obtained for NKX-2311-sensitized TiO_2 solar cells (Hara et al., 2003).

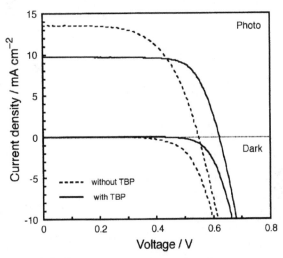

Fig. 19. Current–voltage curves obtained for NKX-2311-sensitized TiO_2 solar cells in an electrolyte of 0.6M DMPImI–0.1M LiI–0.05M I2 in methoxyacetonitrile: (- - -) without TBP, (—) with 0.5M TBP (From Hara et al., 2003).

4. Applications of DSSC

Because of the physical nature of the dye sensitized solar cells, inexpensive, environment-friendly materials, processing, and realization of various colors (kind of the used sensitizing dye); power window and shingles are prospective applications in building integrated photovoltaics BIPV. The Australian company Sustainable Technologies International has produced electric-power-producing glass tiles on a large scale for field testing and the first building has been equipped with a wall of this type (see for example, Figure 20-a). The availability of lightweight flexible dye sensitized cells or modules are attractive for applications in room or outdoor light powered calculators, gadgets, and mobiles. Dye sensitized solar cell can be designed as indoor colorful decorative elements (see Figure 20-b). Flexible dye sensitized solar modules opens opportunities for integrating them with many portable devices, baggage, gears, or outfits (Pagliaro et al., w w w. pv- te ch.org) (see Figure 20-c and Figure 20-d). In power generation, dye sensitized modules with efficiency of 10% are attractive choice to replace the common crystalline Si-based modules. In 2010 Sony announced fabrication of modules with efficiency close to 10% and hence opportunity of commercialization of DSSC modules is attainable.

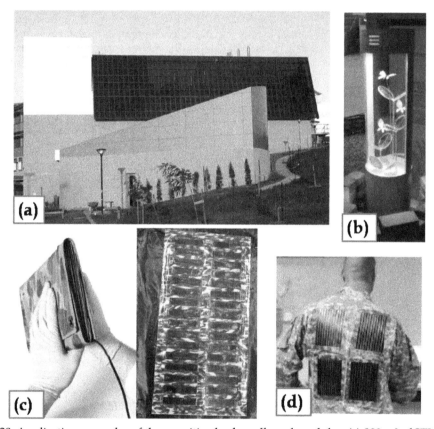

Fig. 20. Application examples of dye sensitized solar cells and modules: (a) 200 m² of STI DSSC panels installed in Newcastle (Australia)– the first commercial DSSC module (http://www.sta.com.au/index.htm), (b) indoor ornament of dye sensitized solar cells leaves (AISIN SEIKI CO.,LTD), (c) flexible DSSC-based solar module developed by Dyesol (http://www.dyesol.com), and (d) jacket commercialized by G24i (http://www.g24i.com).

5. Commercialization of DSSC

Commercialization of dye sensitized solar cells and modules is taking place on almost all continents (Lenzmann & Kroon, 2007). In Asia, specifically in Japan: IMRA-Aisin

Seiki/Toyota, Sharp, Toshiba, Dai Nippon, Peccell Technologies. In Australia: Dyesol. In USA Konarka. G24i in UK, and Solaronix in Switzerland. G24i has announced a DSC module production of 25MW capacity in 2007 in Cardiff, Wales (UK), with extension plans up to 200MW by the end of 2008 (http://www.g24i.com). The success of many labs and companies such as ASIAN and Toyota Central R & D Labs., INC. (see Figure 21) to demonstrate various sizes and colors in a series-connected dye solar cell module in many international exhibitions and conferences reflects the potential role of dye sensitized solar cells systems in the PV technology. In fact, Toyota has installed in their dream house walls a similar kind of DSSCs panels shown in Figure 21-b.

Fig. 21. (a) An example of DSSC module for outdoor application
(Fromhttp://kuroppe.tagen.tohoku.ac.jp/~dsc/cell.html) and (b) Outdoor field tests of DSSC modules produced by Aisin Seiki in Kariya City. Note the pc-Si modules in the second row. (From Gratzel article at
http://rsta.royalsocietypublishing.org/content/365/1853/993.full#ref-3).

Glass substrate is robust and sustains high temperatures, but it is fragile, nonflexible, and pricey when designed for windows or roofs. Flexible DSSCs have been intensively investigated. Miyasaka et. al. (Miyasaka & Kijitori, 2004) used the ITO (indium tin oxide) coated on PET (polyethylene terephatalate) as the substrate for DSSCs. Generally, the conducting glass is usually coated with nanocrystallineTiO$_2$ and then sintered at 450°C-500°C to improve the electronic contact not only between the particles and support but also among the particles. Plastics films have a low ability to withstand heat. The efficiency of plastic-based dye sensitized solar cells is lower than that of using glass substrate (η = 4.1%, J$_{sc}$= 9.0mA/cm^2, V$_{oc}$ = 0.74V, FF = 0.61) because of poor necking of TiO$_2$ particles. Kang et al., (Kang et al.2006), used the stainless steel as the substrate for photoelectrode of DSSCs (see Figure 22). The cell illuminated through the counter electrode due to the non-penetration of light through metal substrate. In their system, the SiOx layer was coated on stainless steel (sheet resistance ~ 1 mΩ per square) and separated ITO from stainless steel, for preventing photocurrent leakage from stainless steel to the electrolyte. The constructed cells resulted in J$_{sc}$ = 12 mA/cm^2, V$_{oc}$ = 0.61 V, FF = 0.66, and η = 4.2 %. Recently, Chang et al. fabricated flexible substrate cell that produced conversion efficiency close to 2.91%. The photoelectrode substrates are flexible stainless steel sheet with thickness 0.07mm and titanium (Ti) sheet with thickness 0.25mm (Chang et al., 2010). Also, the reported approach by Yen et al. in developing a low temperature process for the flexible dye-sensitized solar cells using commercially available TiO$_2$ nanoparticles (such as P25) is interesting since it yielded a conversion efficiency of 3.10% for an incident solar energy of 100 mW/cm^2 (Yen et al. 2010). Because Titanium has extremely high corrosion resistance, compared with stainless steel, Titanium is still the privileged substrate material.

Fig. 22. A prototype of a flexible dye sensitized solar cell using stainless steel substrate (From Kang et al., 2006).

Availability of nonvolatile electrolyte is another issue toward commercialization of single or multi-junction modules. Polymer (solid) electrolyte, hole conductor, and solidified ionic liquids are solvent free choices with high electronic conductivity and chemical stability (Wang et al., 2005.) The key to high power heterojunction DSSC is to increase the effective diffusion length of electron within the nanostructured electrode by increasing the mobility of hole conductor or the extinction coefficient of the sensitizer to ensure more efficient light harvesting action. Since heat and UV light degrade cells performance, development of heat sink and optimized low cost UV coating is a must for outdoor applications. The successes in development of flexible substrate, solid electrolyte, and spectrally broad absorption range inexpensive nontoxic dyes will potentially open the possibility of role-to-role mass production of dye sensitized solar cells and modules (see Figure 23). Molecular engineering of efficient and stable organic sensitizers is an open invitation for many research groups, the successes in this area is expected to advance production and commercialization of DSSC (Kim et al., 2006).

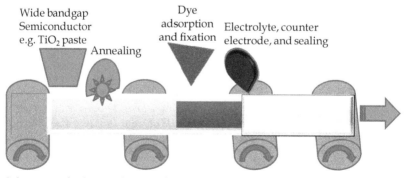

Fig. 23. Schematic of role-to-role manufacturing of flexible dye sensitized solar cells.

6. Conclusions

In This chapter we have discussed one example of the third generation solar cells, called photoelectrochemical cell and now called nanocrystalline dye sensitized solar cells DSSC or Gratzel cell. Nanocrystalline dye sensitized solar cell DSSC is classified as a low cost, environmental friendly, and capable of being highly efficient cell mainly due to materials, charge carriers generation and transport within the cell structure. The nanostructured dye-sensitized solar cell (DSSC) is going to provide economically credible alternative to present day p–n junction photovoltaics. In fact, dye sensitized solar cell are green solar cell mimicking the green leave. In dye sensitized solar cell electricity is generated as a result of electron transfer due to photoexcitation of dye molecules adsorbed to nanostructured wide bandage material photoelectrode. The oxidized dye molecules regenerated by gaining electrons from electrolyte which is reduced by the electrons reaching the counter-electrode of the cell. In dye sensitized solar cells light absorption is separated from carrier transport. From educational point of view, since nanostructured dye sensitized solar cell DSSC is mimicking photosynthesis in plants, it provides an interdisciplinary context for students learning the basic principles of biological extraction, chemistry, physics, environmental science and electron transfer.

The requirements of practical sensitizers are: broad band and high level absorption of visible and near infrared region of the electromagnetic spectrum, exhibit thermal and photochemical stability, definitely chelating to the semiconductor oxide surface and inject electrons into the conduction band with a quantum yield of unity, and owning suitable ground- and excited state redox properties. Investigations of solvent free electrolyte such as polymer based, and ionic liquid are promising. In order to commercialize dye sensitized solar cell in low power applications, flexible DSSCs have been intensively investigated. The solar to electric power conversion efficiency of the DSC in full AM 1.5 sun light validated by accredited PV calibration laboratories has reached over 11 % and modules with efficiency close to 10% has been exhibited in 2010. Nanowires and quantum dots QDSSCs may be a promising solar cell design.

The search for green sources or generators of energy is considered one of the priorities in today's societies and occupies many policy makers' agendas. We at the University of Bahrain are the first to start the investigation of Dye sensitized solar cells DSSCs in the Arabian Gulf. Dye sensitized solar cells using natural organic dies were prepared using low cost materials and natural dyes. Sensitization of wide gap oxides semiconductor materials was accomplished with the growth of nanocrystalline TiO_2. The natural dyes extracted from Henna (*Lawsonia inermis* L.), pomegranate, cherries, and raspberries (Rubus spp.). We found that the nanocrystalline material based solar cell system exhibits an excellent optical absorption parameters for visible and near infrared portion of the electromagnetic spectrum. The performance of natural dye extract sensitized nanocrystalline solar cells can be appreciably enhanced by optimizing preparation technique, using different types of electrolyte, reported additives, and sealing.

In short, compared to Si based solar cells dye sensitized solar cells are of low cost and ease of production, their performance increases with temperature,   possessing bifacial configuration - advantage for diffuse light, have transparency for power windows, color can be varied by selection of the dye, invisible PV-cells based on near-IR sensitizers are feasible, and they are outperforms amorphous Si. Moreover, DSSC shows higher conversion efficiency than polycrystalline Si in diffuse light or cloudy conditions. It is believed that nanocrystalline photovoltaic devices are becoming viable contender for large scale future solar energy converters.

7. Acknowledgments

The author is greatly indebted to the University of Bahrain for financial support. I would like to express my thanks to Prof. Dr. Shawqi Al Dallal for being keen in providing fruitful discussions. Many thanks to Dr. Akil dakil (Department of Physics, University of Bahrain) for facilitating XRD measurements and Dr. Mohammad S. Hussain (National Nanotechnology Center King Abdulaziz City for Science and Technology (KACST)) for providing the SEM image.

8. References

Amao, Y. & Komori, T. (2004). Bio-photovoltaic conversion device using chlorine-e_6 derived from chlorophyll from *Spirulina* adsorbed on a nanocrystalline TiO_2 film electrode. *Biosensors Bioelectronics*, Vol. 19, Issue 8, pp. 843-847.

Baxter, J.B; Walker, A.M; van Ommering, K. & Aydil, E.S. (2006). Synthesis and characterization of ZnO nanowires and their integration into dye-sensitized solar cells . *Nanotechnology*, Vol. 17, Issue 11, pp. S 304-S 312.

Becquerel, A. E. (1839). Mĕmoire sur les effets électriques produits sous l'influence des rayons solaires. *C. R. Acad. Sci. Paris*, Vol. 9, pp. 561-567.

Bergeron, B.V.; Marton, A.; Oskam, G. & Meyer, G.J. (2005). Dye-Sensitized SnO_2 Electrodes with Iodide and Pseudohalide Redox Mediators. *J. Phys. Chem. B*, Vol, 109, pp.935-943.

Boercker, J.E; Schmidt, J.B. & Aydil, E.S. (2009). Transport Limited Growth of Zinc Oxide Nanowires. *Cryst. Growth*, Vol. 9 (6), pp 2783-2789.

Calogero, G.; Di Marco,G.; Caramori, S.; Cazzanti, S.; Roberto Argazzi, R. and Bignozzi, C.A. (2009). Natural dye senstizers for photoelectrochemical cells. *Energy Environ. Sci.*, Vol. 2, Issue 11, pp. 1162-1172.

Chang, H.; Chen, T.L.; Huang, K.D.; Chien, S.H. & Hung, K.C. (2010). Fabrication of highly efficient flexible dye-sensitized solar cells. *Journal of Alloys and Compounds*, Vol. 504, Issue 2, pp. S435-S438.

Cherepy, N.J.; Smestad, G.P.; Gratzel, M. & Zhang, J.Z. (1997). Ultrafast Electron Injection: Implications for a Photoelectrochemical Cell Utilizing an Anthocyanin Dye-Sensitized TiO_2 Nanocrystalline Electrode. *J. Phys. Chem. B*, Vol. 101, pp. 9342-9351.

Chiba, Y.; Islam, A.; Watanabe, Y; Komiya, R.; Koide, N. & Han, L. (2006). Dye-sensitized solar cells with conversion efficiency of 11.1%. *Japanese Journal of Applied Physics, Part 2: Letters & Express Letters* , Vol. 45, pp. 24-28.

Durr M.; Bamedi A.; Yasuda A. & Nelles G. (2004). Tandem dye-sensitized solar cell for improved power conversion efficiencies. *Appl. Phys. Lett.*, Volume 84 (17), pp. 3397-3399.

Ellinson, R.; Beard, M.; Johnson, J.; Yu, P.; Micic, O.; Nozik, A.; Shabaev, A.; and Efros A. (2005). Highly Efficient Multiple Exciton Generation in Colloidal PbSe and PbS Quantum Dots, *Nano Letters*, Vol. 5, No. 5, pp. 865-871.

Farzad, F.; Thompson, D. W.; Kelly, C. A.; and Meyer, G. J. (1999). Competitive Intermolecular Energy Transfer and Electron Injection at Sensitized Semiconductor Interfaces. *J. Am. Chem. Soc.*, Vol. 121, pp. 5577-5578.

Fernando, J.M.R.C. & Senadeera G.K.R. (2008). Natural anthocyanins as photosensitizers for dye-sensitized solar devices. *Current Science*, Vol. 95, No. 5, pp. 663-666.

Garcia, C.G., Polo, A.S. and Iha, N.Y. (2003). Fruit extracts and ruthenium polypyridinic dyes for sensitization of TiO_2 in photoelectrochemical solar cells. *J. Photochem. Photbiol. A(Chem.)*, Vol. 160 (1-2), pp. 87-91.

Gleiter, H. (1989). Nanocrystalline materials. *Prog. Mater. Sci.*, Vol. 33, pp. 223-315.

Gratzel, M. & Hagfeldt, A. (2000). Molecular Photovoltaics. *Acc. Chem. Res*, Vol. 33, pp. 269-277.

Gratzel, M. (2001). Molecular photovoltaics that mimic photosynthesis. *Pure Appl. Chem.*, Vol. 73, No. 3, pp. 459-467.

Gratzel, M., (2003). Dye-sensitized solar cells. *Journal of Photochemistry and Photobiology C: Photochemistry Reviews* 4, pp. 145-153.

Gratzel, M. (2005). Solar Energy Conversion by Dye-Sensitized Photovoltaic Cells. *Inorg. Chem.*, Vol. 44, pp. 6841-6851.

Green, M. A.; Emery, K. Solar Cell Efficiency Tables 19, *Prog. Photovolt.: Res. Appl.* 2002; 10:55-61.

Guo, M.; Diao, P.; Wang, X. & Cai, S. (2005). The effect of hydrothermal growth temperature on preparation and photoelectrochemical performance of ZnO nanorod array films. *Journal of Solid State Chemistry*, Vol., 178, pp. 3210-3215.

Hao, S.; Wu, J.; Huang, Y. & Lin, J. (2006). Natural dyes as photosensitizers for dye-sensitized solar cell. *Sol. Energy*, Vol. 80, Issue 2, pp. 209-214.

Hara, K. & Arakawa, H. (2003). Dye-sensitized Solar Cells, In: Handbook of Photovoltaic Science and Engineering, A. Luque and S. Hegedus, (Ed.), Chapter 15, pp. 663-700, John Wiley & Sons, Ltd, ISBN: 0-471-49196-9.

Hara, K.; Tachibanaa, Y.; Ohgab, Y.; Shinpob, A.; Sugab, S.; Sayamaa,K.; Sugiharaa, H. & Arakawaa, H. (2003). Dye-sensitized nanocrystalline TiO$_2$ solar cells based on novel coumarin dyes. *Solar Energy Materials & Solar Cells*, Vol. 77, pp. 89-103.

Haque,S.A.; Handa,A.; Katja,P.; Palomares,E.; Thelakkat, M. & Durrant, J.R. (2005). Supermolecular control of charge transfer in dye-sensitized nanocrystalline TiO$_2$ films: towards a quantitative structure-function relationship. *Angewandte Chemie International Edition*, Vol. 44, pp. 5740-5744.

Harding, H.E.; Hoke, E.T.; Armistrong, P.B.; Yum, J.; Comte, P.; Torres, T.; Frechet, J.M.J.; Nazeeruddin, M.K.; Gratzel, M. & McGehee, M.D. (2009). Increased light harvesting in dye-sensitized solar cells with energy relay dyes. *Nature Photonics*, Vol. 3, pp. 406-411.

Hasobe, T.; Fukuzumi, S. & Kamat, P.V. (2006). Organized Assemblies of Single Wall Carbon Nanotubes and Porphyrin for Photochemical Solar Cells: Charge Injection from Excited Porphyrin into Single-Walled Carbon Nanotubes. *Journal of Physical Chemistry B*, Vol.110 (50), pp. 25477-25484.

Hasselmann, G. & Meyer, G. (1999). Sensitization of Nanocrystalline TiO2 by Re(I) Polypyridyl Compounds . *J. Phys. Chem.*, Vol. 212, pp. 39-44.

Hirata, N.; Lagref, J.; Palomares, E.J.; Durrant, J.R.; M. Khaja Nazeeruddin, Gratzel,M. & Di Censo, D. (2004). Supramolecular Control of Charge-Transfer Dynamics on Dye-sensitized Nanocrystalline TiO$_2$ Films . *Chem. Eur. J.* Vol. 10, Issue 3, pp. 595-602.

Hoffert, M.I; Caldeira, K.; Jain, A.K.; Haites,E.F.; Harvey, L.D.; Potter, S.D.; Schlesinger, M.E.; Schneider, S.H.; Watts, R.G.; Wigley, T.M. & Wuebbles, D.J. (1998). Energy implications of future stabilization of atmospheric CO$_2$ content. *Nature* , Vol. 395, pp. 881-884.

Horiuchi,T.; Hidetoshi Miura,H.; Sumioka, K. & Satoshi Uchida, S. (2004). High Efficiency of Dye-Sensitized Solar Cells Based on Metal-Free Indoline Dyes. *J. Am. Chem. Soc.*, Vol. 126 (39), pp 12218-12219.

Hoyer, P. & Könenkamp, R., (1995). Photoconduction in porous TiO_2 sensitized by PbS quantum dots. *Appl. Phys. Lett. Vol.* 66, Issue 3, pp. 349-351.

Islam, A.; Hara, K.; Singh, L. P.; Katoh, R.; Yanagida, M.; Murata, S.; Takahashi, Y.; Sugihara, H. & Arakawa, H. (2000) *Chem. Lett.*, pp. 490-491.

Jasim, K. E. & Hassan, A.M. (2009). Nanocrystalline TiO_2 based natural dye sensitised solar cells. *Int. J. Nanomanufacturing*, Vol. 4, Nos. 1/2/3/4, pp.242–247.

Jasim, K. E.; Al Dallal, S. & Hassan, A.M. (2011). Natural dye-sensitised photovoltaic cell based on nanoporous TiO_2, *Int. J. Nanoparticles*,(in press 2011).

Jasim, K. E; Al Dallal, S. & Hassan, A.M. (2011). HENNA (*Lawsonia inermis* L.) DYE-SENSITIZED NANOCRYSTALINE TITANIA SOLAR CELL. *J. Nanotechnology*, submitted for publication.

Jasim, K. E. (2011). Natural Dye-Sensitized Solar Cell Based On Nanocrystalline TiO_2. *Sains Mallaysiana*, submitted for publication.

En Mei Jin; Kyung-Hee Park; Bo Jin; Je-Jung Yun, & Hal-Bon Gu (2010). Photosensitization of nanoporous TiO_2 films with natural dye. Phys. Scr,.Vol. 2010, Issue T139, pp. 014006.

Jiu, J., Wang, F., Isoda, S., and Adachi, M. (2006). *J Phys Chem B Condens Matter Mater Surf Interfaces Biophys.*, Vol. 110(5), pp. 2087-2092

Kang, M.G., Park, N-G., Kim, K-M., Ryu, K.S., Chang S.H. & Kim, K.J. (2003). Highyl efficient polymer gel electrolytes for fye-sensitized solar cells. *3rd World Conference on Phorovolroic Emru Conversion . May 11-18, 2003 Osnh, Japan.*

Kang , M.; Park, N.; Ryu, K.; Chang, S. & Kim, K. (2006). A 4.2% efficient flexible dye-sensitized TiO2 solar cells using stainless steel substrate. *Solar Energy Materials & Solar cells*, Vol. 90, pp. 574-581.

Karami, A. (2010). Synthesis of TiO_2 Nano Powder by the Sol-Gel Method and Its Use as a Photocatalyst. *J. Iran. Chem. Soc.*, Vol. 7, pp. S154-S160.

Kawano, R.; Matsui, H.; Matsuyama, C.; Sato, A.; Susan, Md. A. B.H.; Tanabe, A. & Watanabe, M. (2004). High performance dye-sensitized solar cells using ionic liquids as their electrolytes. *Journal of Photochemistry and Photobiology A*, Vol. 164, no. 1–3, pp. 87–92.

Kelly, C. A.; Thompson, D. W.; Farzad, F. & Meyer, G. J. (1999). Excited State Deactivation of Ruthenium(II) Polypyridyl Chromophores Bound to Nanocrystalline TiO_2 Mesoporous Films. *,Langmuir*, Vol. 15, pp. 731-734.

Kim, K.S.; Kang, Y.S.; Lee, J.H.; Shin, Y.J.; Park, N.G.; Ryu, K.S. & Chang, S.H. (2006). Photovoltaic properties of nano-particulate and nanorod array ZnO electrodes for dye-sensitized solar cell. *Bull Korean Chem. Soc*, Vol. 27, pp. 295-298.

Kim,S.; Lee, J.K.; Kang, S.O.; Ko, J.J.; Yum, J.H.; Fantacci, S.; De Angelis, F.; Di Censo, D.; Nazeeruddinand, Md. K. & Graetzel, M., (2006). Molecular Engineering of Organic Sensitizers for Solar Cell Applications. *J. Am. Chem, Soc*, Vol. 128,pp. 16701-16707.

Kleverlaan, C. J.; Indelli, M. T.; Bignozzi, C. A.; Pavanin, L.; Scandola, F. & Hasselmann, Meyer, G. J. (2000). Stepwise Photoinduced Charge Separation in Heterotriads: Binuclear Rh(III) Complexes on Nanocrystalline Titanium Dioxide. *J. Am. Chem. Soc.*, Vol. 122, pp. 2840-2849.

Kopidakis, N.; Benkstien, K.D.; van de Lagemaat, J. & Frank, A.J. (2003). *J. Phys. Chem. B*, Vol, 107, pp. 11307.

Kuang, D.; Wang, P.; Ito, S.; Zakeeruddin, S.M. & Gratzel, M. (2006). Stable mesoscopic dye-sensitized solar cells based on tetracyanoborate ionic liquid electrolyte. *Journal of the American Chemical Society*, Vol. 128, no. 24, pp. 7732–7733.

Kubo, W.; Murakoshi, K.; Kitamua, T.; Yoshida, S.; Haruki, M.; Hanabusa, K.; Shirai, H.; Wada, Y., & Yanagida, S. (2001). Quasi-Solid-State Dye-Sensitized TiO_2 Solar Cells: Effective Charge Transport in Mesoporous Space Filled with Gel Ele, . *J Phys. Chem. B*, Vol. 105, Issue 51, pp. 12809-12812.

Kubo, W.; Sakamoto, A.; Kitamura, T.; Wada, Y. & Yanagida, S. (2004). Dye-sensitized solar cells: improvement of spectral response by tandem structure, *Journal of Photochemistry and Photobiology A: Chemistry* ,Vol.164, pp. 33–39.

Kumara, G.R.A., Kanebo, S., Okuya, M., Onwaona-Agyeman, B., Konno, A., and Tennakone, K. (2006) . *Sol. Enrgy Mater. Sol. Cells*, Vol. 90, pp.1220.

Lagref, J.J.; Nazeeruddin, M.K., & Graetzel, M. (2008). Artificial photosynthesis based on dye-sensitized nanocrystalline TiO_2 solar cells, *INORGANICA CHIMICA ACTA*, Vol. 361, Issue 3, pp.735-745.

Law, M.; Greene, L. E.; Johnson, J. C.; Saykally, R. & Yang, P. D. (2005). Nanowire dye-sensitized solar cells. *Nature Materials, Vol. 4*, 455-459.

Lenzmann, F.O. & Kroon, J.M. (2007). Recent Advances in Dye-Sensitized Solar Cells, *Advances in OptoElectronics* , Volume 2007, Article ID 65073.

Lewis, N. S. (2007). Toward Cost-Effective Solar Energy Use. *Science*, Vol 315, pp. 798-801.

Lindström, H.; Holmberg, A.; Magnusson, E.; Malmqvist, L. & Hagfeldt, A. (2001). A new method to make dye-sensitized nanocrystalline solar cells at room temperature, *Journal of Photochemistry and Photobiology A: Chemistry*, Vol. 145, pp. 107–112.

Liu, D. and Kamat, P.V. (1993). Photoelectrochemical behavior of thin cadmium selenide and coupled titania/cadmium selenide semiconductor films. *J. Phys. Chem.* Vol. 97, pp. 10769-10773.

Martinson, A. B. F.; Hamann, T. W.; Pellin, M. J. & Hupp, J. T. (2008). New Architectures for Dye-Sensitized Solar Cells. *Chem. – Eur. J.*, Vol. 14, 4458– 4467.

Matsumoto, M.; Wada, Y.; T. Kitamura, T.; Shigaki, K.; Inoue, T.; Ikeda, M. and Yanagida, S. (2001). Fabrication of solid-state dye-sensitized TiO_2 solar cell using polymer electrolyte. *Bull. Chem. Soc. Japan*, vol. 74, pp. 387-393.

Miyasaka, T. & Kijitori, Y. (2004). Low-temperature fabrication of dye-sensitized plastic electrodes by electrophoretic preparation of mesoporous TiO2 layers. *J. Electrochem. Soc.*, Vol. 151(11), pp. A1767-A1773.

Monari, A.; Assfeld, X.; Beley, M. & Gros, P.C. (2011). Theoretical Study of New Ruthenium-Based Dyes for Dye-Sensitized Solar Cells. *J. Phys. Chem. A*, Vol.115 (15), pp. 3596–3603.

Mor, G.K.; Shankar, K.; Paulose, M.; Varghese, O.K. & Grimes, C.A. (2006). Use of Highly-Ordered TiO_2 Nanotube Arrays in Dye-Sensitized Solar Cells. *Nanoletters*, Vol., 6, Issue 2, p.215-218.

Moser, J.-E. (2005). Solar Cells Later rather than sooner, *Nature materials*, Vol. 4, pp. 723-724.

Nansen, R. (1995). S u n Power: The Global Solution for the Coming Energy Crisis s. Ocean Press, ISBN-10: 0964702118, Washington, USA.

Nazerruddin, M.K; Kay, A.; Ridicio, I.; Humphry-Baker, R.; Mueller, E.; Liska, P.; Vlachopoulos, N. & Gratzel, M. (1993). *J. Amer. Chem. Soc.* Vol. 115, pp. 6382-6390.

Nazeeruddin, M. K.; De Angelis, F.; Fantacci, S.; Selloni, A.; Viscardi, G.;Liska, P.; Ito, S.; Takeru, B., & Graetzel, M. (2005). Combined Experimental and DFT-TDDFT Computational Study of Photoelectrochemical Cell Ruthenium Sensitizers. J. Am. Chem. Soc, Vol. 127(48), pp. 16835-16847.

Noack, V., Weller, H., & Eychmuller, A. (2002). J. Phys. Chem. B, Vol. 106, pp. 8514.

Nogueira, A.F.; De Paoli, M.A.; Montanan, I.; Monkhouse, R. & Durrant, I. (2001). J. Phys. Chem. B, Vol. 105, pp. 7417.

Nozik, A.J. (2001). Quantum Dot Solar Cells. Annu. Rev. Phys. Chem. Vol. 52, pp. 193-231.

Nozik, A. J. (2004). Quantum dot solar cells. Next Gener. Photovoltaics, pp. 196-222.

Nozik, A. J. (2005). Exciton multiplication and relaxation dynamics in quantum dots: applications to ultrahigh-efficiency solar photon conversion. Inorg. Chem. Vol. 44,pp. 6893–6899.

O'Regan B. & Gratzel, M. (1991). A low cost, high-efficiency solar cell based on dye-sensitized colloidal TiO2 films. Nature; Vol. 353, pp. 737–739.

Pagliaro, M.; Palmisano, G., & Ciriminna,R. (2008). Working principles of dye-sensitised solar cells and future applications,third print edition of Photovoltaics International journal, w w w.pv- te ch.org.

Pavasupree, S.; Suzuki, Y.; Yoshikawa, S., & Kawahata, R. (2005). Synthesis of Titanate, $TiO_2(B)$, and AnataseTiO$_2$ Nanofibers from Natural Rutile Sand, J. Solid State Chem, Vol. 178 (10), pp. 3110-3116.

Pavasupree, S.; Ngamsinlapasathian, S.; Nakajima, M.; Suzuki, Y., & Yoshikawa, S. (2006). Synthesis, characterization, photocatalytic activity and dye-sensitized solar cell performance of nanorods/nanoparticles TiO2 with mesoporous structure, Journal of Photochemistry and Photobiology A: Chemistry, Vol. 184. pp 163–169.

Plass, R.; Pelet, S.; Krueger, J., & Gratzel, M. (2002). Quantum Dot Sensitization of Organic-Inorganic Hybrid Solar Cells, Phys. Chem. B, Vol. 106 (31), pp. 7578 -7580.

Polo, A.S., and Iha, N.Y. (2006). Blue sensitizers for solar cells: natural dyes from Calafate and Jaboticaba. Sol. Enrgy Mater. Sol. Cells, Vol. 90, pp.1936-1944.

Qu, P.; Thompson, D. W., & Meyer, G. J. (2000). Temperature Dependent, Interfacial Electron Transfer from Ru(II) Polypyridyl Compounds with Low Lying Ligand Field States to Nanocrystalline Titanium Dioxide. Langmuir, Vol. 16, pp. 4662-4671.

Ren, Y.; Zhang, Z.; Gao, E.; Fang, S., & Cai, S. (2001). A dye-sensitized nanoporous TiO$_2$ photoelectrochemical cell with novel gel network polymer electrolyte. J. Appl. Electochem.,Vol. 31, pp. 445-447.

Robel, I.; Bunker, B. A. & Kamat, P. V. (2005). Single-walled carbon nanotube-CdS nanocomposites as light-harvesting assemblies: Photoinduced charge-transfer interactions, Adv. Mater., 17, 20, pp. 2458-2463..

Sayer, R. A.; Hodson, S.L. & Fisher, T.S. (2010). Improved Efficiency of Dye-Sensitized Solar Cells Using a Vertically Aligned Carbon Nanotube Counter Electrode, J. Sol. Energy Eng, Vol. 132, Issue 2, 021007 (4 pages).

Schmidt-Mende, L. & Gratzel, M. (2006). TiO$_2$ pore-filling and its effect on the efficiency of solid-state dye-sensitized solar cells, Thin Solid Films 500, pp. 296–301.

Shen, Q., Katayama, K., Sawada, T., Yamaguchi, M., and Toyoda, T. (2006). Optical Absorption, Photoelectrochemical, and Ultrafast Carrier Dynamic Investigations of TiO$_2$ Electrodes Composed of Nanotubes and Nanowires Sensitized with CdSe Quantum Dots, Japanese Journal of Applied Physics, Vol. 45, No. 6B, pp. 5569-5574.

Shoute, L. C. T., & Loppnow, G. R. (2003). Excited-state Metal-to-Ligand Charge Transfer Dynamics of a Ruthenium(II) Dye in Solution and Adsorbed on TiO2 Nanoparticles from Resonance Raman Spectroscopy. *J. Am. Chem. Soc.*, Vol.125, pp.15636-15646.

Smestad, G.P. (1998). Education and solar conversion: Demonstrating electron transfer, *Sol. Energy Mater. Sol. Cells*, Vol. 55, pp. 157-178.

Smestad, G.P. & Gratzel, M. (1998). Demonstrating Electron Transfer and Nanotechnology: A Natural Dye-Sensitized Nanocrystalline Energy Converter, *Journal of Chemical Education*, Vol. 75 No. 6, pp. 752-756.

Späth, M., Sommeling, P. M., van Roosmalen, J. A. M., Smit, H. J. P., van der Burg, N. P. G., Mahieu, D. R., Bakker, N. J. & Kroon, J. M. (2003). Reproducible Manufacturing of Dye-Sensitized Solar Cells on a Semi-automated Baseline, *Prog.   Photovolt: Res. Appl.* Vol. 11, pp. 207-220.

Suri, P., Panwar, M., & Mehra, R. (2007). Photovoltaic performance of dye-sensitized ZnO solar cell based on Eosin-Y photosensitizer, Materials Science-Poland, Vol. 25, No. 1, pp. 137-144.

SSun, J.Q.; Wang, J.S.; Wu, X.C.; Zhang, G.S.; Wei, J.Y.; Zhang, S.Q.; Li, H. & Chen, D.R. (2006). Novel Method for High-Yield Synthesis of Rutile SnO2 Nanorods by Oriented Aggregation. *Crystal Growth & Design*, Vol., 6, Issue 7, pp.1584-1587.

Suzuki, Y.; Ngamsinlapasathian, S.; Yoshida, R., & Yoshikawa, S. (2006). Partially nanowires-structured porous TiO2 electrode for dye-sensitized solar cell. *Central Euro. J. Chem.*, Vol. 4 (3), pp. 476 -488.

Tennakone, K.; Kumara, G.; Kottegota, I. & Wijayantha, K. (1997). The photostability of dye-sensitized solid state photovoltaic cells: factors determining the stability of the pigment in n-TiO2/Cyanidin/p-CuI cells. *Semicond. Sci. Technol.* Vol. 12, pp. 128.

Tennakone, K.; Perera, V.P.S; Kottegoda, I.R.M. & Kumara, G. (1999). Dye-sensitized solid state photovoltaic cell based on composite zinc oxide/tin (IV) oxide films, *J. Phys. D-App. Phys.*, Vol. 32, No. 4, pp. 372.

Tiwari, A., & Snure, M. (2008). Synthesis and Characterization of ZnO Nano-Plant-Like Electrodes. *Journal of Nanoscience and Nanotechnology*, Vol. 8, pp. 3981-3987.

Vogel, R. & Weller. H. (1994). Quantum-Sized PbS, CdS, AgzS, Sb&, and Bi& Particles as Sensitizers for Various Nanoporous Wide- Bandgap Semiconductors. *J. Phys. Chem.* Vol. 98, pp. 3183-3188.

Wang, P., Zakeeruddin, S.M., Moser, J.,-E., Humphry-Baker, R., and Gratzel, M. (2004). "A solvent-free, SeCN-/(SeCN)3− based ionic liquid electrolyte for high-efficiency dye-sensitized nanocrystalline solar cells," *Journal of the American Chemical Society*, vol. 126, no. 23, pp. 7164–7165.

Wang, P.; Klein, C.; Humphry-Baker, R.; Zakeeruddin, S. M. & Gratzel, M. (2005). Stable >= 8% efficient nanocrystalline dye-sensitized solar cell based on an electrolyte of low volatility. *Appl. Phys. Lett.* Vol. 86 (12), pp. Art. No.123508.

Wongcharee, K.; Meeyoo, V. & Chavadej, S. (2007). Dye-sensitized Solar Cell Using Natural Dye Extract From Rosslea and Blue Pea Flower . *Sol. Enrgy. Mater. Sol. Cells*, Vol. 91 (7), pp. 566-571.

www.g24i.com

www.intertechpira.com

Xiang, J.H., Zhu, P.X., Masuda, Y., Okuya, M., Kaneko, S. & Koumoto, K. (2006). Flexible solar-cell from zinc oxide nanocrystalline sheets self-assembled by an in-situ electrodeposition process. *J. Nanosci. Nanotechnol.* Vol. 6 (6), pp. 1797-1801.

Yanagida, S., (2006). Recent research progress of dye-sensitized solar cells in Japan. *C. R. Chemie.* Vol. 9, pp. 597-604.

Yang, M., Thompson, D., and Meyer, G. (2000). Dual Sensitization Pathways of TiO_2 by $Na_2[Fe(bpy)(CN)_4]$. *Inorg. Chem.* Vol. 39, pp. 3738-3739.

Yanagida, S., Senadeera, G.K.R., Nakamura, K., Kitamura, T., & Wada, Y. (2004). Polythiophene-sensitized TiO_2 solar cells. *J. Photohem. Photbiol. A*, Vol. 166, pp. 75-80.

Yen, W.; Hsieh, C.; Hung, C.; Hong-Wen Wang, H., & Tsui, M. (2010). Flexible TiO2 Working Electrode for Dye-sensitized Solar Cells. *Journal of the Chinese Chemical Society*, Vol 57, pp.1162-1166.

Zaban, A.; Micic, O.I.; Gregg, B.A. & Nozik, A.J. (1998). Photosensitization of Nanoporous TiO_2 Electrodes with InP Quantum Dots. *Langmuir, Vol.* 14 (12), pp. 3153-3156.

Zhang, W.; Zhu, R.; Li, F.; Qing Wang, Q. & Bin Liu, B. (2011). High-Performance Solid-State Organic Dye Sensitized Solar Cells with P3HT as Hole Transporter. J. Phys. Chem. C, Vol. 115 (14), pp 7038–7043.

Zhu, K.; Neale, N.R.; Halverson, A.F.; Kim, J.Y. & Frank, A.J. (2010). Effects of Annealing Temperature on the Charge-Collection and Light-Harvesting Properties of TiO_2 Nanotube-Based Dye-Sensitized Solar Cells. J. Phys. Chem. C, Vol. 114 (32), pp 13433–13441.

Zhao, J.; Wang, A. & Green, M.A. (1999). 24.5% Efficiency Silicon PERT Cells on MCZ Substrates and 24.7% Efficiency PERL Cells on FZ Substrates. *Progress in Photovoltaics*, Vol. 7, pp. 471-474.

Zweibel, K. & Green, M.A. (ed.) (2000). *Progress in Photovoltaics: Research and Applications,* Volume 8, Issue 1, pp. 171 – 185, John Wiley & Sons, Ltd;

9

Dye-Sensitized Solar Cells Based on Polymer Electrolytes

Mi-Ra Kim, Sung-Hae Park, Ji-Un Kim and Jin-Kook Lee
Department of Polymer Science & Engineering, Pusan National University,
Jangjeon-dong, Guemjeong-gu, Busan,
South Korea

1. Introduction

Dye-sensitized solar cells (DSSCs) using organic liquid electrolytes have received significant attention because of their low production cost, simple structure and high power conversion efficiency [1-5]. Recently, the power conversion efficiencies of DSSCs using Ruthenium complex dyes, liquid electrolytes, and Pt counter electrode have reached 10.4 % (100 mW/cm², AM 1.5) by Grätzel group [6]. However, the important drawback of DSSCs using liquid electrolyte is the less long-term stability due to the volatility of the electrolyte contained organic solvent. In the viewpoint for commercialize, durability is a crucial component. Then, gel electrolytes are being investigated to substitute the liquid electrolytes [7-10]. One way to make a gel electrolyte is to add organic or inorganic (or both) materials. In the past decades, many studies have been carried out on this kind of gel electrolyte, and great progress has been achieved [11-12]. The advantages of them include; limited internal shorting, leakage of electrolytes and non-combustible reaction products at the electrode surface existing in the gel polymer electrolytes [13-14]. However, because of their complicated preparing technology and poor mechanical strength, they cannot be used in commercial production [15-16]. To overcome this problem, the polymer membrane is soaked in an electrolyte solution that has been examined [17-19].

To prepare the polymer membrane for polymer electrolyte, a number of processing techniques such as drawing [20], template synthesis [21-22], phase separation [23], electrospinning [24], etc. have been used. Among of these, the electrospinning technology is a simple and low-cost method for making ultra-thin diameter fibers. This technique, invented in 1934, makes use of an electrical field that is applied across a polymer solution and a collector, to force a polymer solution jet out from a small hole [25]. When the diameters of polymer fiber materials are shrunk of micrometers to submicrons or nanometers, several amazing characteristics appear such as a very large surface area to volume ratio, flexibility in surface functionalities, and superior mechanical performance compared to any other known forms of this material [26]. In recent years, the electrospinning method has gained greater attention. A vastly greater number of synthetic and natural polymer solutions were prepared with electrospun fibers, such as poly(ethylene oxide) (PEO) in distilled water [27], polyurethane in N,N-dimethylformamide (DMF) [28], poly(ε-caprolactone) (PCL) in acetone [29], PVDF in acetone/ N,N-dimethylacetamide (DMAc) (7:3 by weight) [30], and regenerated cellulose in 2:1(w/w) acetone/DMAc [31].

Many applications of electrospun fibers were also studied. In addition, this technique is highly versatile and allows the processing of not only many different polymers into polymeric nanofibers, but also the co-processing of polymer mixtures, mixtures of polymers, and low molecular weight nonvolatile materials, etc [13,32].

2. Principle

2.1 Dye-sensitized solar cells (DSSCs)
2.1.1 History of DSSCs
The history of the sensitization of semiconductors to light of wavelength longer than that corresponding to the band gap has been presented elsewhere [33,34]. It is an interesting convergence of photography and photo-electrochemistry, both of which rely on photo-induced charge separation at a liquid–solid interface. The silver halides used in photography have band gaps of the order of 2.7–3.2 eV, and are therefore insensitive to much of the visible spectrum, just as is the TiO_2 now used in these photo-electrochemical devices.

The material has many advantages for sensitized photochemistry and photo-electrochemistry: it is a low cost, widely available, non-toxic and biocompatible material, and as such is even used in health care products as well as domestic applications such as paint pigmentation. The standard dye at the time was *tris*(2,2'-bipyridyl-4,4'-carboxylate) ruthenium(II), the function of the carboxylate being the attachment by chemisorption of the chromophore to the oxide substrate. Progress thereafter, until the announcement in 1991 of the sensitized electrochemical photovoltaic device with a conversion efficiency at that time of 7.1% under solar illumination, was incremental, a synergy of structure, substrate roughness and morphology, dye photophysics [35] and electrolyte redox chemistry. That evolution has continued progressively since then, with certified efficiency now over 10%.

2.1.2 Structure and working principles of DSSCs
The DSSC consists of the following staffs (Fig. 1). (1) transparent conductive oxide glass (F-doped SnO_2 glass (FTO glass), (2) Nanoporous TiO_2 layers (diameter ; 15-20 nm), (3) dye monolayer bonded to TiO_2 nano-particles, (4) electrolytes consisting of I- and I_3^- redox species, (5) platinum, (6) a counter electrode.

A schematic presentation of the operating principles of the DSSC is given in Fig. 2. At the heart of the system is a mesoscopic oxide semiconductor film, which is placed in contact with a redox electrolyte or an organic hole conductor. The choice of material has been TiO_2 (anatase) although alternative wide hand gap oxides such as ZnO, and Nb_2O_5 have also been investigated. Attached to the surface of the nanocrystalline film is a monolayer of the sensitizer. Photo-excitation of the latter results in the injection of an electron into the conduction band of the oxide. The dye is regenerated by electron donation from the electrolyte, usually an organic solvent containing a redox system, such as the iodide/triiodide couple. The regeneration of the sensitizer by iodide intercepts the recapture of the reduction of the conduction band electron by the oxidized dye. The iodide is regenerated in turn by triiodide at the counter electrode when the circuit is completed via electron migration through the external load.

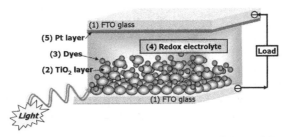

Fig. 1. A schematic presentation of a cross-section structure of the DSSC.

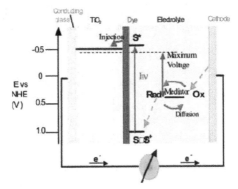

Fig. 2. A schematic presentation of the operating principles of the DSSC.

$$TiO_2/S + h\nu \rightarrow TiO_2/S^* \tag{I}$$

$$TiO_2/S^* \rightarrow TiO_2/S^+ + e_{cb} \tag{II}$$

$$TiO_2/S^+ + e_{cb} \rightarrow TiO_2/S \tag{III}$$

$$TiO_2/S^+ + (3/2)I^- \rightarrow TiO_2/S + (1/2)I_3^- \tag{IV}$$

$$(1/2)I_3^- + e_{(pt)} \rightarrow (3/2)I^-,\ I_3^- + 2e_{cb} \rightarrow 3I^- \tag{V}$$

Light absorption is performed by a monolayer of dye (S) adsorbed chemically at the semiconductor surface and excited by a photon of light (Eq. (I)). After having been excited (S*) by a photon of light, the dye-usually a transition metal complex whose molecular properties are specifically for the task is able to transfer an electron to the semiconductor (TiO₂) by the injection process (Eq. (II)). The efficiency of a DSSC in the process for energy conversion depends on the relative energy levels and the kinetics of electron transfer processes at the liquid junction of the sensitized semiconductor/electrolyte interface. For efficient operation of the cell, the rate of electron injection must be faster than the decay of the dye excited state. Also, the rate of rereduction of the oxidized dye (dye cation) by the electron donor in the electrolyte (Eq. (IV)) must be higher than the rate of back reaction of the injected electrons with the dye cation (Eq. (III)), as well as the rate of reaction of injected electrons with the electron acceptor in the electrolyte (Eq. (V)). Finally, the kinetics of the reaction at the counter electrode must also guarantee the fast regeneration of charge mediator (Eq. (V)), or this reaction could also become rate limiting in the overall cell performance [36-39].

2.1.3 Present DSSC research and development

2.1.3.1 Sensitizer (Dye)

The ideal sensitizer for a single junction photovoltaic cell converting standard global AM 1.5 sunlight to electricity should absorb all light below a threshold wavelength of about 920 nm. In addition, it must also carry attachment groups such as carboxylate or phosphonate to firmly graft it to the semiconductor oxide surface. Upon excitation it should inject electrons into the solid with a quantum yield of unity. The energy level of the excited state should be well matched to the lower bound of the conduction band of the oxide to minimize energetic losses during the electron transfer reaction.

Its redox potential should be sufficiently high that it can be regenerated via electron donation from the redox electrolyte or the hole conductor. Finally, it should be stable enough to sustain about 108 turnover cycles corresponding to about 20 years of exposure to natural light.

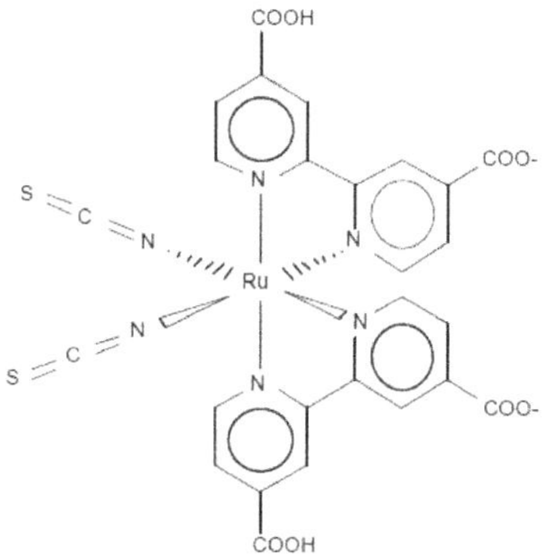

Fig. 3. Chemical structure of the N3 ruthenium complex used as a charge transfer sensitizer in DSSCs.

The best photovoltaic performance both in terms of conversion yield and long-term stability has so far been achieved with polypyridyl complexes of ruthenium and osmium. Sensitizers having the general structure $ML_2(X)_2$, where L stands for 2,2'-bipyridyl-4,4'-dicarboxylic acid M is Ru or Os and X presents a halide, cyanide, thiocyanate, acetyl acetonate, thiacarbamate or water substituent, are particularly promising. Thus, the ruthenium complex cis-$RuL_2(NCS)_2$, known as N3 dye, shown in Fig. 3 has become the paradigm of heterogeneous charge transfer sensitizer for mesoporous solar cells.

2.1.3.2 Mesoporous oxide film development

When the dye-sensitized nanocrystalline solar cell was first presented, perhaps the most puzzling phenomenon was the highly efficient charge transport through the nanocrystalline TiO_2 layer. The mesoporous electrodes are very much different compared to their compact analogs because (i) the inherent conductivity of the film is very low; (ii) the small size of the nanocrystalline particles does not support a built-in electrical field; and (iii) the electrolyte penetrates the porous film all the way to the back-contact making the semiconductor/electrolyte interface essentially three-dimensional. Charge transport in mesoporous systems is under keen debate today and several interpretations based on the Montrol Scher model for random displacement of charge carriers in disordered solids [40] have been advanced. However, the "effective" electron diffusion coefficient is expected to depend on a number of factors such as trap filling and space charge compensation by ionic motion in the electrolyte. Therefore, the theoretical and experimental effort will continue as there is a need for further in depth analysis of this intriguing charge percolation process. The factors controlling the rate of charge carriers percolation across the nanocrystalline film are presently under intense scrutiny. Intensity modulated impedance spectroscopy has proved to be an elegant and powerful tool [41,42] to address important questions related to the characteristic time constants for charge carrier transport and reaction dynamics in DSSCs. On the material science side, future research will be directed towards synthesizing structures with a higher degree of order than the random fractal-like assembly of nanoparticles shown in Fig. 4. A desirable morphology of the films would have the mesoporous channels or nanorods aligned in parallel to each other and vertically with respect to the transparent conducting oxide (TCO) glass current collector. This would facilitate pore diffusion, give easier access to the film surface avid grain boundaries and allow the junction to be formed under better control.

Fig. 4. Scanning electron micrograph of a sintered mesoscopic TiO_2 film supported on an FTO glass. The average particle size is 20 nm.

One approach to fabricate such oxide structures is based on surfactant templates assisted preparation of TiO_2 nanotubes as described in recent paper by Adachi et al. [43]. The hybrid nanorod-polymer composite cells developed by Huynh et al. [44] have confirmed the superior photovoltaic performance of such films with regards to random particle networks.

2.1.3.3 Polymer electrolytes

The polymer-based material is generally produced in a thin-film configuration by casting or spin-coating techniques. Polymer electrolytes are composed by alkaline salts (e.g. lithium or sodium salts) dissolved in a high molar mass polyether host (e.g. poly(ethylene oxide) (PEO) or poly(propylene oxide) (PPO)) [45]. In polymer electrolytes, the polymer matrix should be an efficient solvent for the salt, capable of dissociating it and minimizing the formation of ion pairs. The solubility of the salt relies on the ability of the electron donor atoms in the polymer chain to coordinate the cation through a Lewis type acid–base interaction. This interaction also depends on the lattice energy of the salt and the structure of the host polymer. The mechanism for ionic motion in polymer electrolytes results from a solvation-desolvation process along the chains that occurs predominantly in the amorphous polymer phase. Since the ionic motion is strictly correlated with the segmental motion of the polymer chains, the ionic conductivity increases with increasing chain mobility. The ionic conductivity is also a function of the number of charge carriers in the polymer matrix. However, above a limiting high salt concentration the segmental motion of the polymer chains is reduced due to an "ionic cross-linking" which decreases ionic conductivity [46].

2.1.4 Characteristics of DSSCs

There are several factors for characterization of DSSCs below; (See Fig. 5).

2.1.4.1 Incident photon-to current efficiency (IPCE)

The Incident Photon-to-Current Efficiency (IPCE), also called external quantum efficiency, is defined as the number of electrons generated by light in the external circuit divided by the number of incident photons as a function of excitation wavelength. It is expressed in Eq. (1) [47]. A high IPCE is a prerequisite for high-power photovoltaic applications, which depends on the sensitizer photon absorption, excited state electron injection. And electrons transport to the terminals.

$$\text{IPCE} = [(1.25 \times 10^3) \times \text{photocurrent density} (mAcm^{-2})] \times [\text{Wavelength} (nm) \times \text{photon flux } (Wm^{-2})]^{-1} \quad (1)$$

2.1.4.2 Open-circuit voltage (V_{oc})

If there is no external circuit, the incoming photons will still create hole-electron pairs and they will still travel downhill to the layers, but they will pile up there because there is no external wire. The number of carriers leaking back is equal to the number being generated by the incoming light, an equilibrium voltage has been reached. This is called the open-circuit voltage (V_{oc}). In DSSCs, V_{oc} is defined by the difference between Fermi level of TiO_2 and redox potential of electrolyte [10].

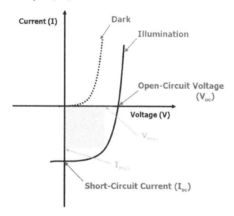

Fig. 5. Several factors for characterization of DSSC.

2.1.4.3 Short-circuit current (I_{sc}) and Short-circuit current density (J_{sc})

If the external circuit is simply a wire and has no appreciable resistance, the current that flows is the short-circuit current (I_{sc}) and is directly related to the number of photons of light being absorbed by the cell. Short-circuit current density (J_{sc}) is short-circuit current per active area.

2.1.4.4 Fill factor (FF)

The fill factor (FF) is obtained by diving the product of current and voltage measured at the power point (maximum output power P_{max}) by the product of short-circuit current and the open-circuit voltage. The power point is the maximum product of the cell voltage and the photocurrent obtained on the I-V curve.

$$P_{max} = I_{max} \times V_{max} = FF \times I_{sc} \times V_{oc} \quad (2)$$

$$FF = P_{max} / I_{sc} \times V_{oc} \qquad (3)$$

2.1.4.5 Power conversion efficiency

The power conversion efficiency (η) of the dye-sensitized solar cell is determined by the photocurrent density measured at short-circuit, V_{oc}, the FF of the cell, and the intensity of the incident light (P_i) as shown in follow equation.

$$\eta = I_{sc} \times V_{oc} \times FF / P_i \qquad (4)$$

Sometimes the use of the ratio of the maximum power output (P_{max}) to the incident power input (P_{in}), defined as

$$\eta = P_{max} / P_{in} = I_{sc} \times V_{oc} \times FF / P_{in} \qquad (5)$$

2.1.5 Advantages and disadvantages of DSSCs

The major advantage of the concept of dye sensitization is the fact that the conduction mechanism is based on a majority carrier transport as opposed to the minority carrier transport of conventional inorganic cells. This means that bulk or surface recombination of the charge carriers in the TiO_2 semiconductor cannot happen. Thus, impure starting materials and a simple cell processing without any clean room steps are permitted, yet resulting in promising power conversion efficiencies of 7 - 11% and the hope of a low-cost device for photoelectrochemical solar energy conversion. On the other hand impure materials can result in a strongly reduced lifetime of the cells. The most important issue of the dye-sensitized cells is the stability over the time and the temperature range which occurs under outdoor conditions. Although it could be shown, that intrinsic degradation can considerably be reduced, the behavior of the liquid electrolyte under extreme conditions is still unknown. For a successful commercialization of these cells, the encapsulation/sealing, the coloration and the electrolyte filling has to be transferred into fully automated lines including the final closure of the filling openings. Therefore, a significant effort is taken in order to replace the liquid electrolyte by a gel electrolyte, a solid-state electrolyte or a p-type conducting polymer material.

2.2 Electrospinning
2.2.1 History of electrospinning method

The process of using electrostatic forces to form synthetic fibers, known as electrospinning, has been known for over 100 years. From 1934 to 1944, Formhals published a series of patents [25,48,49,50], describing an experimental setup for the production of polymer filaments using an electrostatic force. A polymer solution, such as cellulose acetate, was introduced into the electric field. It was not until 1934, when Formhals patented a process, that electrospinning truly surfaced as a valid technique for sinning small-diameter fibers. In 1952, Vonnegut and Neubauer were able to produce streams of highly electrified uniform droplets of about 0.1mm in diameter [51]. They invented a simple apparatus for the electrical atomization. In 1955, Drozin investigated the dispersion of a series of liquids into aerosols under high electric potentials [52]. Formhals studied for a better understanding of the electrospinning process; however, it would be nearly 30years before Taylor would publish work regarding the jet forming process. In 1969, Taylor published his work examining how the polymer droplet at the end of a capillary behaves when an electric field is applied [53]. In his studies he found that the pendant droplet develops into a cone (now called the Taylor cone) when the surface tension is balanced by electrostatic forces. In 1971, Baumgarten made an apparatus to electrospun acrylic fibers with diameters in the range of 0.05-1.1 microns [54]. Since 1980s and especially in recent years, the electrospinning process has regained more attention probably due to interest in nanotechnology, as ultrafine fibers or fibrous structures of various polymers with diameters down to submicrons or nanometers can be easily fabricated with this process. One of the most important applications of traditional micro-size fibers, especially engineering fibers such and carbon, glass, and Kevlar fivers is to be used as reinforcements in composite developments [55]. In addition to composite reinforcement, other application fields based on electrospun polymer nanofibers, such as medical prosthesis [55-57], filtration, cosmetics [59], tissue engineering [60], liquid crystal device [61], electromagnetic shielding and photovoltaic device [62], have been steadily extended especially in recent years.

2.2.2 Principles of electrospinning method
Fig. 6 shows the schematic diagram of the electrospinning method. There are basically three components to fulfill this process: a high voltage supplier, a capillary tube with a pipette or needle of a small diameter, and a metal collecting screen. A typical electrospinning setup consists of a capillary through which the liquid to be electrospun is forced; a high voltage source with positive or negative polarity, which injects charge into the liquid; and a grounded collector. Once it has been ejected out of the metal needle with a small hole, the polymer solution was introduced into the electric field. The polymer filaments formed between two electrodes bearing electrical charges of opposite polarity. Before the polymer solution reach the collector, the polymer solution jet evaporates or solidifies, and is collected as an interconnecting web of small fibers. One of these electrodes was placed into the solution and the other onto a collector (Fig. 7).

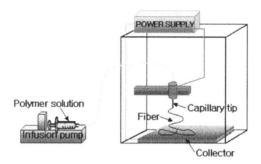

Fig. 6. Schematic diagram of the electrospinning method.

2.2.3 Ideal targets of electrospun nanofibers
As long as a polymer can be electrospun into nanofibers, ideal targets would be in that: (1) the diameters of the fibers be consistent and controllable, (2) the fiber surface be defect-free or defect-controllable, and (3) continuous single nanofibers be collectable. However researches so far have shown that there three targets are by no means easily achievable [26].

2.2.4 Processing parameters
There are a number of processing parameters that can greatly affect fiber formation and structure. The parameters are polymer concentration, applied voltage, flow rate, tip to

collector distance (TCD) and Solvent volatility. Furthermore, all parameters can influence the formation of bead defects.

2.2.4.1 Polymer concentration

The polymer concentration determines the spinnability of a solution, namely whether a fiber forms or not. The solution must have a high enough polymer concentration for chain entanglements to occur; however the solution cannot be either too dilute or too concentrated. If the solution is too dilute then the fibers break up into droplets before reaching the collector due to the effects of surface tension. However, if the solution is too concentrated then fibers cannot be formed due to the high viscosity, which makes it difficult to control the solution flow rate through the capillary. Thus, an optimum range of polymer concentrations exists in which fibers can be electrospun when all other parameters are held constant. A higher viscosity results in a larger fibers diameter [53, 63]. When polymers dissolve in a solvent, the solution viscosity is proportional to the polymer concentration. Thus, the higher the polymer concentration the larger the resulting nanofiber diameters will be.

2.2.4.2 Applied voltage

The strength of the applied electric field controls formation of fibers from several microns in diameter to tens of nanometers. Deitzel et al. examined a polyethylene oxide (PEO)/water system and found that increases in applied voltage altered the shape of the surface at which the Taylor cone and fiber jet were formed [64]. At lower applied voltages the Taylor cone formed at the tip of the drop; however, as the applied voltage was increased the volume of the drop decreased until the Taylor cone was formed at the tip of the capillary, which was associated with an increase in bead defects seen among the electrospun fibers (Fig.8). Moreover, another parameter which affects the fiber diameter to a remarkable extent is the applied voltage. In general, a higher applied voltage ejects more fluid in a jet, resulting in a larger fiber diameter [27].

2.2.4.3 Flow rate

Polymer flow rate also has an impact of fiber size, and additionally can influence fiber porosity as well as fiber shape. In the Taylor's work, they realized that the cone shape at the tip of the capillary cannot be maintained if the flow of solution through the capillary is insufficient to replace the solution ejected as the fiber jet [53]. They demonstrated that both fiber diameter and pore size increase with increasing flow rate. Additionally, at high flow rates significant amounts of bead defects were noticeable, due to the inability of fibers to dry completely before reaching the collector [60].

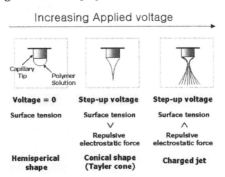

Fig. 7. Effect of varying the applied voltage on the formation of the Taylor cone.

2.2.4.4 Tip to collector distance (TCD)

The distance of between capillary tip and collector can also influence fiber size by 1-2 orders of magnitude. Additionally, this distance can dictate whether the end result is electrospinning or electrospraying. Doshi and Reneker found that the fiber diameter decreased with increasing distances from the Taylor cone [65].

2.2.4.5 Solvent volatility

Choice of solvent is also critical as to whether fibers are capable of forming, as well as influencing fiber porosity. In order for sufficient solvent evaporation to occur between the capillary tip and the collector a volatile solvent must be used. As the fiber jet travels through the atmosphere toward the collector a phase separation occurs before the solid polymer fibers are deposited, a process that is greatly influenced by volatility of the solvent. Zhao et al. examined the structural properties of 15 wt % of poly(Vinylidene Fluoride) nanofibers with different volume ratios in DMF/Acetone [66]. When DMF was used as the solvent without acetone, bead-fibers were found. When 9:1 DMF/acetone was used a s the solvent in the polymer solution, beads in the electrospun almost disappeared. Furthermore, the ultafine fibers without beads demonstrated clearly when the acetone amount in the solution increased to 20 %. Acetone is more volatile than DMF. Furthermore, the changes of solution properties by the addition of acetone could probably improve the electrospun membrane morphology and decrease the possibility of bead formation.

3. Results

3.1 Preparation of electrospun poly(vinylidene fluoride-hexafluoropropylene) (PVDF-HFP) nanofibers

Generally, in the electrospinning method, the changing of the parameters had a great effect on fiber morphology. To prepare the electrospun PVDF-HFP nanofiber films with the suitable morphology, we prepared the electrospun PVDF-HFP nanofiber films by several parameters such as the applied voltage(voltage supplier: NNC-ESP100, Nano NC Co., Ltd.), the tip-to-collector distance (TCD), and the concentration of the PVDF-HFP. First, the PVDF-HFP was dissolved in acetone/DMAc (7/3 weight ratio) for 24 hours at room temperature. Then, we prepared the electrospun PVDF-HFP nanofibers by the electrospinning method with different parameters. The applied voltage was ranged from 8 to 14 kV, TCD was varied from 13 to 21 cm, and the concentration of PVDF-HFP varied from 11 to 17 wt %. On all occasions, we used a syringe pump (781100, Kd Scientific) to control the flow rate of the polymer solution, the solution flow rate was 2 ml/h.

In the electrospinning method, the changing of the polymer concentration had a great effect on fiber morphology. To investigate the influence of polymer concentrations on the electrospun PVDF-HFP nanofibers, we prepared the PVDF-HFP nanofibers. When the polymer concentration were varied from 11 wt% to 17 wt%, TCD and applied voltages were 15 cm and 14 kV, respectively. Over the polymer solution of 19 wt% and below the polymer solution of 9 wt%, the nanofibers did not form. Fig. 8 shows the surface images of the electrospun PVDF-HFP nanofibers observed by FE-SEM and the diameter distributions of nanofibers. The increase of the polymer concentration resulted in an increase of the average fiber diameter of the electrospun PVDF-HFP nanofibers. In particular, the PVDF-HFP nanofiber, which was prepared from 15 wt% of polymer concentration showed a highly regular morphology with an average diameter of 800 - 1000 nm.

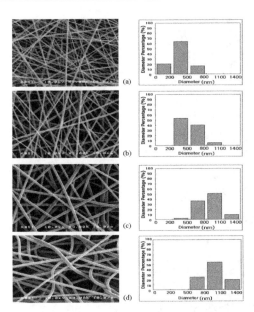

Fig. 8. FE-SEM images of electrospun PVDF-HFP nanofibers with different polymer concentrations (Applied voltage = 14 kV, TCD = 15 cm, flow rate = 2 ml/h) and their diameter distributions: (a) 11 wt%, (b) 13 wt%, (c) 15 wt%, (d) 17 wt%.

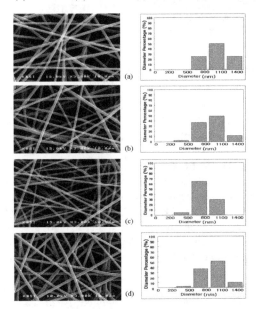

Fig. 9. FE-SEM images of electrospun PVDF-HFP nanofiber with different applied voltages (TCD = 15 cm, polymer concentration = 15 wt%, flow rate = 2 ml/h) and their diameter distributions: (a) 8 kV, (b) 10 kV, (c) 12 kV, (d) 14 kV.

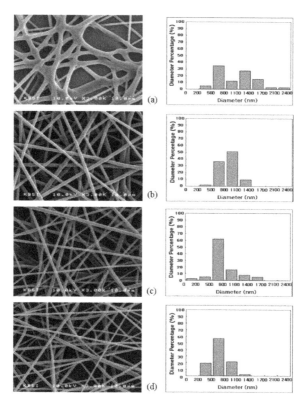

Fig. 10. FE-SEM images of electrospun PVDF-HFP nanofibers with different TCDs (Applied voltage = 14 kV, polymer concentration = 15 wt%, flow rate = 2 ml/h) and their diameter distributions: (a) 13 cm, (b) 15 cm, (c) 17 cm, (d) 19 cm.

To investigate the effect of applied voltage, experiments were carried out when the applied voltage was varied from 8 kV to 14 kV, TCD and polymer concentrations were held at 15 cm and 15 wt%, respectively. The morphologies of electrospun PVDF-HFP nanofibers prepared are shown in Fig. 9.

In addition, we prepared the electrospun PVDF-HFP nanofibers when the TCD was varied from 13 cm to 19 cm, applied voltage and polymer concentrations were held at 14 kV and 15 wt%, respectively. The morphologies of prepared electrospun PVDF-HFP nanofibers prepared are shown in Fig. 10. When the TCD was just close below 13 cm, irregular fiber morphology was formed, because the polymer jet arrived at the collector before the solidification. Therefore, we were able to optimize the preparation condition at an applied voltage of 14 kV, a polymer concentration of 15 wt%, and TCD of 15 cm to obtain the regular PVDF-HFP nanofibers.

As the changing of such parameters in the electrospinning method, the diameter and the morphology of the nanofibers fabricated were changed. At the condition of the 15 wt% of PVDF-HFP polymer solution, 14 kV of the applied voltage, 15 cm of the TCD and 2 ml/h of the flow rate, the nanofibers of the electrospun PVDF-HFP films showed extremely regular morphology with diameter of average 0.8 ~ 1.0 μm.

3.2 Characterizations of PVDF-HFP nanofibers

The pore size, the volume fraction and interconnectivity of pore domain, and the type of porous polymer matrix will determine the uptake and the ion conductivity of the electrolyte [63]. To investigate the effect of porous polymer matrix, the spin-coated PVDF-HFP film was also fabricated by using conventional spin-coating method, and measured the ionic conductivity under the same condition. The ionic conductivity of the spin-coated PVDF-HFP film was 1.37×10^{-3} S/cm, and this value showed lower value than the electrospun PVDF-HFP nanofiber film.

To measure the uptake and the porosity of the electropsun PVDF-HFP nanofiber films from electrolyte solution, the electropsun PVDF-HFP nanofiber films were taken out from the electrolyte solution after activation and excess electrolyte solution on the film surface was wiped.

The electrolyte uptake (U) was evaluated according to the following formula:

$$U = [(m-m_0)/m_0] \times 100\%$$

where m and m_0 are the masses of wet and dry of the electrospun nanofiber films, respectively.

The porosity (P) of the electrospun nanofibers was calculated from the density of electrospun PVDF-HFP nanofibers (ρ_m, g/cm^3) and the density of pure PVDF-HFP (ρ_p= 1.77g/cm^3):

$$P \text{ (vol.\%)} = (1 - \rho_m/\rho_p) \times 100$$

The density of the electrospun PVDF-HFP nanofibers was determined by measuring the volume and the weight of the electrospun PVDF-HFP nanofibers. The uptake and the porosity of the electrospun PVDF-HFP nanofiber film was obtained 653±50 % and 70±2.3 %, respectively, regardless the diameter and the morphology of nanofibers prepared with various parameters.

3.3 Fabrications of DSSCs devices using PVDF-HFP nanofibers

We prepared the DSSC devices, sandwiched with working electrode using TiO_2 impregnated dyes and counter electrode using a platinum (Pt, T/SP) electrode as two electrodes. The DSSC device was fabricated using this following process. The TiO_2 pastes (Ti-Nanoxide, HT/SP) were spread on a FTO glass using the doctor blade method and calcinated at 500 °C. The sensitizer Cis-di(thiocyanato)-N,N-bis(2,2'-bypyridil-4.4'-dicarboxylic acid)ruthenium (II) complex (N3 dye) was dissolved in pure ethanol in a concentration of 20 mg per 100 ml of solution. The FTO glass deposited TiO_2 was dipped in an ethanol solution at 45 °C for 18 hours. The electospun PVDF-HFP nanofibers or the spin-coated PVDF-HFP film were cut by 0.65 cm × 0.65 cm after drying, and put on the TiO_2 adsorbed the dyes, the electrolyte solution was dropped above them, and dried in a dry oven at 45 °C for 2 hours to evaporate wholly the solvent. To compare with the electrospun PVDF-HFP nanofiber films, the conventional spin-coating method was used for making a spin-coated PVDF-HFP film. In all cases, the thickness of the electrospun PVDF-HFP nanofibers and spin-coated PVDF-HFP film were 30±1 µm by using digimatic micrometer. The electrolyte was consisted of 0.10 M of iodine (I_2), 0.30 M of 1-propyl-3-methylimidazolium iodide (PMII), and 0.20 M of tetrabutylammonium iodide (TBAI) in the solution of ethylene carbonate (EC)/ propylene carbonate (PC)/ acetonitrile (AN) (8:2:5

v/v/v). The Pt pastes were spread on a FTO glass using the doctor blade method and calcinated at 400 °C.

3.4 Photovoltaic properties of the DSSC devices using PVDF-HFP nanofibers

The DSSC devices using several different electrospun PVDF-HFP nanofibers on various parameters were fabricated and their photovoltaic characteristics are summarized in Table 1 – 3. I-V curves of the DSSC devices using them are shown in Fig. 11. The concentration of the PVDF-HFP solution was 15 wt% in acetone/DMAc (7/3 by weight ratio). The photovoltaic characteristics of the DSSC devices were measured by using Solar Simulator (150 W simulator, PEC-L11, PECCELL) under AM 1.5 and 100 mW/cm² of the light intensity.

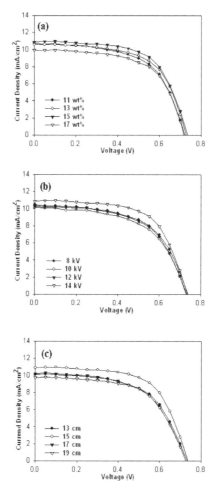

Fig. 11. I-V curves of DSSC devices using electrospun PVDF-HFP nanofibers under illumination at AM 1.5 condition: (a) different polymer concentrations, (b) different applied voltages, (c) different TCDs.

Polymer concentration (wt.%)	V_{OC} (V)	J_{SC} (mA/cm²)	FF	η (%)
11	0.74	10.88	0.60	4.78
13	0.73	10.57	0.62	4.78
15	0.74	10.89	0.63	5.02
17	0.72	9.92	0.62	4.41

Table 1. Photovoltaic performances of DSSC devices using electrospun PVDF-HFP nanofibers on different polymer concentrations

Applied voltage (kV)	V_{OC} (V)	J_{SC} (mA/cm²)	FF	η (%)
8	0.74	10.50	0.57	4.41
10	0.73	10.10	0.56	4.17
12	0.74	10.30	0.58	4.35
14	0.74	10.88	0.63	5.02

Table 2. Photovoltaic performances of DSSC devices using electrospun PVDF-HFP nanofibers on different applied voltages

TCD (cm)	V_{OC} (V)	J_{SC} (mA/cm²)	FF	η (%)
13	0.73	10.10	0.58	4.30
15	0.74	10.88	0.63	5.02
17	0.73	10.20	0.57	4.23
19	0.73	9.72	0.60	4.21

Table 3. Photovoltaic performances of DSSC devices using electrospun PVDF-HFP nanofibers on different TCDs

Type	V_{OC} (V)	J_{SC} (mA/cm²)	FF	η (%)
Electrospun PVDF-HFP nanofiber film	0.75	12.3	0.57	5.21
Spin-coated PVDF-HFP film	0.67	3.87	0.55	1.43

Table 4. Photovoltaic characteristics of DSSC devices using electrospun PVDF-HFP nanofiber film and spin-coated PVDF-HFP film in polymer electrolytes

The active area of the DSSC devices measured by using a black mask was 0.25 cm². The V_{OC}, J_{SC}, FF, and η of the DSSC device using the spin-coated PVDF-HFP film were 0.67 V, 3.87 mA/cm², 0.56, and 1.43 %, respectively. The η of DSSC device using the spin-coated PVDF-HFP film was lower than it of the DSSC device using electrospun PVDF-HFP nanofiber films, because of the decrease of J_{SC}, and all data are summarized in Table 4 and their I-V curves are shown in Fig. 12. This result seemed that because the porosity of the electrospun PVDF-HFP nanofibers is higher than it of the spin-coated PVDF-HFP film, ion transfer occurred well and regular nanofiber morphology helped to transfer ion produced by redox mechanism, therefore, overall power conversion efficiency of DSSC devices using the electrospun PVDF-HFP nanofiber films was higher than that of the DSSC device using spin-coated PVDF-HFP film. However, the minute change of nanofibers diameter was influenced little on power conversion efficiency.

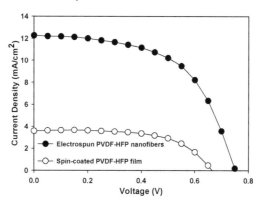

Fig. 12. I-V curves of the DSSC devices using electrospun PVDF-HFP nanofibers and spin-coated PVDF-HFP film.

3.5 Effect of electrolyte in the electrospun PVDF-HFP nanofibers on DSSC

The photovoltaic performance of DSSC devices using the electrospun PVDF-HFP nanofibers showed remarkable improved results compared to DSSC devices using the spin-coated PVDF-HFP film. To prove these results, the interfacial charge transfer resistances were investigated by the EIS measurement. The EIS data were measured with impedance analyzer at same condition using $FTO/TiO_2/electrolyte/Pt/FTO$ cells, and fitted by Z-MAN software (WONATECH) and Echem analyst (GAMRY). The Nyquist plots of the $FTO/TiO_2/electrolyte/Pt/FTO$ cells and charge transfer resistances are shown in Fig. 13 and Table 5, respectively. The equivalent circuit of DSSC devices is shown in Fig. 14. The R_S, $R1_{CT}$ and $R2_{CT}$ were series resistance, the charge transfer resistance of Pt/electrolyte interface, and the charge transfer resistance of TiO_2/electrolyte interface, respectively. The $R2_{CT}$ of the DSSC device using the spin-coated PVDF-HFP film was similar to that of the DSSC device using the electrospun PVDF-HFP nanofibers. However, the R_S and $R1_{CT}$ of the DSSC device using the spin-coated PVDF-HFP film were higher than those of the DSSC device using the electrospun PVDF-HFP nanofibers. These results showed that the spin-coated film has a higher resistance than the electropun nanofibers, and poor I^-/I^{3-} activity between Pt and electrolyte affected to the low value of the J_{SC}. As a result, the η of the DSSC device using the spin-coated PVDF-HFP film showed low value.

Type	R_S (Ω)	$R1_{CT}$ (Ω)	$R2_{CT}$ (Ω)
Electrospun PVDF-HFP nanofibers	21.70	11.01	11.07
Spin-coated PVDF-HFP film	31.87	25.02	14.37

Table 5. The series resistances (R_S), the charge transfer resistance of the Pt/electrolyte ($R1_{CT}$) and TiO_2/electrolyte ($R2_{CT}$) in the DSSC devices under AM 1.5 by the EIS measurement

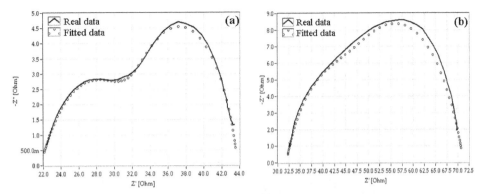

Fig. 13. Nyquist plots the FTO/TiO₂/electrolyte/Pt/FTO device using (a) electrospun PVDF-HFP nanofiber film electrolyte, and (b) spin-coated PVDF-HFP film electrolyte.

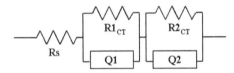

Fig. 14. The equivalent circuit of the DSSC device. (R_S: Series resistance, $R1_{CT}$: charge transfer resistance of Pt/electrolyte, $R2_{CT}$: charge transfer resistance of TiO₂/electrolyte, Q1 and Q2: constant phase element)

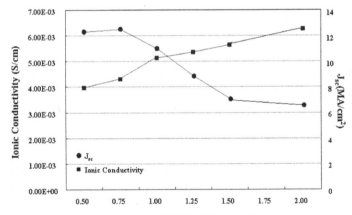

Fig. 15. Ionic conductivities of electrospun PVDF-HFP nanofiber films and J_{sc} of DSSC devices using electrospun PVDF-HFP nanofiber films with mole ratio of iodine to TBAI.

In addition, to investigate the photovoltaic effect of I_2 concentrations on DSSC using the electrospun PVDF-HFP nanofiber, we prepared FTO/TiO₂/Dye/Electrolyte/Pt/FTO devices with various mole ratios of I_2 to TBAI in electrolyte solutions. In Table 6, as the increase of the I_2 concentration in electrolyte, the ionic conductivity of the electrospun

PVDF-HFP nanofiber films increased, while the photocurrent density of the DSSC devices using the electrospun PVDF-HFP nanofibers electrolyte decreased. The relationship between the ionic conductivity the electrospun PVDF-HFP nanofiber films and the photocurrent density of the DSSC devices are illustrated in Fig. 15 and I-V curves are shown in Fig. 16. In general, the photocurrent density of DSSC using the liquid electrolyte is proportionate to the ionic conductivity in electrolyte. From these results, we found that the photocurrent density and the efficiency on DSSC using the electrospun PVDF-HFP nanofibers electrolyte are not necessarily proportionate to the ionic conductivity in electrolyte.

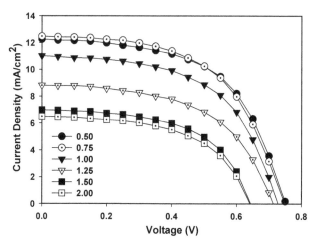

Fig. 16. The I-V curves of the DSSC devices using electrospun PVDF-HFP nanofibers with mole ratio of iodine to TBAI.

4. Future outlooks

During the rebirth of polymer electrospining over the past decade the applicability of electrospun fibers has become apparent across many fields. This highly adaptable process allows the formation of functional fibrous membranes for applications such as tissue engineering, drug delivery, sensor, cosmetic and photovoltaic devices. Electrospun nanofibers offer an unprecedented flexibility and modularity in design. Improvements in strength and durability, and their incorporation in composite membranes, will allow there scaffolds to compete with existing membrane technology. Currently, the research field of electrospnning is ripe with functional materials from resorbable cells to ceramic solid-phase catalyst and continued research interest is expected to improve most areas of full cells and photovoltaic cells.

5. Acknowledgement

This research was supported by the Converging Research Center Program through the National Research Foundation of Korea (NRF) funded by the Ministry of Education, Science and Technology (20090082141).

6. References

Adachi, M.; Murata, Y.; Okada, I. & Yoshikawa, S. (2003). Formation of Titania Nanotubes and Applications for Dye-Sensitized Solar Cells. *Journal of the Electrochemical Society*, Vol. 150, No. 8, pp. G488-G493, ISSN 0013-4651

Amadelli, R.; Argazzi, R.; Bignozzi, C. A. & Scandola, F. (1990). Design of antenna-sensitizer polynuclear complexes. Sensitization of titanium dioxide with [Ru(bpy)2(CN)2]2Ru(bpy(COO)2)22. *Journal of the American Chemical Society*, Vol. 112, No. 20, pp. 7099-7103, ISSN 0002-7863

Anton, Formhals (1934). Process and apparatus for preparing artificial threads. US Patent 1,975,504.

Anton, Formhals (1939). Method and apparatus for spinning. US Patent 2160962

Anton, Formhals (1940). Artificial thread and method of producing same. US Patent 2187306

Anton, Formhals (1944). Method and apparatus for spinning. US Patent 2349950

Armand, M. (1990). Polymers with Ionic Conductivity. *Advanced Materials*, Vol. 2, No. 6-7, pp. 278-286, ISSN 1521-4095

Asano, T.; Kubo, T. & Nishikitani, Y. (2004). Electrochemical properties of dye-sensitized solar cells fabricated with PVDF-type polymeric solid electrolytes. *Journal of Photochemistry and Photobiology A: Chemistry*, Vol. 164, No. 1-3, pp. 111-115, ISSN 1010-6030

Bach, U.; Lupo, D.; Comte, P.; Moser, J. E.; Weissortel, F.; Salbeck, J.; Spreitzer, H. & Gratzel, M. (1998). Solid-state dye-sensitized mesoporous TiO2 solar cells with high photon-to-electron conversion efficiencies. *Nature*, Vol. 395, No. 6702, pp. 583-585, ISSN 0028-0836

Baumgarten, P. K. (1971). Electrostatic spinning of acrylic microfibers. *Journal of Colloid and Interface Science*, Vol. 36, No. 1, pp. 71-79, ISSN 0021-9797

Berry, John P. (Wirral, GB2) (1990). Electrostatically produced structures and methods of manufacturing. US Patent 4965110

Bornat, A. L., GB2) (1987). Production of electrostatically spun products. US Patent 4689186

Cao, J. H.; Zhu, B. K. & Xu, Y. Y. (2006). Structure and ionic conductivity of porous polymer electrolytes based on PVDF-HFP copolymer membranes. *Journal of Membrane Science*, Vol. 281, No. 1-2, pp. 446-453, ISSN 0376-7388

Caruso, R. A.; Schattka, J. H. & Greiner, A. (2001). Titanium Dioxide Tubes from Sol–Gel Coating of Electrospun Polymer Fibers. *Advanced Materials*, Vol. 13, No. 20, pp. 1577-1579, ISSN 1521-4095.

Chand, S. (2000). Review Carbon fibers for composites. *Journal of Materials Science*, Vol. 35, No. 6, pp. 1303-1313, ISSN 0022-2461

Deitzel, J. M.; Kleinmeyer, J.; Harris, D. & Tan, N. C. B. (2001). The effect of processing variables on the morphology of electrospun nanofibers and textiles. *Polymer*, Vol. 42, No. 1, pp. 261-272, ISSN 0032-3861

Deitzel, J. M.; Kleinmeyer, J.; Harris, D. & Tan, N. C. B. (2001). The effect of processing variables on the morphology of electrospun nanofibers and textiles. *Polymer*, Vol. 42, No. 1, pp. 261-272, ISSN 0032-3861.

Demir, M. M.; Yilgor, I.; Yilgor, E. & Erman, B. (2002). Electrospinning of polyurethane fibers. *Polymer*, Vol. 43, No. 11, pp. 3303-3309, ISSN 0032-3861

Dloczik, L.; Ileperuma, O.; Lauermann, I.; Peter, L. M.; Ponomarev, E. A.; Redmond, G.; Shaw, N. J. & Uhlendorf, I. (1997). Dynamic Response of Dye-Sensitized Nanocrystalline Solar Cells: Characterization by Intensity-Modulated Photocurrent

Spectroscopy. *The Journal of Physical Chemistry B*, Vol. 101, No. 49, pp. 10281-10289, ISSN 1520-6106

Doshi, J. & Reneker, D. H. (1995). Electrospinning process and applications of electrospun fibers. *Journal of Electrostatics*, Vol. 35, No. 2-3, pp. 151-160, ISSN 0304-3886

Drozin, V. G. (1955). The electrical dispersion of liquids as aerosols. *Journal of Colloid Science*, Vol. 10, No. 2, pp. 158-164, ISSN 0095-8522

Feng, L.; Li, S.; Li, H.; Zhai, J.; Song, Y.; Jiang, L. & Zhu, D. (2002). Super-Hydrophobic Surface of Aligned Polyacrylonitrile Nanofibers. *Angewandte Chemie International Edition*, Vol. 41, No. 7, pp. 1221-1223, ISSN 1521-3773

Grätzel, M. (2004). Conversion of sunlight to electric power by nanocrystalline dye-sensitized solar cells. *Journal of Photochemistry and Photobiology A: Chemistry*, Vol. 164, No. 1-3, pp. 3-14, ISSN 1010-6030

Hagfeldt, A. & Gratzel, M. (1995). Light-Induced Redox Reactions in Nanocrystalline Systems. *Chemical Reviews*, Vol. 95, No. 1, pp. 49-68, ISSN 0009-2665

Hagfeldt, A. & Grätzel, M. (2000). Molecular Photovoltaics. *Accounts of Chemical Research*, Vol. 33, No. 5, pp. 269-277, ISSN 0001-4842

Hohman, M. M.; Shin, M.; Rutledge, G. & Brenner, M. P. (2001). Electrospinning and electrically forced jets. II. Applications. *Physics of Fluids*, Vol. 13, No. 8, pp. 2221-2236, ISSN 1070-6631

Hou, H. Q.; Jun, Z.; Reuning, A.; Schaper, A.; Wendorff, J. H. & Greiner, A. (2002). Poly(p-xylylene) nanotubes by coating and removal of ultrathin polymer template fibers. *Macromolecules*, Vol. 35, No. 7, pp. 2429-2431, ISSN 0024-9297

Hu Y. J.; Chen B.; Yuan Y. (2007). Preparation and Electrochemical Properties of Polymer Li-ion Battery Reinforced by non-woven Fabric. *J. Cent. South Univ.Technol*, Vol. 14, No. 1, pp. 47-49, ISSN 1005-9784

Huang, H. T. & Wunder, S. L. (2001). Ionic conductivity of microporous PVDF-HFP/PS polymer blends. *Journal of the Electrochemical Society*, Vol. 148, No. 3, pp. A279-A283, ISSN 0013-4651

Huang, Z.-M.; Zhang, Y. Z.; Kotaki, M. & Ramakrishna, S. (2003). A review on polymer nanofibers by electrospinning and their applications in nanocomposites. *Composites Science and Technology*, Vol. 63, No. 15, pp. 2223-2253, ISSN 0266-3538

Huynh, W. U.; Dittmer, J. J. & Alivisatos, A. P. (2002). Hybrid Nanorod-Polymer Solar Cells. *Science*, Vol. 295, No. 5564, pp. 2425-2427, ISSN 0036-8075

Jeong, Y.-B. & Kim, D.-W. (2004). Cycling performances of Li/LiCoO2 cell with polymer-coated separator. *Electrochimica Acta*, Vol. 50, No. 2-3, pp. 323-326, ISSN 0013-4686

Kalyanasundaram, K. & Grätzel, M. (1998). Applications of functionalized transition metal complexes in photonic and optoelectronic devices. *Coordination Chemistry Reviews*, Vol. 177, No. 1, pp. 347-414, ISSN 0010-8545

Kelly, C. A. & Meyer, G. J. (2001). Excited state processes at sensitized nanocrystalline thin film semiconductor interfaces. *Coordination Chemistry Reviews*, Vol. 211, No. 1, pp. 295-315, ISSN 0010-8545

Kim, D. W. & Sun, Y. K. (2001). Electrochemical characterization of gel polymer electrolytes prepared with porous membranes. *Journal of Power Sources*, Vol. 102, No. 1-2, pp. 41-45, ISSN 0378-7753

Kim, D. W.; Kim, Y. R.; Park, J. K. & Moon, S. I. (1998). Electrical properties of the plasticized polymer electrolytes based on acrylonitrile-methyl methacrylate copolymers. *Solid State Ionics*, Vol. 106, No. 3-4, pp. 329-337, ISSN 0167-2738

Kim, J. R.; Choi, S. W.; Jo, S. M.; Lee, W. S. & Kim, B. C. (2005). Characterization and properties of P(VdF-HFP)-based fibrous polymer electrolyte membrane prepared by electrospinning. *Journal of the Electrochemical Society*, Vol. 152, No. 2, pp. A295-A300, ISSN 0013-4651

Komiya, R.; Han, L.; Yamanaka, R.; Islam, A. & Mitate, T. (2004). Highly efficient quasi-solid state dye-sensitized solar cell with ion conducting polymer electrolyte. *Journal of Photochemistry and Photobiology A: Chemistry*, Vol. 164, No. 1-3, pp. 123-127, ISSN 1010-6030

Liu, H. & Hsieh, Y.-L. (2003). Surface methacrylation and graft copolymerization of ultrafine cellulose fibers. *Journal of Polymer Science Part B: Polymer Physics*, Vol. 41, No. 9, pp. 953-964, ISSN 1099-0488

M. Armand, in: J.R. MacCallum, C.A. Vincent (Eds.). (1987). Current state of PEO-based electrolyte *Polymer Electrolyte Reviews-1*, Elsevier Applied Science, London

Ma P. X.; Zhang R. (1999). *Synthetic nano-scale fibrous extracellular matrix*. pp. 60-72, John Wiley

Martin, C. R. (1996). Membrane-Based Synthesis of Nanomaterials. *Chemistry of Materials*, Vol. 8, No. 8, pp. 1739-1746, ISSN 0897-4756

McEvoy, A. J. & Grätzel, M. (1994). Sensitisation in photochemistry and photovoltaics. *Solar Energy Materials and Solar Cells*, Vol. 32, No. 3, pp. 221-227, ISSN 0927-0248

Meyer, G. J. (1997). Efficient Light-to-Electrical Energy Conversion: Nanocrystalline TiO2 Films Modified with Inorganic Sensitizers. *Journal of Chemical Education*, Vol. 74, No. 6, pp. 652l, ISSN 0021-9584

Michot, T.; Nishimoto, A. & Watanabe, M. (2000). Electrochemical properties of polymer gel electrolytes based on poly(vinylidene fluoride) copolymer and homopolymer. *Electrochimica Acta*, Vol. 45, No. 8-9, pp. 1347-1360, ISSN 0013-4686

Nazeeruddin, M. K.; Kay, A.; Rodicio, I.; Humphry-Baker, R.; Mueller, E.; Liska, P.; Vlachopoulos, N. & Graetzel, M. (1993). Conversion of light to electricity by cis-X2bis(2,2'-bipyridyl-4,4'-dicarboxylate)ruthenium(II) charge-transfer sensitizers (X = Cl-, Br-, I-, CN-, and SCN-) on nanocrystalline titanium dioxide electrodes. *Journal of the American Chemical Society*, Vol. 115, No. 14, pp. 6382-6390, ISSN 0002-7863

Nelson, J. (1999). Continuous-time random-walk model of electron transport in nanocrystalline TiO_{2} electrodes. *Physical Review B*, Vol. 59, No. 23, pp. 15374, ISSN 1098-0121

Nogueira, A. F. & De Paoli, M.-A. (2000). A dye sensitized TiO2 photovoltaic cell constructed with an elastomeric electrolyte. *Solar Energy Materials and Solar Cells*, Vol. 61, No. 2, pp. 135-141, ISSN 0927-0248

Ondarçuhu, T. & Joachim, C. (1998). Drawing a single nanofibre over hundreds of microns. *EPL (Europhysics Letters)*, Vol. 42, No. 2, pp. 215, ISSN 0295-5075

O'Regan, B. and M. Gratzel (1991). A low-cost, high-efficiency solar cell based on dye-sensitized colloidal TiO2 films. *Nature*, Vol.353, No.6346, pp. 737-740, ISSN 0028-0836

Péchy, P.; Renouard, T.; Zakeeruddin, S. M.; Humphry-Baker, R.; Comte, P.; Liska, P.; Cevey, L.; Costa, E.; Shklover, V.; Spiccia, L.; Deacon, G. B.; Bignozzi, C. A. & Grätzel, M. (2001). Engineering of Efficient Panchromatic Sensitizers for Nanocrystalline TiO2-Based Solar Cells. *Journal of the American Chemical Society*, Vol. 123, No. 8, pp. 1613-1624, ISSN 0002-7863

Reneker, D. H.; Kataphinan, W.; Theron, A.; Zussman, E. & Yarin, A. L. (2002). Nanofiber garlands of polycaprolactone by electrospinning. *Polymer*, Vol. 43, No. 25, pp. 6785-6794, ISSN 0032-3861

Senecal, Kris (N. Smithfield, RI, US); Samuelson, Lynne (Marlborough, MA, US); Sennett, Michael (Sudbury, MA, US); Schreuder-gibson, Heidi (Holliston, MA, US) (2006). Conductive (electrical, ionic, and photoelectric) polymer membrane articles, and method for producing same. US Patent 7109136

Sill, T. J. & von Recum, H. A. (2008). Electrospinning: Applications in drug delivery and tissue engineering. Biomaterials, Vol. 29, No. 13, pp. 1989-2006, ISSN 0142-9612

Smith, Daniel J. (Stow, OH); Reneker, Darrell H. (Akron, OH); Mcmanus, Albert T. (San Antonio, TX); Schreuder-gibson, Heidi L. (Holliston, MA); Mello, Charlene (Rochester, MA); Sennett, Michael S. (Sudbury, MA) (2004). Electrospun fibers and an apparatus therefor. US Patent 6753454

Stephan, A. M.; Nahm, K. S.; Anbu Kulandainathan, M.; Ravi, G. & Wilson, J. (2006). Poly(vinylidene fluoride-hexafluoropropylene) (PVdF-HFP) based composite electrolytes for lithium batteries. European Polymer Journal, Vol. 42, No. 8, pp. 1728-1734, ISSN 0014-3057

Stergiopoulos, T.; Arabatzis, I. M.; Katsaros, G. & Falaras, P. (2002). Binary Polyethylene Oxide/Titania Solid-State Redox Electrolyte for Highly Efficient Nanocrystalline TiO2 Photoelectrochemical Cells. Nano Letters, Vol. 2, No. 11, pp. 1259-1261, ISSN 1530-6984

Sze S M. (1981). Physics of Semiconductor Devices (New York : Wiley) p 264

Taylor, G. (1969). Electrically Driven Jets. Proceedings of the Royal Society of London. Series A, Mathematical and Physical Sciences, Vol. 313, No. 1515, pp. 453-475, ISSN 0080-4630

van de Lagemaat, J.; Park, N. G. & Frank, A. J. (2000). Influence of Electrical Potential Distribution, Charge Transport, and Recombination on the Photopotential and Photocurrent Conversion Efficiency of Dye-Sensitized Nanocrystalline TiO2 Solar Cells: A Study by Electrical Impedance and Optical Modulation Techniques. The Journal of Physical Chemistry B, Vol. 104, No. 9, pp. 2044-2052, ISSN 1520-6106

Vonnegut, B. & Neubauer, R. L. (1952). Production of monodisperse liquid particles by electrical atomization. Journal of Colloid Science, Vol. 7, No. 6, pp. 616-622, ISSN 0095-8522

Wang, P.; Zakeeruddin, S. M. & Grätzel, M. (2004). Solidifying liquid electrolytes with fluorine polymer and silica nanoparticles for quasi-solid dye-sensitized solar cells. Journal of Fluorine Chemistry, Vol. 125, No. 8, pp. 1241-1245, ISSN 0022-1139

Wang, P.; Zakeeruddin, S. M.; Moser, J. E.; Nazeeruddin, M. K.; Sekiguchi, T. & Gratzel, M. (2003). A stable quasi-solid-state dye-sensitized solar cell with an amphiphilic ruthenium sensitizer and polymer gel electrolyte. Nat Mater, Vol. 2, No. 6, pp. 402-407, ISSN 1476-1122

Watanabe, M.; Kanba, M.; Matsuda, H.; Tsunemi, K.; Mizoguchi, K.; Tsuchida, E. & Shinohara, I. (1981). High lithium ionic conductivity of polymeric solid electrolytes. Die Makromolekulare Chemie, Rapid Communications, Vol. 2, No. 12, pp. 741-744, ISSN 0173-2803

Waters, Colin M. (Tattingstone, GB2); Noakes, Timothy J. (Selbourne, GB2); Pavey, Ian (Fernhurst, GB2); Hitomi, Chiyoji (Tsokuba, JP) (1992). Liquid crystal devices. US Patent 5088807

Zhizhen Zhao, Jingquing Li, Xiaoyan Yuan Xing Li, Yuanyuan Zhang, Jing Sheng. (2005). journal of Applied Polmer Scence, 97, 466-474. Zhao, Z. Z.; Li, J. Q.; Yuan, X. Y.; Li, X.; Zhang, Y. Y. & Sheng, J. (2005). Preparation and properties of electrospun poly(vinylidene fluoride) membranes. Journal of Applied Polymer Science, Vol. 97, No. 2, pp. 466-474, ISSN 0021-8995

10

Development of Dye-Sensitized Solar Cell for High Conversion Efficiency

Yongwoo Kim[1] and Deugwoo Lee[2]
[1]Korea Industrial Complex Corporation
[2]Pusan National University
Korea

1. Introduction

The Solar cell energy is presently promising because of oil inflation, fuel exhaustion, global warming, and space development. Many advanced countries rapidly develop the solar cell energy under a nation enterprise. Particularly, dye-sensitized solar cell (DSC), the 3rd generation solar cell, has low-cost of manufactures about 1/3~1/5 times compared with the silicon solar cell, which encourages the research globally.

Dye-sensitized Solar Cell (DSC) is evaluated to be low-cost technology as the manufacturing DSC is more inexpensive 5 times than producing Silicon Solar Cell. Currently, the best conversion efficiency is 11%, the tile-shaped modules are being produced in STI, Austria. Moreover the efficiency to increase over 15% and the process of fabricating DSC for commercialization are attempted to be highly researched.

Recently, production of nano-particles becomes available due to development of nano-technologies. Since they have broad contact area comparing to the existing compound materials being generally used and increased mechanical, thermal and electrical characteristics, etc., they are attracting public attention as a new material to implement various functions. Especially, nano-tube has more excellent mechanical and electrical characteristics than normal particle type materials. And it is known that the smaller its diameter, i.e. aspect ratio, is, the better its characteristics are. Accordingly a lot of researches related to nano-compound materials have been being progressed nationally and internationally (Gojny et al., 2003; Jijima, 1991; Chang et al., 2001).

It is new methods to improve light conversion efficiency using several approaches such as nanocrystalline CNT/TiO$_2$ hybrid material, reflect mirror with micro pyramid structure, and concentrating light with Fresnel lens.

Figure.1 shows the operational principle and structure of dye sensitized cell. If visible rays are absorbed by n-type nano particles TiO$_2$ that dye molecules are chemically absorbed on the surface, the dye molecules generate electron-hole pairs, and the electron were injected into the conduction band of semiconductor's oxides. These electrons that are injected into the semiconductor's oxide electrode generate current through each nano particles' interfaces. The holes that are made from dye molecules are deoxidized by receiving electrons, thus causing the dye-sensitized cells begin to work (Zhang et al., 2010).

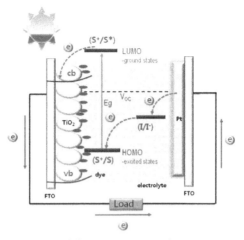

Fig. 1. A schematic representation of the construction of a DSSC

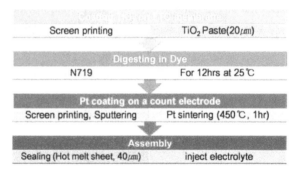

Fig. 2. Process of manufacturing DSSC

We applied screen printing method on FTO membrane to TiO₂ paste in 20μm. Coated working electrode membrane was sintered at 450°C and digested them into dye (N719) for about 12 hours.

Fig. 3. Measurement system and program

In order to understand efficiency increase of solar cells due to coating method, we compared individual efficiencies using solar cell simulator. Measuring efficiency of solar cells had been progressed under AM (air mass) 1.5 conditions (1sun, 100mW/cm²).

2. A study of photocatalyst of the TiO₂ thin film with acid dispersed CNT

As one of major variables of dye-sensitized solar cell, the dye absorbed light energy transits from ground state to excited state and gets electron injection. Electrons are injected in a very fast speed of femtosecond or picoseconds unit and oxidized dye is renewed within several nanoseconds. On the other hand, since rejoining speed that the electron becomes dissolved into electrolyte via surface state is slow such as micro seconds or milliseconds and most of photons are injected semiconductor conduction band, those electrons that are not injected meet holes again to be restored and decrease efficiency of solar cell. If we utilize CNT that is advantageous for electrical and thermal conduction as a compound material, we can increase electron injection speed than rejoining speed and increase efficiency of dye-sensitized solar cells through movement of more electrons.

Since multi walled carbon nanotube (MWCNT) is a material to well transfer electricity and heat, it can function as a basic electrode. Nanotubes function as electrodes to penetrate into broad TiO₂ surface mutually and assist extracting charge carriers efficiently from dye layer. Since these electrodes are very clear under longer wavelength, they are advantageous for solar light spectrums.

This study utilizes CNT that is advantageous for electrical and thermal conduction as a compound material so that surface resistance between clear electrodes and dye layer decreases, electron injection speed can be increased more than the rejoining speed of electrons and efficiency of dye-sensitized solar cell can be also increased through movement of more electrons. In order to perform the task, this study will composite nano-crystal TiO₂ with sol-gel technique (Tracey et al., 1998), establish optimal thin film formation conditions through identification of correlation between permeability and conductivity by manufacturing distributed CNT through acid treatment and compound thin film, and perform characteristics assessment by applying it to solar cells upon surface resistance between dye layer and FTO.

Dye-sensitized solar cell is a device to apply photosynthesis principle of the plant by combining pigments to absorb light energy within chloroplast with polymer and applying them to the solar cell. Dye-sensitized solar cell basically consists of dye polymer to absorb solar light, semiconductor compound having broad band gap serving as n-type semiconductor, electrolyte serving as p-type semiconductor, relative electrode for catalysis, and clear electrode to permeate solar light. Dye-sensitized solar cell is composed of semiconductor nanoparticles where dye absorbs solar energy and separation/transfer of generated electrons are diffused upon electron concentration difference. Processes to generate and transfer electrons play an essential role to determine performance of the cells. Firstly entering time of excited electrons from the dye into TiO₂ should be shorter than that of joining with holes and being exterminated. Normally entering time of electron is very fast in femtoseconds through picoseconds and oxidated dye is renewed within several nanoseconds (Tachibana et al., 1996).

2.1 Composition of TiO₂

Manufacturing methods of TiO₂ for photocatalyst generally includes hydro thermal method (Chen et al., 1995), sedimentation method (Ellis et al., 1989; Lee et al., 2000), sol-gel technique (Ding et al., 1995; Johnson, 1985; Hwang & Kim, 1995), CVD method (Lee et al., 1999), etc. Hydro thermal method that may get mainly powder materials has complex equipment and difficult for continuous work, while sedimentation method allows easy production but it has disadvantages cohesion between particles, powder with possibility of uneven composition, and difficulties for fineness as bulk materials from sintering and harmonization of crystalline. Comparing to them, sol-gel technique has advantages to allow easy acquisition of TiO_2 powder for photocatalyst with even composition relatively simply and low temperature composition available.

2.2 CNT dispersion
Carbon nanotube exists as a form of bundle or cohesion body due to strong van der Waals forces between tubes as like interaction between graphite plates. This cohesion phenomenon obstructs formation of 3-dimensional network structure in manufacturing compound to increase electrical or mechanical properties. It is an important technology to disperse carbon nanotube and release individual strand, as it increases functional efficiency. Dispersion methods of carbon nanotube can be classified into mainly mechanical dispersion, dispersion using strong acid, dispersion with solution, etc. Firstly, dispersion using strong acids is to agitate with composition of nitric acid and hydrochloric acetone at 130°C for 6h or using composition of sulfuric acid and hydrochloric acid. Dispersing with solution is to melt carbon nanotubes through surface active agents such as SDS (sodium dodecyl sulfate), Triton X-100, LDS, etc. The most various mechanical dispersion methods include supersonic treatment by mixing acetone and methanole, or using ball milling method to minimize length and diameter distribution of carbon nanotubes. Others include dispersions with grinding process using mortars, abrasion process, high shear strength using liquid, etc.

2.3 Experiment method and conditions
2.3.1 Production and coating of CNT/TiO$_2$
TTIP that is used for this study generates photocatalyst TiO_2 under low temperature heating conditions but it has a disadvantage fast hydrolysis with air or small moisture due to very string reactivity. So we had made even chelate compound by adding AcAc into the ethanol solution and severely agitating it for 30 minutes at room temperature using agitator in order

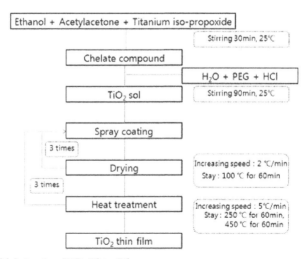

Fig. 4. Process of fabricating TiO_2 Thin Film

to control hydrolysis speed of TTIP. And we have prepared water solution by melting HCl 0.15 mole catalyst for stability of sol and PEG 0.5 mole adhesion to increase osculation with plate to be coated into 50mole distilled water and drop it into the chelate compound. At the moment, it is possible to manufacture yellow clear sol by severely agitating and reacting chelate compound at room temperature for 90min.

For dispersion of CNT, we mixed 60% nitric acid and distilled water in 1:3 volume ratio and severely agitated it at 130°C for 6h. After acid treatment, we mixed a lot of distilled water, apply ultra sonication and continued to neutralize it using PTFE thin film filter with 1um in diameter until it has been neutralized.

In order to understand permeability and electrical characteristics of composited TiO_2 sol and dispersed CNT by concentrations, we have mage mixture solution by establishing variables with Volume% concentrations 0.5, 2.5, 5 and 10.

This study with TiO_2 thin film manufacturing method using sol-gel technique has an advantage to allow applying various processes such as dip-coating method, spinning method, Spray, etc. As spray coating out of them has advantages such as uniform thickness, various compositions to manufacture films by mixing solutions, no big influence from viscosity of solution comparing to spinning method, and available coating without influences from patterned plate, plate surface energy or roughness. It is considered that it is the most appropriate for commercialization as it allows easy manufacture and application of broad area plate.

2.3.2 Manufacture of Dye-sensitized solar cell

We applied composited CNT/TiO_2 to screen printing on FTO plate to coat about 3μm in thickness and coated TiO_2 paste from Dyesol in 15μm. Then we could increase the whole efficiency of cells by decreasing surface resistance between TiO_2 layer absorbed with dye and FTO plate and assisting movement of electrons. Fig. 3 shows coated plated by CNT concentrations, from which we can see that these plates with high content of CNT have thicker colours. We sintered them at 450°C and digested them into dye (N719) for about 24 hours. We coated platinum onto FTO plate with relative electrode and sintered it at 450°C. We sealed prepared 2 plates with Hot melt sheet having about 60μm in thickness and completed them by injecting electrolyte.

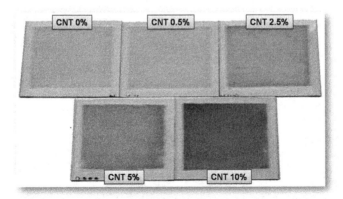

Fig. 5. Feature of CNT/TiO_2 thin film by screen-printing

2.4 Conclusion and considerations
2.4.1 Change of permeability upon CNT concentration

In order to check permeability and electrical characteristics of CNT by concentrations, We coated glass plate with about 5μm in thickness using spray technique, and produced TiO₂ thin film by increasing temperature and sintering it at 450°C for 1 hour. Fig. 6(a) shows composited CNT/TiO₂ thin film surface with very small and even particle. In case of (b), it shows thin film surface of TiO₂ paste from Dyesol with uneven large particles but very excellent porosity. From the results of measurements for permeability of each thin film, as CNT content increases, permeability drops straight and reaches only 10% when concentration becomes 5, 10%.

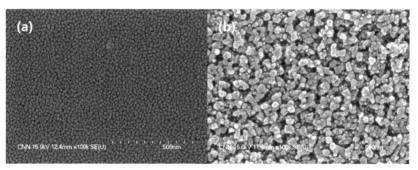

Fig. 6. SEM images of CNT/TiO₂ (a) and TiO₂ paste(b)

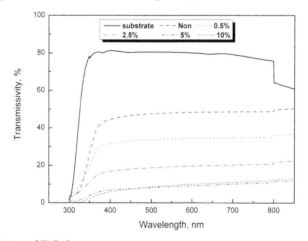

Fig. 7.Transmissivity of TiO₂ layers

2.4.2 Change of surface resistance upon CNT concentration

Fig. 8 shows measurement results of surface resistances upon CNT concentrations. It shows a graph acquired upon each condition by increasing voltage from -1V to 1V by 50mV interval. As CNT concentration with excellent electrical characteristics increases and surface resistance becomes low, it is considered that it would be advantageous for transfer of electrons separated from the dye.

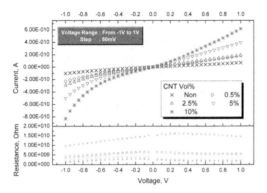

Fig. 8. I-V curve and resistance of TiO$_2$ layers

2.4.3 Efficiency change of Dye-sensitized solar cell upon CNT concentration

Fig. 9 shows I-V and P-V curve of CNT 0.5% indicating the highest energy efficiency out of CNT concentrations. Efficiency can be calculated from output power by estimating voltage (V$_{mp}$) and current (I$_{mp}$) to achieve the maximum output from I-V curve when strength of solar light becomes input power.

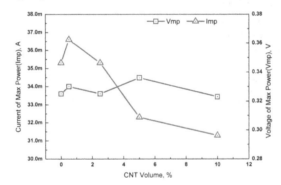

Fig. 9. I-V and P-V curve of DSC with CNT 0.5%

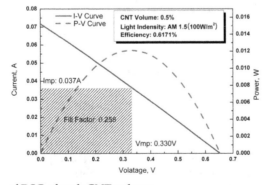

Fig. 10. I$_{mp}$ and V$_{mp}$ of DSC of each CNT volume

In order to confirm if movement of electrons increases according to content of CNT within CNT/TiO$_2$ compound material, we compared V$_{mp}$ and Imp values upon each condition on Fig. 10. V$_{mp}$ values refer to voltage values when solar cell performs the maximum power, and CNT contents show relatively even values throughout the whole areas. In case of Imp, we can see that it shows high value on very low CNT content such as 0.5%. It means that a lot of electrons are transferred comparing to no CNT, in spite of low generation of electrons due to low permeability. Under high CNT content, such as 5, 10%, we can see that generated electrons are small as surface resistance is low but permeability is low.

2.5 Conclusions

This study has confirmed efficiency of solar cells upon impact of permeability and surface resistance by manufacturing CNT/TiO$_2$ compound material to increase transfer of electrons separated from dye layer using sol-gel technique as a variable having effect on efficiency of dye-sensitized solar cell and applying in to the solar cells.

- It was possible to develop CNT/TiO$_2$ compound material to remarkably reduce surface resistance of solar cells and increase efficiency of them and acquire the highest efficiency at 0.5% Vol concentration of CNT.
- It is concluded that as CNT concentration increases, surface resistance becomes low with excellent electrical characteristics, but the solar cells reacting against solar light receive large influence from permeability.
- High permeability increases voltage of solar cells and CNT/TiO$_2$ compound material with low surface resistance effectively transfers electrons generated from the dye, so they contribute to increase of the current.

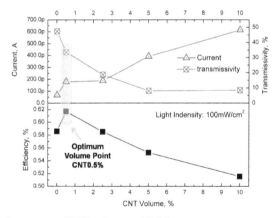

Fig. 11. Determine of optimum CNT volume of DSC

3. The recovery of light with a reflector of microstructures

We have studied a reflector recovering the loss light for improvement conversion efficiency of DSC. One of methods to increase efficiency of dye-sensitized solar cell is to expand surface area of semiconductor oxide such as TiO$_2$. Since dye polymer has high efficiency when it is absorbed on the semiconductor as a single molecule layer, solar light absorption becomes larger as the surface area of the semiconductor on which dye polymer is absorbed is wide. Consequently efficiency of the cell is improved as TiO_2 particles are small and porousness is high. This study has researched that the dye produces electrons as much as possible using a method to lengthen scattering distance by reflecting entered solar light rather than expanding surface area of the oxide.

The reflector angle was determined by optical analysis program. Micro pyramid patterns with the 112.6° were processed using the ultra precision shaping machine in order to maximize the conversion efficiency due to increasing light distance. In addition, a comparative study carried out about the conversion efficiency. We made the DSC that is attached reflector with mirror angle 112.6° below. We measured conversion efficiency of solar cell by solar simulator that can irradiate 100mW/cm² (1Sum, AM 1.5).

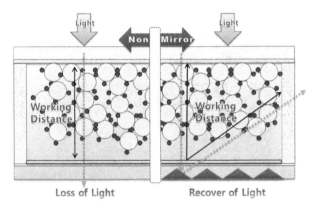

Fig. 12. Scheme of DSC with reflector

As a result of this experiment, the DSC with micro pyramid mirror improves efficiency about 2% against the DSC with black plate. Reflected light can cross more dye of TiO_2 Layer. Therefore, Voltage of maximum power (V_{mp}) increases other reflector with different mirror angle.

We propose a new method to improve light conversion efficiency using a reflector to collect light being lost from permeation through the dye-sensitized solar cell. The goals of this study are to

• Fabrication of reflector with microstructure
• Improve working distance of scattering light
• Control mirror angle by optimum design
• Improve photoelectric efficiency with micro pyramid arrayed mirror

3.1 Light simulation

We designed a micro reflector to lengthen scattering distance of reflected solar light on the dye layer of dye-sensitized solar cell using light analysis program.

In order to maximize light scattering distance of dye layer, we allowed value of Light Angle (θ_l) to be the maximum so that reflected light can be spread out widely on the dye layer. Light Angle and Mirror Angle (θ_m) are indicated on Figure 13. We measured Light Angle according to aspect ratio of height and width of micropattern of the reflector, and determined Mirror Angle when this value became the maximum. Result values of analysis according to Aspect Ratio values are indicated on Table 1 and Figure 14.

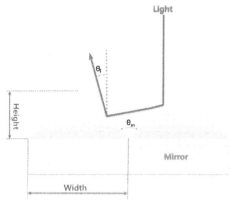

Fig. 13. Scheme of light path and mirror angle

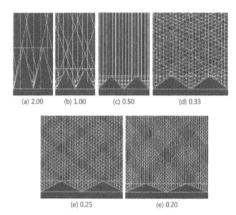

Fig. 14. Simulation results pursuant to aspect ratio

Aspect Ratio	Height [μm]	Width [μm]	Mirror Angle [degree]	Light Angle [degree]	Working distance [μm]
2	50	25	28.8	11.6	40.8
1	50	50	53.2	21.2	42.9
0.5	50	100	90.0	0.0	40
0.33	50	150	112.6	67.5	104.5
0.25	50	200	126.9	53.6	67.4
0.2	50	250	136.4	43.8	55.4

Table 1. Results of light simulation

The results of light analysis said that when Aspect Ratio was 0.33 and Mirror Angle was 112.6°, Light Angle had max value of 67.5°. At the moment, considering that thickness of dye layer of dye-sensitized solar is 40μl, solar light can meet the largest amount of dye with about 104.5μm of scattering distance.

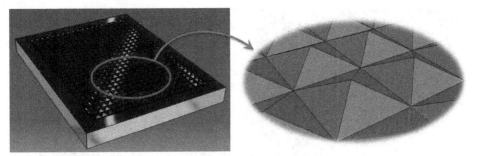

Fig. 15. Modelling of microstructure arrayed mirror

3.2 Fabrication of a reflector

In case of creating micropatterns on the material through micro cutting using single crystal diamond tool, pattern dimensions such as height, width and length of the micropattern depend on shape, cutting depth and feeding amount of the machining tool. In case of shaping machining of material using single crystal diamond having a certain tool angle, its sectional shape has 3-dimensional structure of pyramid shape. In order to machine pyramid shape, process consisted of machining one axis with diamond shaping machining and machining other axis by rotating table by 90° (Kim, 2005).

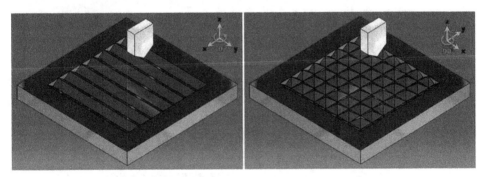

Fig. 16. Tool path of Pyramid shape

For machining sample, Oxygen Free Copper (OFHC copper) to be mostly used for general reflector mirror, etc. has been used. From the results of light analysis, a single crystal diamond bite with s112.6° of tool angle to maximize Light Angle has been used. Cutting length was total 50μm with 5μm steps, and feeding interval was 100μm. Height of mound of machined material was about 33.35μm. After machining, material was washed with supersonic wave to remove micro burr or chip, etc. generated from machining, treated with acid for about 15 seconds with fluoric acid and coated with gold through electrolytic plating. Micropattern machining experiment had been performed using commercial super precision machine. Shape of micro machined sample had been measured using SEM. Figure 18 shows pyramid shaped reflector machined by shaping machine and has 100μm width and 33.35μm height with tetrahedron shape. There was burr generated at the bottom of micro machined work that would be created during 90°rotation machining. It is considered to study machining conditions of cutting amount and feeding speed.

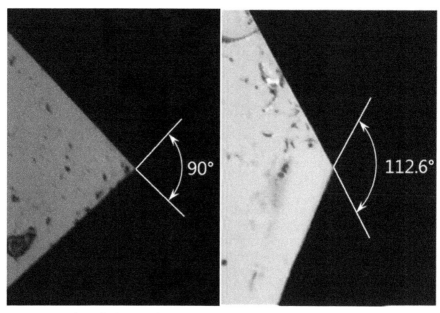

Fig. 17. Images of single diamond tool

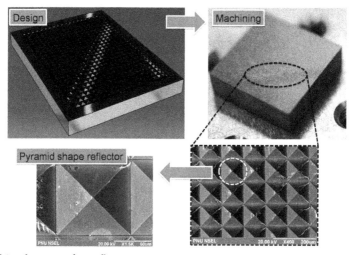

Fig. 18. Machined pattern for reflector

3.3 Results
We made DSC by handmade which of size is 11mm x 4mm. It is smaller area than my precedent study that obtains about 0.5% of efficiency. And size of mirror also becomes smaller before. Figure 19 shows I-V and P-V line diagram in order to measure efficiency of the cell by attaching a black plate at the bottom of dye-sensitized solar cell, pass-through of light in figure 20, and pyramid shaped reflector in figure 21.

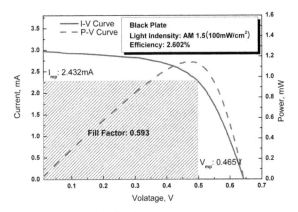

Fig. 19. I-V and P-V curve based on black plate

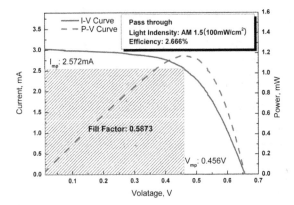

Fig. 20. I-V and P-V curve without a reflector

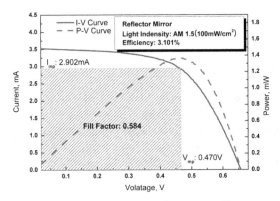

Fig. 21. I-V and P-V curve based on micro pyramid pattern reflector

3.4 Conclusions

This study machined micropatterned micro reflector and measured conversion efficiency of solar cell in order to increase efficiency of dye-sensitized solar cell. Its results are summarized as follows:

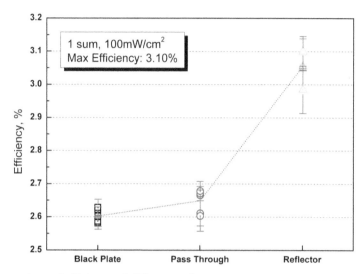

Fig. 22. Comparison of efficiency of different reflector

It was possible to get the optimal reflection angle 112.6° to increase conversion efficiency through reflector to collect lost light passing through dye-sensitized solar cell by performing light analysis.

When using reflector with pyramid shape, it was confirmed that it could get higher energy conversion efficiency by about 17% than a black plate

In case of micro pyramid reflector, maximized inclination to lengthen distance to meet dye layer of TiO_2 had maximized max power current (I_{mp}) and brought about entire efficiency improvement, different from vertical reflection.

4. Photovoltaic performance of Dye-sensitized solar cell by concentrating sunlight

One of method that can be improved efficiency of solar cell is concentrating light. Various factors influence the production of electricity from solar cells such as solar radiation, solar cell installation angle, direction, shade, solar cell module temperature. Among these, solar cell installation angle, direction, and shade have almost no influence on the system performance once the solar cell system is installed unless artificial external effects are given because they are determined when the solar cell system is designed and installed. After installation, the performance of a solar cell system varies greatly by the solar radiation reaching the module surface and the surface temperature. The higher the solar radiation, the higher the efficiency of the solar cell becomes. Due to the nature of the solar cell module, the power production increases in proportion to the solar radiation, and the power generation increases as the surface temperature of the solar cell module increases. Therefore, we can improve the performance of solar cell modules by compulsorily increasing the solar cell module temperature through solar concentration.

This study intended to develop solar cell module that can maximize the efficiency of unit cells of dye-sensitized solar cell (DSSC) by maximizing solar concentration and minimizing solar loss while analyzing and improving the factors that influence the efficiency of DSSC.

In this study, and was concentrated by Fresnel lens. High temperature heat on concentration can decrease efficiency of solar cell so as cooling radiator was installed. Maximum concentrating ratio was 26 times of 1sun (2.6W/cm²). When the solar energy of High density was illuminated on a DSC, It was confirmed that temperature and concentrating ratio affect efficiency of DSC.

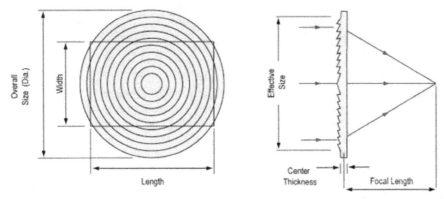

Fig. 23. Structure of Fresnel lens

Contents	Value
Grooves/Inch	100
Size H x L (inches)	5.0 x 5.0
Effective Size (inches)	Φ 4.0
Effective Focal Length (inches)	2.8
Center Thickness (inches)	0.06
Transmission (%)	92(wavelength 400~1100nm)

Table 2. Specifications of Fresnel lens

When high density light is illuminated in the DSC using concentrating lens, conversion efficiency is reached up to 16.11%. The enhancement in overall device efficiency is a result of increased open circuit potential and short circuit current. If coolant system is used, it can help guarantee of stable performance of a high efficiency of DSC at 45°C.

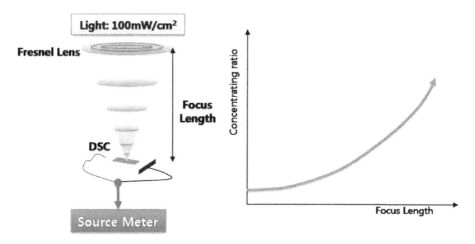

Fig. 24. Schematic of concentrating DSC

4.1 Evaluation of solar cell performance by changing temperature

High temperature is generated when solar energy is concentrated to improve energy conversion efficiency. Performance drops due to sealing problems such as the leakage and evaporation of electrolytes resulting from the changes in the volume of volatile electrolytes and the increase of vapour tension (Kim, 2007). To solve this problem, many efforts are being made to achieve performance reliability such as replacement of liquid electrolytes with solid electrolytes, development of new materials for sealing, and the performance of thermal stability tests (Fischer et al., 1997). Performance varies greatly by the surface temperature and the solar radiation that reaches the cell surface of a solar cell. The higher the solar radiation, the higher power production becomes, and the performance of a solar cell varies by surface temperature. In this study, the effects of the changing cell temperature on the efficiency of solar cells, and the optimum conditions for thermal stability in solar concentration and the production of solar cell module were investigated.

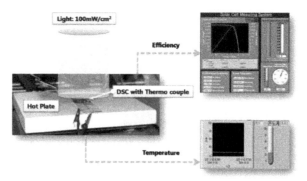

Fig. 25. Schematic design of measurements

Figure 25 shows a schematic diagram of a device for measuring the changes in the efficiency of DSSC according to changing temperature. A hot plate was used as a heat source, and a resin epoxy was used for sealing to prevent the leakage and evaporation of electrolytes due to exposure to high temperature. To examine the cell efficiency under changing temperature, a thermocouple for measuring temperature was attached to the DSSC. For this thermocouple, the K-type from Omega was used. The change in the efficiency of the solar cell was measured while the temperature was varied from 35°C to 65°C in 5°C steps.

4.2 Performance evaluation of the solar cell by solar concentration rate

The solar cell device was fabricated in such a way to obtain high efficiency by increasing the energy density through solar concentration. The lens for solar concentration was a Fresnel lens with the conventional curved surface of the lens replaced by concentric grooves, and fine patterns were formed on the thin, light plastic surface. Each groove has a refracting surface like a very small prism with a fixed focal distance and a low aberration. Because the lens is thin, it has a low loss from light absorption. A high groove density provides high image quality and a low groove density increases efficiency.

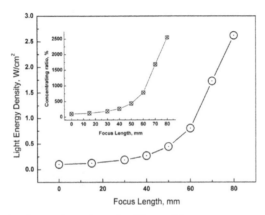

Fig. 26. Energy density due to focus length of Fresnel lens

The focal distances of the Fresnel lens were defined as 15, 30, 40, 50, 60, 70, and 80mm. A power meter was used to measure the concentrated energy density to determine the solar concentration rate for each focal distance. If was found that the energy density increased exponentially as the focal distance increased. As shown in Figure 26, the solar concentration rate at the highest focal distance was approx. 26 times ($2.619W/cm^2$) the 1sun ($100mW/cm^2$) condition.

4.3 Results

Figure 27 shows the results of the efficiency of the DSSC measured by different cell temperatures with the solar intensity of 1sun (AM 1.5, $100mW/cm^2$). The cell efficiency increased as the cell temperature increased and abruptly dropped from 45°C.

Figure 28 shows the maximum output, maximum output current (I_{mp}) and voltage (V_{mp}) at various temperatures as percentages of the values at 35°C to determine the factors influencing cell efficiency and output. It shows I-V line diagrams comparing the changes of I_{SC} and V_{OC} at different cell temperature. I_{SC} increased as the cell temperature increased and dropped from 55°C while V_{OC} decreased as the temperature increased.

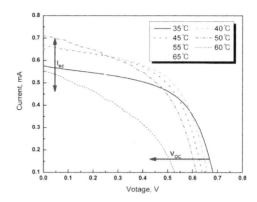

Fig. 27. Comparison of I-V curve due to temperature change

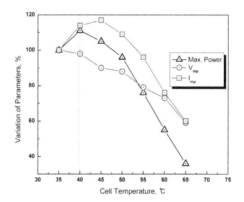

Fig. 28. Performance changes due to temperature change

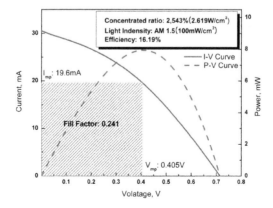

Fig. 29. I-V curves of DSC due to Focus length

The changing efficiency of the DSSC by solar concentration rate was measured at varying focal distances with the prepared lens and stage. Figure 29 shows the I-V line diagrams for each solar concentration rate. When the focal distance was 80mm and the solar concentration was at the maximum of 2,543%, the cell efficiency was 16.2%.

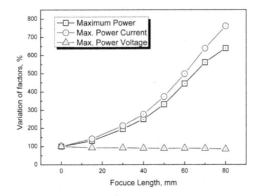

Fig. 30. Performance changes due to focus length

Figure 30 shows the maximum output for each focal distance and the voltage and current changes in percentages at the maximum output to determine the factors influencing efficiency improvement. The maximum output increased as the solar concentration rate increased, indicating cell efficiency improvement. It was found that the increase of current (I_{mp}) by solar concentration had a direct influence.

4.4 Conclusions

This study investigated the changes in efficiency when concentrated solar radiation with high energy density was applied to DSSC to determine the factors influencing efficiency.

- Imp increased as the cell temperature increased and dropped from 45°C while V_{mp} decreased as temperature increased.
- The efficiency of DSSC at changing temperatures was investigated when high heat was generated by solar concentration, and the highest efficiency was obtained at 45°C. As temperature increased over this value, the cell efficiency dropped sharply. Thus, a cooling device is essential when manufacturing a power generation system using solar concentration.
- The high energy density obtained by solar concentration increased the efficiency of DSSC by 6.4 times on average and up to 16.1% by absolute value. Because current density can be increased by solar concentration, it is possible to implement solar cells with a high output.

5. Concentrating system of Dye-sensitized solar cell with a heat exchanger

Conversion efficiency of solar cell is the key point for reducing price to manufacture products. The efficiency is expected to be improved by using the concentrator system, because lost energy density of concentrator system increase in proportion to quantity of concentration.

In this study, the conversion efficiency is expected to be improved by concentrating light which has high energy density through the concentrating lens. In this process, DSC will emit heat at high temperature and make defection like evaporation and leaks of an electrolyte. To protect this problem, we have discussed the way to ensure steady cells by developing the system available to return the heat of the high temperature.

Cell of temperature was maintained 30°C at 1sun(100mW/cm²) condition, Concentrated light density was 2.6W/cm² that is about 26suns. The cell is measured for 480 minutes because it is generally running for 8 hours during a day. On average, the conversion efficiency of the cell is 13. 24%. Finally we conform that the solar cell using concentration system with a heat exchange is available to steady and highly improve the conversion efficiency.

5.1 Concentrating system with a heat exchanger
The dye-sensitized solar cell with concentrated light generates high heat from concentrated light with high density and results in defections such as leakage of electrolyte, evaporation, etc. In order to prevent them, the researcher has installed a cooler under the solar cell and executed stability test. The stability test has a meaning to confirm efficiency change and ensure performance reliability of the cell when they have been exposed to the light for a long time. Figure 31 shows apparatus and conditions used for this test. Efficiencies have been acquired for a certain time period by keeping temperature of the cell at 30°C using the cooler and radiating light with 2.6W/cm² that is 25.4 times of the maximum light concentration under 1sun. Measurement time was 480 minutes considering that the number of hours when the solar cell can be operated during daytime on clean weather us 8 hours.

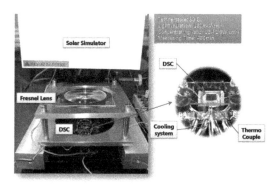

Fig. 31. Equipment for thermal stability test

5.2 Results
In order to measure efficiency change of the dye-sensitized solar cell upon change of light concentration coefficient, efficiencies have been measured according to focal distances using the prepared lens and stage. Figure 32 shows I-V curve of the solar cell upon light concentration coefficients. When the light concentration coefficient is a maximum of 2,543% at 80mm of focal distance, efficiency of the cell showed 16.2%.

Figure 33 shows efficiency changes of the dye-sensitized solar cell for 480 minutes in a graph. As the measurement was started and time passed, the efficiency was linearly reduced and showed 11.5% after 480 minutes, reduced by 25.6% comparing to 15.4% of initial efficiency. It is considered that it could perform 13.2% of average efficiency over the entire time period. Consequently, it is possible to realize a stable and high efficient solar cell with light concentration utilizing the cooler.

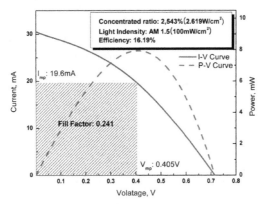

Fig. 32. I-V curves of DSC on Focus length 80mm

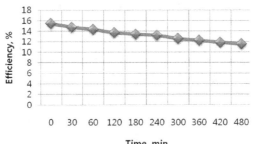

Fig. 33. Efficiency change of DSC due to time

5.3 Conclusion
When concentrating light through the Fresnel lens that has less light loss with thinner than normal lens and may increase energy density with small aberration against focus, it was possible to confirm a maximum light concentration coefficient at 80mm of focal distance.

When keeping a certain temperature (about 30°C) using the cooler, it was possible to get average 13.2% efficiency for 8 hours using the condenser lens. This shows that it would be possible to realize the high efficient dye-sensitized solar cell by making light concentration and cooling system in a module.

Light concentration is mostly advantageous as a practical technology of the high efficient dye-sensitized solar cell. In addition, it is possible to increase comprehensive energy use rate by progressing power generation and heating at the same time as a cogeneration pattern using the high heat generated from light concentration. This has applied light concentration and cooling on the basis of a single cell, but it would be possible to get the higher efficiency from fabrication cost per unit area and operation of the circulation system such as motor, etc. if it will be extended to a large area in a form of a power plant.

6. Acknowledgment

This chapter is composed to be based on my thesis of doctorate and proceedings of conferences.

7. References

Gojny, F. H., Nastalczyk, J., Roslaniec Z., & Sculte, K. (2003). Surface Modified Multi-walled Carbon Nanotubes in CNT/Epoxy-composites, *Chemistry Physical Letters*, Vol. 370, Issues 5-6, pp. 820-824, ISSN:0009-2614

Jijima, S. (1991). Helical Microtubules of Graphitic Carbin. *Nature*, Vol. 354, pp. 56-58 , ISSN:0028-0836

Chang, H., Lee, J., Lee, S., & Lee, Y.(2001). Adsorption of NH_3 and NO_2 Molecules on Carbon-nanotubes, *Applied Physical Letters*, Vol. 79, No. 23, pp. 3863-3865, ISSN:0003-6951

Zhang, J., Yang, G., Sun, Q., Zheng, J., Wang, P., Zhu, Y., & Zhao, X. (2010). The improved performance of dye sensitized solar cells by bifunctional aminosilane modified dye sensitized photoanode. *Journal of Renewable and Sustainable Energy*, Vol. 2, Issue 1, p. 10 , ISSN:1941-7012

Tracey, S., M. Hodgson, S. N. B., Ray, A. K., & Ghassernlooy, Z. (1998). The Role and Interaction of Process Parameters on The Nature of Alkoxide Derived Sol-gel Films. *Journal of Materials Processing Technology*, Vol. 77, pp. 86-94, ISSN:0924-0136

Tachibana, Y. Moser, J. E. Graltzel, M. Klug, D. R. and Durrant ,J. R. (1996). Subpicosecond Interfacial Charge Separation in Dye-Sensitized Nanocrystalline Titanium Dioxide Films. *J. Phys. Chem.* Vol. 100, pp.20056-20062, ISSN: 0022-3654

Chen, Q., Qian, Y., Chen, Z., Zhou, G., & Zhang, Y. (1995). Preparation of TiO_2 Powder with Different Morphologies by An Oxidation-hydrothermal Combination Method. *Materials Letters*, Vol. 22, Issues 1-2, pp.77-80, ISSN: 0167-577X

Ellis, S. K., & McNamara, E. P. Jr. (1989). Powder Synthesis Research at CAMP. *American Ceramic Society bulletin*, Vol. 68, No. 5, pp. 988-991 , ISSN: 0002-7812

Lee, B. M., Shin, D. Y., & Han, S. M. (2000). Synthesis of Hydrous TiO_2 Powder by Dropping Precipitant Method and Photocatalytic Properties. *Journal of Korean Ceramic Society*, Vol. 37,pp. 308-313.

Ding, X. Z., Qi, Z. Z., & He, Y. Z. (1995). Study of the room temperature ageing effect on structural evolution of gel-derived nanocrystalline titania powders. *Journal of Materials Science Letters*, Vol. 15, No. 4, pp.320-322 , ISSN:0059-1650

Johnson, D. W. Jr. (1985). Sol-gel Processing of Ceramics and Glass. *American Ceramic Society bulletin*, Vol. 64, No. 12, pp.1597-1602, ISSN: 0002-7812

Hwang, K. S., & Kim, B. H. (1995). A Study on the Characteristics of TiO_2 Thin Films by Sol-gel Process. *Journal of Korean Ceramic Society*, Vol. 32, pp.281-288

Lee, H. Y., Park, Y. H., & Ko, K. H. (1999). Photocatalytic Characteristics of TiO_2 Films by LPMOCVD. *Journal of Korean Ceramic Society*, Vol. 36, pp.1303-1309

Kim, S. W. (2005). *Die machining with micro tetrahedron patterns array using the ultra precision shaping machine* , PhD. Thesis of Pusan National University

Kim, J. H. (2007). Dye-Sensitized Solar Cell. *News & Information for Chemical Engineers*, Vol. 25, No. 4, p.390

Fischer, J. E., Dai, H., Thess, A., Lee, R., Hanjani, N. M., Dehaas, D. L., & Smalley, R. E. (1997). Metallic resistivity in crystalline ropes of single-wall carbon nanotubes. Physical Review B, Vol. 55, No. 8, pp.4921-4924, ISSN: 0163-1829

11

Shape Control of Highly Crystallized Titania Nanorods for Dye-Sensitized Solar Cells Based on Formation Mechanism

Motonari Adachi[1,4], Katsuya Yoshida[2], Takehiro Kurata[2],
Jun Adachi[3], Katsumi Tsuchiya[2], Yasushige Mori[2] and Fumio Uchida[4]
*[1]Research Center of Interfacial Phenomena, Faculty of Science and Engineering,
Doshisha University, 1-3 Miyakodani, Tatara, Kyotanabe,
[2]Department of Chemical Engineering and Materials Science,
Doshisha University, 1-3 Miyakodani, Tatara, Kyotanabe,
[3]National Instituite of Biomedical Innovation, 7-6-8 Asagi Saito, Ibaraki,
[4]Fuji Chemical Co., Ltd., 1-35-1 Deyashikinishi-machi, Hirakata,
Japan*

1. Introduction

Utilization of solar energy - the part transmitted to the earth in the form of light- relies on how effectively it can be converted into the form of electricity. In this regard, dye-sensitized solar cells have attracted recent attention as they are expected to offer the possibility of inexpensive yet efficient solar energy conversion. The performance of dye-sensitized solar cells depends critically on a constituent nanocrystalline wide-band-gap semiconductor (usually titania, TiO_2, nanoparticles) on which a dye is adsorbed. The electrical and optical properties of such nanoparticles are often dependent on their morphology and crystallinity in addition to size, and hence, it is essential to be able to control the particle size, shape, their distributions and crystallinity (Empedocles et al., 1999; Nirmal & Brus, 1999; Manna et al., 2000), which requires an in-depth understanding of the mechanisms of nucleation and growth as well as such processes as aggregation and coarsening.

Among the unique properties exhibited by nanomaterials, the movement of electrons and holes in semiconductor materials is dominated mainly by the well-known quantum confinement, and the transport properties related to phonons and photons are largely affected by the size, geometry, and crystallinity of the materials (Alivisatos, 1996a, 1996b; Murray et al., 2000; Burda et al., 2005). Up to now, various ideas for morphological control were introduced (Masuda & Fukuda, 1995; Masuda et al., 1997; Lakshmi et al., 1997a, 1997b; Penn & Banfield, 1998; Banfield et al., 2000; Peng et al., 2000; Puntes et al., 2001; Pacholski et al., 2002; Tang et al., 2002, 2004; Peng, 2003; Scher et al., 2003; Yu et al., 2003; Cao, 2004; Cheng et al., 2004; Cui et al., 2004; Garcia & Tello, 2004; Liu et al., 2004; Pei et al., 2004; Reiss et al., 2004; Song & Zhang, 2004; Wu et al., 2004; Yang et al., 2004; Zhang et al., 2004) based on: (1) a mixture of surfactants used to bind them selectively to the crystallographic faces for CdS (Scher et al., 2003), (2) monomer concentration and ligand effects for CdSe (Peng et al., 2000), (3) growth rate by controlling heating rate for $CoFe_2O_4$ (Song & Zhang, 2004), (4) biological routes in peptide sequence for FePt (Reiss et al., 2004), (5) controlled removal of protecting organic stabilizer for CdTe (Yu et al., 2003; Tang et al., 2002, 2004), (6) anodic alumina used as a template (Masuda & Fukuda, 1995; Masuda et al., 1997), and (7) the "oriented attachment" mechanism for nanoparticles (Penn & Banfield, 1998; Banfield et al., 2000). A number of methods have been developed to control the shape of nanocrystals on the basis of these ideas.

Titanium dioxide has a great potential in alleviating the energy crisis through effective utilization of solar energy with photovoltaics and water splitting devices, and is believed to be the most promising material for the electrode of dye-sensitized solar cells (Fujishima & Honda, 1972; Fujishima et al., 2000; Hagfeldt & Grätzel, 2000; Grätzel, 2000, 2001, 2004, 2005; Nazeeruddin et al., 2005). To further pursue this potential in terms of its morphology in dispersion, we have synthesized highly crystallized nanoscale "one-dimensional" titania materials such as titania nanowires having network structure (Adachi et al., 2004) and titania nanorods (Jiu et al., 2006), which were confirmed to provide highly efficient dye-sensitized solar cells (Adachi et al., 2007, 2008; Kurata et al., 2010).

Extremely high crystalline features of nanorods can be perceived in the images of high-resolution transmission electron microscopy (Yoshida et al., 2008; Kurata et al., 2010) as shown in Fig. 1. A highly magnified, high-resolution transmission electron microscopy image (Fig. 1b) demonstrates a well-regulated alignment of titanium atoms in crystalline anatase structure with essentially no lattice defects. The TiO_2 anatase (101) face, (-101) face, and (001) face are clearly observed; a specific feature definitely captured and to be noted is that the nanorod edge is sharply demarcated by the kinks consisting of (101) and (-101) planes. Such bare anatase crystal with atomic alignment - anatase TiO_2 crystals not covered with amorphous or additional phases around the edge or rim - is extremely important, when used as the materials for the electrodes, to achieve high performance for electrons transport and dye adsorption in the dye-sensitized solar cells. The longitudinal direction of the nanorod is along the c-direction, and the lattice spacing of 0.95 nm for the (001) plane and that of 0.35 nm for the {101} plane agree quite well with the corresponding values recorded in JCPDS. Such visual evidence strongly supports that the electron transport rate in the titania nanorods is expected to be very rapid, bringing highly efficient dye-sensitized solar cells through the use of the titania nanorods as the materials for the electrodes.

So far we have attained the power conversion efficiency ranging from 8.52% (Kurata et al., 2010) to 8.93% (Yoshida et al., 2008) using these nanorods as the electrode of dye-sensitized solar cells. In order to realize further improvement in conversion efficiency, we need to investigate the ways to control the shape as well as size of these nanorods by maintaining the extremely high crystalline feature of the nanorods. To accomplish the proper control of size and shape of nanorods, we examined the formation processes of nanorods under the most suitable condition for making nanorods, which is called "standard condition" hereafter, the results of which were detailed in a published work (Kurata et al., 2010).

In this chapter we first present the formation processes of titania nanorods under the standard condition in reasonable depth (Kurata et al., 2010). We then present the effects of both the concentrations of reactants, especially ethylenediamine, and the temperature-change strategy on the formation processes of nanorods. Based on all these findings, shape and size control of highly crystallized titania nanorods was proposed and carried out, leading to high-aspect-ratio, longer titania nanorods with highly crystallized state being successfully synthesized. We finally present that high dispersion of titania nanorods having highly crystallized state can be attained with the help of acetylacetone.

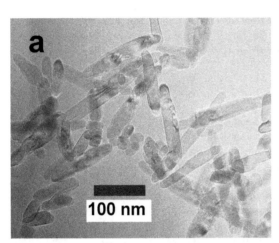
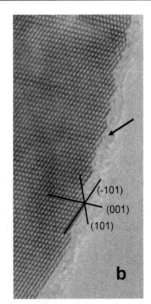

Fig. 1. Transmission electron microscopy images of highly crystallized titania nanorods covered with dye: (a) low-magnification image of titania nanorods, and (b) high-resolution image near the edge of a titania nanorod with dye coverage indicated by the arrow.

2. Experimental

The experimental procedure under the standard condition has been described in detail in our previous papers (Jiu et al., 2006; Kurata et al., 2010). Here, we summarize the essential part of the standard procedure and describe the modifications made on it. First, a 10-wt% aqueous solution of blockcopolymer F127 [(PEO)$_{106}$-(PPO)$_{70}$-(PEO)$_{106}$] was prepared using deionized pure water (Millipore Milli-Q). Cetyltrimethylammonium bromide was dissolved in the F127 solution at 308 K with a fixed concentration of cetyltrimethylammonium bromide, 0.055 M. In some modified cases the synthesis was carried out under no cetyltrimethylammonium bromide conditions. Ethylenediamine was added as a basic catalyst and also as a shape director (Sugimoto et al., 2003). The concentration of ethylenediamine was 0.25 M in the standard condition; in the modified conditions, the ethylenediamine concentration was varied from 0 to 0.5 M in order to examine its effects. After a transparent solution was obtained, tetraisopropyl orthotitanate (0.25 M) was added into the solution with stirring. This solution was stirred for half a day in the standard condition. The solution including white precipitates obtained by hydrolysis and condensation reactions of tetraisopropyl orthotitanate was then transferred into a Teflon autoclave sealed with a crust made of stainless steel, and reacted at 433 K for a desired period.

In the modified cases with temperature strategy, the reaction temperature was reduced during the preparation from 433 to 413 K to investigate its effects on the reaction mechanism. When acetylacetone was used to modify tetraisopropyl orthotitanate by binding acetylacetone to Ti atoms of tetraisopropyl orthotitanate, the transparent solutions were obtained after one-week stirring before hydrothermal reaction. The reaction product obtained under the hydrothermal condition at a desired time was washed by isopropyl alcohol and deionized pure water, followed by separating the reaction product by centrifugation (Kokusan H-40F). After the washing, the obtained sample was dried in vacuum for 24 h (EYELA Vacuum Oven VOS-450-SD). To gain additional insight into the underlying mechanism for the transition from amorphous-like structure to titania anatase crystalline structure in the early stage of the reaction, changes in shape and crystalline structure of reaction products upon calcination at 723 K for 2 h were observed and measured.

3. Results and discussion

3.1 Formation processes under standard condition

First of all, the formation processes under the standard condition are described prior to comparing the experimental results and discussing the effects of various modifications on those under the modified conditions. Typical transmission electron microscopy images of reaction products at 0.5, 2, 3.5, 4, 6, and 24 h under the standard condition (Kurata et al., 2010) are shown in Fig. 2. At 0.5 h, only a film-like structure was observed. At 2 h, the shape of reaction products was still mostly film-like, while some deep-black wedge-shaped structure partly appeared. At 3.5 h, the main structure was still film-like, with uneven light and dark patches recognized. At 4 h, however, only rod-shaped products were observable, signifying that the film-like shape with amorphous-like structure changed to nanorod-shaped titania in a time interval between 3.5 and 4 h. After 6 h, only nanorod shape was observed. The morphology was observed to change very slowly with time after 6 h.

Fig. 2. Transmission electron microscopy images of reaction products at 0.5, 2, 3.5, 4, 6, and 24 h under standard condition (Kutata et al., 2010).

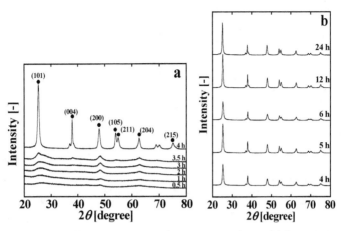

Fig. 3. Variation in X-ray diffraction spectra of reaction products: (a) from 0.5 to 4 h, (b) from 4 to 24 h.

Fig. 3 shows the variation in X-ray diffraction spectra (a) from 0.5 to 4 h and (b) from 4 to 24 h under the standard condition (Kurata et al., 2010), i.e., 0.25-M tetraisopropyl orthotitanate, 10-wt% F127, 0.055-M cetyltrimethylammonium bromide, 0.25-M ethylenediamine, and at 433 K. In the initial stage of reaction, X-ray diffraction spectra showed almost no clear peak, indicating the TiO_2 formed was amorphous. From 2 to 3.5 h, tiny and broad anatase peaks appeared, but the main structure of titania was still amorphous-like. During 3.5 to 4 h interval, a drastic change in the X-ray diffraction spectrum was detected, signifying the evolution from amorphous-like to clear anatase crystalline structure. From 4 to 24 h, X-ray diffraction spectra showed no appreciable changes.

In order to investigate the underlying process for the transition from amorphous-like structure to titania anatase crystalline structure in the early stage of the reaction, variations in shape and crystalline structure of reaction products upon calcination at 723 K for 2 h were utilized by Kurata et al. (2010). Fig. 4 shows the structural change from amorphous to anatase phase at 0.5 h after calcination, and the amorphous-like structure at 2 and 3.5 h also changing to anatase phase. At 4 h, the anatase crystalline structure was already formed before calcination. After 6 h, the X-ray diffraction patterns obtained before calcination almost completely coincided with those after calcination, indicating that crystalline structure before calcination did not change upon calcination owing to the highly crystallized state already achieved prior to calcination.

Transmission electron microscopy images of reaction products at reaction times of 0.5, 2, 3.5, 4, 6, and 24 h after calcination at 723 K for 2 h (Kurata et al., 2010) are shown in Fig. 5. Titania anatase nanoparticles with diameter around 10 nm were identifiable for the reaction products obtained at 0.5 h upon the calcination. While the product obtained at 1 h also changed to nanoparticles, the product obtained at 2 h changed to a mixture of nanoparticles and nanorods on the calcination. Similarly, a mixture of nanoparticles and nanorods were obtained for the product of 3.5 h upon the calcination. The fraction of rods at 3.5 h increased in comparison with that at 2 h. The nanorods formation could thus be claimed to be attributed to the growth of nuclei with anatase-like structure on the calcination. X-ray diffraction spectra before the calcination at 2 and 3.5 h were quite different from those of highly crystallized titania anatase at 6 and 24 h.

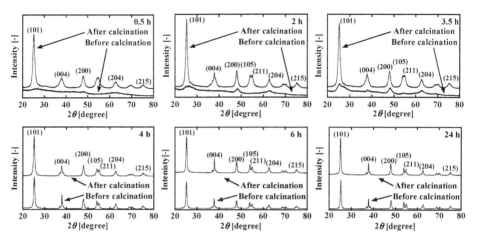

Fig. 4. Variation in X-ray diffraction patterns of reaction products upon calcination at 723 K for 2 h for the samples obtained at reaction times of 0.5, 2, 3.5, 4, 6, and 24 h.

The peak at 48.3 deg corresponding to (200) plane (2θ = 48.1 deg) in anatase phase was clearly observable and larger than those at 37.7 and 63 deg corresponding to (004) and (204) planes. Furthermore, no peak is observable at 38.6 deg, which corresponds to characteristic peak of (11) plane of Lepidocrocite (two-dimensional titania crystal). Therefore, the crystalline structure generated from film-like amorphous phase is inferred to be very thin two-dimensional anatase crystal.

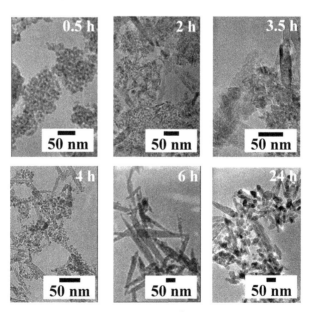

Fig. 5. Transmission electron microscopy images of reaction products obtained at 0.5, 2, 3.5, 4, 6, and 24 h after calcination at 723 K for 2 h.

The intensity ratio of (004) peak to (200) peak in X-ray diffraction spectra is shown in Fig. 6 as a function of the reaction time before and after calcination (Kurata et al., 2010). The almost zero ratio was obtained from 0.5 to 2 h in the absence of calcination, indicating that ordering of amorphous titania from random connection to crystal, evidenced partly in Fig. 2 and 3, occurred only in the film with no growth in the c-axis. The ratio, then, had increased progressively up to 0.38 at 3.5 h revealing slight growth in the c-axis, until the ratio attained a maximum rate of increase between 3.5 and 4 h duration, corresponding to the drastic change in the shape and crystalline structure of reaction products. Such an overwhelming increase in the intensity ratio (004)/(200) indicates that the phase transition from amorphous-like phase to anatase phase can bring about significant growth in the c-axis. The highest value was obtained at 4 h and slightly decreased with time, asymptotically approaching a constant of ~1.2 after 6 h. After calcination, the ratio gradually increased from 0.5 up to 6 h, and then reached a constant value, which was identical to the value obtained before calcination. These two distinctive trends shown in Fig. 6 signify that the crystalline structure of nanorods did not change on calcination, maintaining the intensity ratio at the same asymptotic level (~1.2) before and after calcination.

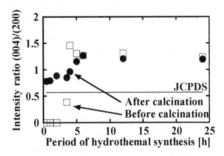

Fig. 6. Intensity ratio of (004) peak to (200) peak in X-ray diffraction spectra with reaction time under conditions before and after calcination.

3.2 Effects of ethylenediamine concentration and temperature change on the formation processes of nanorods

We investigated the effects of both ethylenediamine concentration and temperature change on the formation processes of nanorods. In particular, their mechanistic contributions to size and shape control of highly crystallized titania nanorods were inferred, together with the results of formation processes under standard condition mentioned above.

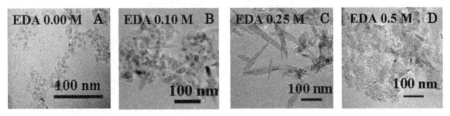

Fig. 7. Transmission electron microscopy images of reaction products synthesized at different ethylenediamine concentrations by hydrothermal method. Reaction conditions: 433 K, 6 h. "EDA" designates ethylenediamine.

Fig. 7 shows the effects of ethylenediamine concentration on the morphology of reaction products at 433 K for 6 h. When the ethylenediamine concentration was 0, titania particles with aspect ratio of roughly unity were formed. As the concentration was changed from 0 to 0.1 M, the morphology of titania shifted from particulate to a mixture of particles and rods. As the concentration reached 0.25 M (i.e., the value used under the standard condition, and thus as expected), only nanorods were observed to form, while at an ethylenediamine concentration as high as 0.5 M the observed products appeared to be unexpectedly film-like titanate. The corresponding X-ray diffraction spectra for the given series of samples are shown in Fig. 8. When the ethylenediamine concentration was 0 M, typical anatase peaks were obtained where (004) peak has a lower height than (200) peak, matching the spherical shape observed in Fig. 7A. When the ethylenediamine concentration was 0.25 M, a clear anatase spectrum was observed with higher (004) peak in comparison to (200) peak, signifying the formation of titania nanorods. For 0.1-M ethylenediamine concentration an intermediate spectrum between those of 0 and 0.25 M was observed due to the formation of particle-rod mixture as discussed above (see Fig. 7B). When the concentration became 0.5 M, a weak amorphous-like spectrum was obtained, corresponding to the observation of film-like structure in Fig. 7D. All these results signify that there should exist an optimum ethylenediamine concentration for controlling the rate of formation of titania nanorods at ≈ 0.25 M, above which - specifically at as high as 0.5 M - the reaction rate tends to slow down; that is, the morphological transition would be delayed. Such inference could be made by referring the morphology transformation as depicted in Fig. 2 under the standard condition with 0.25-M ethylenediamine.

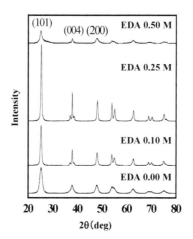

Fig. 8. X-ray diffraction spectra of the same samples shown in Fig. 7.

To further investigate the formation processes of titania nanorods at this high ethylenediamine concentration of 0.5 M, we carried out a series of experiments for evaluating the time course of the formation processes; the results are shown in Fig. 9. Film-like structure was observed up to 6 h as stated above; after 8 h, however, only rod shape was identifiable, signifying that the transformation from the amorphous film-like structure to the anatase titania nanorods has been almost completed by this time. Fig. 10 shows transmission electron microscopy and high-resolution transmission electron microscopy images of the reaction product obtained at 45 h. Ordered alignment of titania atoms in anatase structure can be clearly perceived, indicating the formation of highly crystallized titania anatase nanorods.

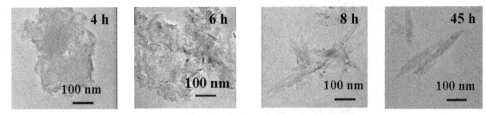

Fig. 9. Transmission electron microscopy images of reaction products obtained at different reaction times for 0.5-M ethylenediamine concentration. The rest of reaction conditions are the same as in the standard condition.

The effect of temperature change/reduction was examined by obtaining the time course of the formation processes at 413 K based on scanning electron microscopy images and X-ray diffraction measurements; the results are shown in Figs. 11 and 12, respectively. As shown in Fig. 11, the film-like structure was still observed at even 36 h. The X-ray diffraction spectrum obtained at 36 h shows no significant peaks, i.e., amorphous-like phase formation, which was observed under the standard condition at only up to 3.5 h (see Fig. 3a). Therefore, the reaction rate at 413 K became significantly slower. From scanning electron microscopy images, coexistence of titania nanorods and film was observed until 56 h, which was never recognized at the standard reaction temperature 433 K. It was after 64 h that only titania nanorods were finally observed. The scanning electron microscopy image obtained at 64 h shows a wide distribution in length of nanorods from roughly 10 to 600 nm, implying that nucleation and growth of nanorods would proceed concurrently because of the slow reaction rate at 413 K. The X-ray diffraction spectrum at 48 h, on the other hand, shows anatase peaks, though each peak height is not high. The peak height increases gradually with time up to 64 h. This observation suggests again the coexistence of amorphous-like films and titania nanorods. The peak height becomes higher with an increase in the fraction of titania nanorods up to 64 h.

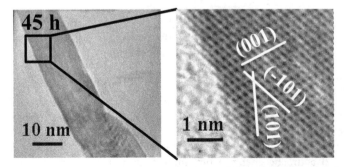

Fig. 10. Transmission electron microscopy and high-resolution transmission electron microscopy images of reaction product obtained at 45 h under 0.5-M ethylenediamine.

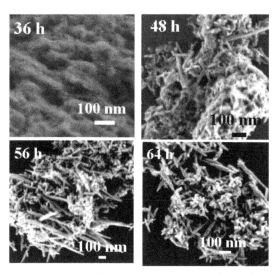

Fig. 11. Scanning electron microscopy images of reaction products obtained under the condition of 413 K at various times.

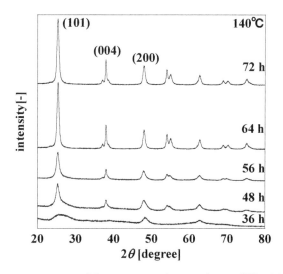

Fig. 12. X-ray diffraction patterns of the same reaction products of Fig. 11.

The observations and measurements made on temperature change described above are summarized in Fig. 13. At 413 K, the reaction is slow, resulting in concurrence of nucleation and growth of nanorods. At 433 K, on the other hand, the reaction occurs rapidly, resulting in 1) the prevalence of nucleation almost exclusively in the amorphous phase in the early reaction stage, 2) a drastic change from amorphous phase to crystalline titania anatase nanorods, and 3) no concurrence of nucleation and growth of nanorods. These findings should give some hints for the strategy for size and shape control.

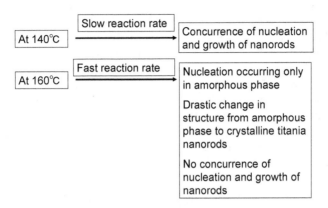

Fig. 13. Effects of reaction temperature on characteristics of formation processes at 413 K and 433 K.

3.3 Strategy for shape and size control of highly crystallized titania nanorods

The proposed strategy is given in Fig. 14. Nuclei are to be generated at a higher temperature 433 K in the early stage of reaction. These nuclei formed coincidently are to be reacted at a reduced temperature 413 K under hydrothermal conditions without further nucleation. Then, growth of rather uniform-sized and shaped nanorods is expected, in the aid of high concentration of ethylenediamine in effectively reducing their nucleation rate. In addition, we can also use the effectiveness of acetylacetone in obtaining good dispersion of nanorods (to be discussed later).

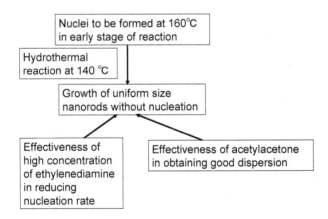

Fig. 14. Strategy for size and shape control of titania nanorods.

The specific preparation procedures are as follows. The even formation of nuclei was attempted at 433 K under the standard condition via the hydrothermal reaction for 2 h before being cooled down to room temperature. Ethylenediamine was then added to have its concentration be 0.5 M for selectively reducing nucleation rate. Formation reaction with these precursory nuclei was successively carried out at 413 K under the hydrothermal condition for 52 h; the reaction conditions are in the following: 0.25-M tetraisopropyl orthotitanate, 10-wt% F127, 0.055-M cetyltrimethylammonium bromide and 0.5-M ethylenediamine. A transmission electron microscopy image of thus obtained nanorods is shown in Fig. 15. Over 800-nm long, high-aspect-ratio nanorods were indeed obtained. In comparison to the nanorods images for 433 K at 24 h shown in Fig. 2 and those for 413 K at 64 h in Fig. 11, the nanorods obtained based on the proposed shape-control strategy were certainly improved in terms of morphological uniformity, despite the presence of some shorter nanorods, which stems from the not completely avoidable occurrence of nucleation during the formation reaction.

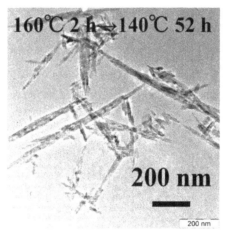

Fig. 15. Transmission elecron microscopy image of shape-controlled nanorods prepared based on the strategy given in Fig. 14.

3.4 Highly dispersed titania nanorods obtained with the help of acetylacetone

The effect of addition of acetylacetone was examined separately. In the experiments the same moles of acetylacetone and tetraisopropyl orthotitanate were mixed with each other to make a 1:1 complex. The complex was added to an aqueous solution of 10-wt% F127 containing 0.3-0.5 M ethylenediamine but no cetyltrimethylammonium bromide. The solution was stirred for one week at room temperature. The solution became transparent after 1-week stirring, which was never observed in the absence of acetylacetone. Adding acetylacetone thus must have a critical effect on particle dispersion. An example of nanorods thus obtained is shown in Fig. 16 (top) under the condition of 0.3-M ethylenediamine. Very good dispersion of titania nanorods was attained, and highly crystallized state is obvious as demonstrated in the high-resolution image in Fig. 16 (bottom). Since acetylacetone is known to adsorb on the surface of titania anatase crystal (Connor et al., 1995), adsorbed acetylacetone molecules could prevent aggregation of titania nanorods, resulting in such good dispersion. Also, since acetylacetone is expected to affect the formation mechanism, utilization of acetylacetone might improve the shape-control scheme of nanorods.

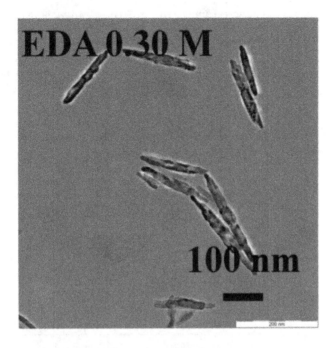

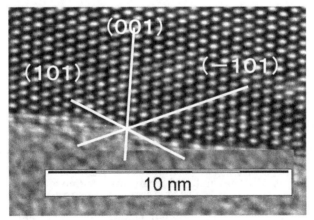

Fig. 16. Highly dispersed titania nanorods obtained with the help of acetylacetone (top) and highly crystallized feature of the nanorods demonstrated by high-resolution transmission electron microscpy image (bottom).

3.5 Application for dye-sensitized solar cells

The application of highly crystallized titania nanorods for making dye-sensitized solar cells was already reported (Yoshida et al., 2008; Kurata et al., 2010). A titania electrode made of titania nanorods was successfully fabricated as follows. The complex electrodes were prepared by the repetitive coating-calcining process: 3 layers of titania nanoparticles (Jiu et al., 2004, 2007) were first coated on FTO conducting glass, followed by 7 layers of mixed gel composed of titania nanorods and P-25. High light-to-electricity conversion efficiencies of 8.52 to 8.93% were achieved as exemplified in Fig. 17. We are now trying to get much higher power conversion efficiency by utilizing the shape-controlled, highly crystallized titania nanorods with high dispersion as a titania electrode of dye-sensitized solar cells.

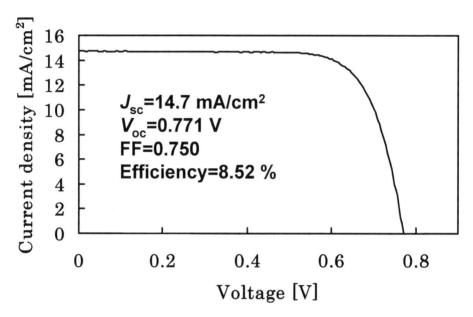

Fig. 17. I-V curve for complex dye-sensitized solar cell electrode consisting of highly crystallized titania nanorods, P-25, and titania nanoparticles.

4. Conclusions

The formation processes of highly crystallized titania nanorods were revealed in detail under 10-wt% F127, 0.25-M tetraisopropyl orthotitanate, 0.055-M cetyltrimethylammonium bromide, 0.25-M ethylenediamine, and 433 K (standard) conditions.

Strategy for shape and size control of highly crystallized titania nanorods was proposed through the findings obtained by examining the effects of both ethylenediamine concentration and temperature change on the formation processes of titania nanorods. Over 800-nm long and high-aspect-ratio, highly crystallized titania nanorods were successfully synthesized following the proposed strategy.

5. References

Adachi, M. Jiu, J. & Isoda, S. (2007). Synthesis of morphology-controlled titania nanocrystals and application for dye-sensitized solar cells. *Current Nanoscience*, 3, pp. 285-295

Adachi, M. Jiu, J. Isoda, S. Mori, Y. & Uchida, F. (2008). Self-assembled nanoscale architecture of TiO$_2$ and application for dye-sensitized solar cells. *Nanotechnology. Science and Applications*, 1, pp. 1-7

Adachi, M. Murata, Y. Takao, J. Jiu, J. Sakamoto, M. & Wang, F. (2004). Highly efficient dye-sensitized solar cells with titania thin film electrode composed of network structure of single-crystal-like TiO$_2$ nanowires made by "Oriented Attachment" mechanism. *J. Am. Chem. Soc.*, 126, pp. 14943-14949

Adachi, M. Sakamoto, M. Jiu, J. Ogata, Y. & Isoda, S. (2006). Determination of parameters of electron transport in dye-sensitized solar cells using electrochemical impedance spectroscopy. *J. Phys. Chem. B* 110, pp. 13872-13880

Alivisatos, A. P. (1996a). Semiconductor Clusters, Nanocrystals, and Quantum Dots. *Science*, 271, pp. 933-937

Alivisatos, A. P. (1996b). Perspectives on the physical chemistry of semiconductor nanocrystals. *J. Phys. Chem.*, 100, pp. 13226-13239

Banfield, J. F. Welch, S. A. Zhang, H. Ebert, T. T. & Penn, R. L. (2000). Aggregation-based crystal growth and microstructure development in natural iron oxyhydroxide biomineralization products. *Science*, 289, pp. 751-754

Burda, C. Chen, X. Narayanan, R. & El-Sayed, M. A. (2005). Chemistry and properties of nanocrystals of different shapes. *Chem. Rev.*, 105, pp. 1025-1102

Cao Y. C. (2004) Synthesis of square gadolinium-oxide nanoplates. *J. Am. Chem. Soc.*, 126, pp. 7456-7457

Cheng, B. Rusell, J. M. Shi, W. Zhang, L. & Samulski, E. T. (2004). Large-scale, solution-phase growth of single-crystalline SnO$_2$ nanorods. *J. Am. Chem. Soc.*, 126, pp. 5972-5973

Connor, P. A.. Dobson, K. D & McQuillan: A. J. (1995). New sol-gel attenuated total reflection infrared spectroscopic method for analysis of adsorption at metal oxide surface in aqueous solution. Chelation of TiO2, ZrO2, and Al2O3 surfaces by catechol, 8-quinolinol, and acetylacetopne. *Langmuir*, 11, pp. 4193-4195

Cui, Y. Bjork, M. T. Liddle, J. A. Sonnichsen, C. Boussert, B. & Alivisatos, A. P. (2004). Integration of colloidal nanocrystals into lithographically patterned devices. *Nano Lett.*, 4, pp. 1093-1098

Empedocles, S. A. Neuhauser, R. Shimizu, K. & Bawendi, M. G. (1999). Photoluminescence from single semiconductor nanostructures. *Adv. Mater.*, 11, pp. 1243-1256

Fabregat-Santiago, F. Bisquert, J. Palomares, E. Kuang, D. Zakeeruddin, S. M. & Grätzel, M. (2007). Correlation between photovoltaic performance and impedance spectroscopy of dye-sensitized solar cells based on ionic liquid: *J. Phys. Chem. C*, 111, pp. 6550-6560

Fujishima, A. & Honda, K. (1972). Electrochemical photolysis of water at a semiconductor electrode. *Nature*, 238, pp. 37-38

Fujishima, A. Rao, T. N. & Tryk, D. A. (2000). Titanium dioxide photocatalysis. *J. Photpchem. Photobio. C*, 1, pp. 1-21

Fukuda, K. Ebina, Y. Shibata, T. Aizawa, T. Nakai, I. & Sasaki, T. (2007) Unusual crystallization behaviors of anatase nanocrystallites from a molecularly thin titania nanosheet and its stacked forms: Increase in nucleation temperature and oriented growth. *J. Am. Chem. Soc.*, 129, pp. 202-209.

Garcia, R. & Tello, R. (2004). Size and shape controlled growth of molecular nano-structures on silicon oxide templates. *Nano Lett.*, 4, pp. 1115-1119

Grätzel, M. (2000). Photovoltaic performance and long-term stability of dye-sensitized meosocopic solar cells. *C. R. Chemie.*, 9, pp. 578-583

Grätzel, M. (2001). Photoelectrochemical cells. *Nature*, 414, pp. 338-344

Grätzel, M. (2004). Conversion of sunlight to electric power by nanocrystalline dye-sensitized solar cells. *J. Photpchem. Photobio. A: Chemistry*, 164, pp. 3-14

Grätzel, M. (2005). Solar Energy Conversion by Dye-Sensitized Photovoltaic Cells. *Inorg. Chem.*, 44, pp. 6841-6851

Hagfeldt, A. & Grätzel, M. (2000). Molecular photovoltaics. *Acc. Chem. Res.*, 33, pp. 269-277

Jiu, J. Isoda, S. Adachi, M. & Wang, F. (2007). Preparation of TiO2 nanocrystalline with 3-5 nm and application for dye-sensitized solar cell. *J. Photochem. Photobio. A: Cemistry*, 189, pp. 314-321

Jiu, J. Isoda, S. Wang, F. & Adachi, M. (2006). Dye-sensitized solar cells based on a single-crystalline TiO2 nanorod film. *J. Phys. Chem. B*, 110, pp. 2087-2092

Jiu, J. Wang, F. Sakamoto, M. Takao, J. & Adachi, M. (2004). Preparation of nanocrystaline TiO$_2$ with mixed template and its application for dye-sensitized solar cells. *J. Electrochem. Soc.*, 151, pp. A1653-A1658

Kurata, T. Mori, Y. Isoda, S. Jiu, J. Tsuchiya, K. Uchida, F. & Adachi, M. (2010). Characterization and formation process of highly crystallized single crystalline TiO$_2$ nanorods for dye-sensitized solar cells. *Current Nanoscience*, 6, pp. 269-276

Lakshmi, B. B. Dorhout, P. K. & Martin, C. R. (1997a). Sol-gel template synthesis of semiconductor nanostructures. *Chem. Mater.*, 9, pp. 857-862

Lakshmi, B. B. Dorhout, P. K. & Martin, C. R. (1997b). Sol-gel template synthesis of semiconductor oxide micro- and nanostructures. *Chem. Mater.*, 9, pp. 2544-2550

Liu, C. Wu, X. Klemmer, T. Shukla, N. Yang, X. Weller, D. Roy, A. G. Tanase, M.& Laughlin, D. (2004). Polyol process synthesis of monodispersed fept nanoparticles. *J. Phys. Chem. B*, 108, pp. 6121-6123

Manna, L. Scher, E. C. & Alivisatos, A. P. (2000). Synthesis of soluble and processable rod-, arrow-, teardrop-, and tetrapod-shaped CdSe nanocrystals. *J. Am. Chem. Soc.*, 122, pp. 12700-12706

Masuda, H. & Fukuda, K. (1995). Ordered Metal nanohole arrays made by a two-step replication of honeycomb structures of anodic alumina. *Science.* 268, pp. 1466-1468

Masuda, H. Yamada, H. Satoh, M. Asoh, H. Nakao, M. & Tamamura, T. (1997) Highly ordered nanochannel-array architecture in anodic alumina. *Appl. Phys. Lett.*, 71, pp. 2770-2772

Murray, C. B. Kagan, C. R. & Bawendi, M. G. (2000). Synthesis and characterization of monodisperse nanocrystals and close-packed nanocryatal assemblies. *Annu. Rev. Mater. Sci.*, 30, pp. 545-610

Nazeeruddin, M. K. De Angelis, F. Fantacci, S. Selloni, A. Viscardi, G. Liska, P. Ito, S. Bessho, T. & Graetzel, M. (2005) Combined experimental and DFT-TDDFT computational study of photoelectrochemical cell ruthenium sensitizers. *J. Am. Chem. Soc.*, 127, pp. 16835-16847

Nirmal, M. & Brus, L. (1999). Luminescence photophysics in semiconductor nanocrystals. *Acc. Chem. Res.*, 32, pp. 407-414

Pacholski, C. Kornowski, A. & Weller, H. (2002). Self-assembly of ZnO: From nanodots to nanorods. *Angew. Chem. Int. Ed.*,. 41, pp. 1188-1191

Pei, L. Mori, K. & Adachi, M. (2004). Formation process of two-dimensional networked gold nanowires by citrate reduction of AuCl₄⁻ and the shape stabilization. *Langmuir*, 20, pp. 7837-7843

Peng, X. (2003) Mechanisms for the shape-control and shape-evolution of colloidal semiconductor nanocrystals. *Adv. Mater.*, 15, pp. 459-463

Peng, X. G. Manna, L. Yang, W. D. Wickham, J. Scher, E. Kadavanich, A. & Alivisatos, A. P. (2000). Shape control of CdSe nanocrystals. *Nature*, 404, pp. 59-61

Penn, R. L. & Banfield, J. F. (1998). Morphology development and crystal growth in nanocrystalline aggregates under hydrothermal conditions: insights from titania. *Science*, 281, pp. 969-971

Puntes, V. F. Krishnan, K. M. & Alivisatos, A. P. (2001). Colloidal nanocrystal shape and size control: The case of cobalt. *Science*, 291, pp. 2115-2117

Reiss, B. D. Mao, C. Solis, D. J. Ryan, K. S. Thomson, T. & Belcher. (2004). A. M. Biological routes to metal alloy ferromagnetic nanostructures. *Nano Lett.*, 4, pp. 1127-1132

Scher, E. C. Soc, R. Manna, L.& Alivisatos, A. P. (2003). Shape control and applications of nanocrystals. *Phil. Trans. R. Soc.* Lond. A, 361, pp. 241-257

Song, Q. & Zhang, Z. J. Shape control and associated magnetic properties of spinel cobalt ferrite nanocrystals, J. (2004). *Am. Chem. Soc.*, 126, pp. 6164-6168

Sugimoto, T. Zhou, X. & Muramatsu, A. (2003). Synthesis of uniform anatase TiO₂ nanoparticles by gel–sol method 4. Shape control. *J. Colloid Interface Sci.*, 259, pp. 53-61

Tang, Z. Kotov, N. A. & Giersig. M. (2002). Spontaneous Organization of single CdTe nanoparticles into luminescent nanowires. *Science*, 297, pp. 237-240

Tang, Z. Ozturk, B. Wang, Y. & Kotov. N. A. (2004). Simple preparation strategy and one-dimensional energy transfer in CdTe nanoparticle chains. *J. Phys. Chem. B*, 108, pp. 6927-6931

Wu, G. Zhang, L. Cheng, B. Xie, T. & Yuan, X. (2004). Synthesis of Eu₂O₃ nanotube arrays through a facile sol-gel template approach. *J. Am. Chem. Soc.*, 126, pp. 5976-5977

Yang, D. Wang, R. Zhang, J. & Liu, Z. (2004). Synthesis of nickel hydroxide nanoribbons with a new phase: A solution chemistry approach. *J. Phys. Chem. B*, 108, pp. 7531-7533

Yoshida, K. Jiu, J. Nagamatsu, D. Nemoto, T. Kurata, H. Adachi, M. & Isoda, S. (2008). Structure of TiO2 Nanorods formed with doiuble surfactants, *Molecular crystals and liquid crystals*. 491, pp. 14-20

Yu, W. W. Wang, Y. A. & Peng, X. (2003). Formation and stability of size-, shape-, and structure-controlled cdte nanocrystals: Ligand effects on monomers and nanocrystals. *Chem. Mater.*, 15, pp. 4300-4308

Zhang, H. Sun, J. Ma, D. Bao, X. Klein-Hoffmann, A. Weinberg, G. Su, D. & Schlogl, R.
(2004). Unusual mesoporous SBA-15 with Parallel channels running along the short
axis. *J. Am. Chem. Soc.*, 126, pp. 7440-7441

Permissions

The contributors of this book come from diverse backgrounds, making this book a truly international effort. This book will bring forth new frontiers with its revolutionizing research information and detailed analysis of the nascent developments around the world.

We would like to thank Professor, Doctor of Sciences, Leonid A. Kosyachenko, for lending his expertise to make the book truly unique. He has played a crucial role in the development of this book. Without his invaluable contribution this book wouldn't have been possible. He has made vital efforts to compile up to date information on the varied aspects of this subject to make this book a valuable addition to the collection of many professionals and students.

This book was conceptualized with the vision of imparting up-to-date information and advanced data in this field. To ensure the same, a matchless editorial board was set up. Every individual on the board went through rigorous rounds of assessment to prove their worth. After which they invested a large part of their time researching and compiling the most relevant data for our readers. Conferences and sessions were held from time to time between the editorial board and the contributing authors to present the data in the most comprehensible form. The editorial team has worked tirelessly to provide valuable and valid information to help people across the globe.

Every chapter published in this book has been scrutinized by our experts. Their significance has been extensively debated. The topics covered herein carry significant findings which will fuel the growth of the discipline. They may even be implemented as practical applications or may be referred to as a beginning point for another development. Chapters in this book were first published by InTech; hereby published with permission under the Creative Commons Attribution License or equivalent.

The editorial board has been involved in producing this book since its inception. They have spent rigorous hours researching and exploring the diverse topics which have resulted in the successful publishing of this book. They have passed on their knowledge of decades through this book. To expedite this challenging task, the publisher supported the team at every step. A small team of assistant editors was also appointed to further simplify the editing procedure and attain best results for the readers.

Our editorial team has been hand-picked from every corner of the world. Their multi-ethnicity adds dynamic inputs to the discussions which result in innovative outcomes. These outcomes are then further discussed with the researchers and contributors who give their valuable feedback and opinion regarding the same. The feedback is then collaborated with the researches and they are edited in a comprehensive manner to aid the understanding of the subject.

Apart from the editorial board, the designing team has also invested a significant amount of their time in understanding the subject and creating the most relevant covers. They scrutinized every image to scout for the most suitable representation of the subject and create an appropriate cover for the book.

The publishing team has been involved in this book since its early stages. They were actively engaged in every process, be it collecting the data, connecting with the contributors or procuring relevant information. The team has been an ardent support to the editorial, designing and production team. Their endless efforts to recruit the best for this project, has resulted in the accomplishment of this book. They are a veteran in the field of academics and their pool of knowledge is as vast as their experience in printing. Their expertise and guidance has proved useful at every step. Their uncompromising quality standards have made this book an exceptional effort. Their encouragement from time to time has been an inspiration for everyone.

The publisher and the editorial board hope that this book will prove to be a valuable piece of knowledge for researchers, students, practitioners and scholars across the globe.

List of Contributors

Yang Jiao, Fan Zhang and Sheng Meng
Beijing National Laboratory for Condensed Matter Physics and Institute of Physics, Chinese Academy of Sciences, Beijing, China

Seigo Ito
Department of Electrical Engineering and Computer Sciences, Graduate School of Engineering, University of Hyogo, Hyogo, Japan

Y. Chergui, N. Nehaoua and D. E. Mekki
Physics Department, LESIMS Laboratory, Badji Mokhtar University, Annaba, Algeria

Ying-Hung Chen, Chen-Hon Chen, Shu-Yuan Wu, Chiung-Hsun Chen, Ming-Yi Hsu, Keh-Chang Chen and Ju-Liang He
Department of Materials Science and Engineering, Feng Chia University, Taichung, Taiwan, R.O.C.

Zhigang Chen, Qiwei Tian, Minghua Tang and Junqing Hu
State Key Laboratory for Modification of Chemical Fibers and Polymer Materials, College of Materials Science and Engineering, Donghua University, China

Sadia Ameen, Young Soon Kim and Hyung-Shik Shin
Energy Materials & Surface Science Laboratory, Solar Energy Research Center, School of Chemical Engineering, Chonbuk National University, Jeonju, Republic of Korea

M. Shaheer Akhtar
New & Renewable Energy Material Development Center (NewREC), Chonbuk National University, Jeonbuk, Republic of Korea

N. Stem and S. G. dos Santos Filho
Universidade de São Paulo/ Escola Politécnica de Engenharia Elétrica (EPUSP), Brazil

E. F. Chinaglia
Centro Universitário da FEI/ Departamento de Física, Brazil

Khalil Ebrahim Jasim
Department of Physics, University of Bahrain, Kingdom of Bahrain

Mi-Ra Kim, Sung-Hae Park, Ji-Un Kim and Jin-Kook Lee
Department of Polymer Science & Engineering, Pusan National University, Jangjeon-dong, Guemjeong-gu, Busan, South Korea

Yongwoo Kim
Korea Industrial Complex Corporation, Korea

Deugwoo Lee
Pusan National University, Korea

Motonari Adachi
Research Center of Interfacial Phenomena, Faculty of Science and Engineering, Doshisha University, 1-3 Miyakodani, Tatara, Kyotanabe, Japan

Katsuya Yoshida, Takehiro Kurata, Katsumi Tsuchiya and Yasushige Mori
Department of Chemical Engineering and Materials Science, Doshisha University, 1-3 Miyakodani, Tatara, Kyotanabe,

Jun Adachi
National Instituite of Biomedical Innovation, 7-6-8 Asagi Saito, Ibaraki, Japan

Motonari Adachi and Fumio Uchida
Fuji Chemical Co., Ltd., 1-35-1 Deyashikinishi-machi, Hirakata, Japan

Printed in the USA
CPSIA information can be obtained
at www.ICGtesting.com
JSHW011449221024
72173JS00004B/1005